ENVIRONMENTAL HEALTH - PHYSICAL, CHEMICAL AND BIOLOGICAL FACTORS

PARASITES

ECOLOGY, DISEASES AND MANAGEMENT

Environmental Health - Physical, Chemical and Biological Factors

Additional books in this series can be found on Nova's website
under the Series tab.

Additional e-books in this series can be found on Nova's website
under the e-book tab.

ENVIRONMENTAL HEALTH - PHYSICAL, CHEMICAL AND BIOLOGICAL FACTORS

PARASITES

ECOLOGY, DISEASES AND MANAGEMENT

GILMAR S. ERZINGER
EDITOR

NOVA BIOMEDICAL
New York

For permission to use material from this book please contact us:
Telephone 631-231-7269; Fax 631-231-8175
Web Site: http://www.novapublishers.com

Additional color graphics may be available in the e-book version of this book.

LIBRARY OF CONGRESS CATALOGING-IN-PUBLICATION DATA

ISBN: 978-1-62257-692-0

Published by Nova Science Publishers, Inc. † New York

Contents

PREFACE

Knowledge of parasitology has increased in recent years, due to the development of new technologies in molecular biology and real image analysis. The effects of globalization, and even those of low-grade heat, in conjunction with the setbacks caused by air, land and water pollution, may create new dangers for humans and animals worldwide. Thus, all disciplines of veterinary biology and medicine must be enriched with knowledge of parasitology.

Parasitology is an interdisciplinary science that combines biology, medicine, pharmacology and animal medicine in the study of parasites in terrestrial and aquatic human and animal diseases. Research efforts over the past year have made important findings regarding the life cycles of parasites and, more importantly, about their pathogenicity, physiology, metabolism, epidemiology and transmission. These discoveries have provided insights into the biosecurity required to prevent the transmission of these diseases.

The scientist Heinz Mehlhorn (2011) stated that the recent pandemics of several emerging diseases have proven that we "*all live in very thin ice*" or "*on an already rumbling volcano.*"

This book comprises research from internationally renowned scientists and provides a solid foundation from which to explore various aspects of parasitic diseases, including clinical, laboratory, and differential diagnostic options for understanding the control and containment of various parasites. All of the topics covered in this book are relevant to parasitologists, zoologists, biologists and veterinarians, who are general practitioners and examine many patients early in the course of parasitic diseases. Many details of less common parasitic diseases have deliberately been omitted, and the book instead emphasizes epidemiology and effective new methods of diagnosis, treatment and prevention. For practitioners, this book provides concise guidance and substantial information on the diagnosis and treatment of a number of important parasitic diseases to facilitate patient care; thus, this book can be used as a continuing education resource for human and veterinary medicine.

A building-block approach provides the logic that students and researchers need in an easy-to-use, memorable format. Many illustrations are included for the information of the student and general practitioner and may also be used in client education.

The material is presented in stages, extending from basic principles and concepts in parasitology to the systematic description of major parasitic diseases that affect humans and animals on land and in water. This chapter concludes with a discussion of the principles of diagnosis and control of parasitic diseases.

Prof. Gilmar Sidnei Erzinger. Ph.D.
University of Joinville Region
Joinville, Brazil
February, 2012

In: Parasites
Editor: Gilmar S. Erzinger

ISBN: 978-1-62257-692-0

Chapter 1

Malaria and Dengue: Epidemiology, Control, Life Cycle of the Vector and Biosafety in Parasitology Labs

Gilmar Sidnei Erzinger[1,3,*], Ozair Souza[2] and Marco Fabio Mastroeni[3]

[1]Department of Pharmacy, University of Joinville Region, Brazil
[2]Department of Chemical Engineering, University of Joinville Region, Brazil
[3]Department of Medicine, University of Joinville Region, Brazil

Abstract

More than 300 parasitic worm diseases and 70 protozoan-elicited diseases have been described in humans. An overview of parasites is presented by Cox and in the literature cited therein [1]. A number of human parasites live in aquatic ecosystems at some point in their developmental cycle. The global distribution of these organisms is not limited to subtropical and tropical countries, and certain medically important parasites have been found in mid-latitude bodies of water, such as in Europe. One of the probable effects of global warming is the presence of malaria in Germany, which has a temperate climate [2].

According to the World Health Organization (WHO), malaria is one of the world's most deadly diseases, killing approximately 880,000 people, primarily children in sub-Saharan Africa. Several approaches to pest control have been taken in the past, including chemical pesticides (e.g., nematicides, molluscicides, insecticides). More recent approaches have employed specific chemicals (e.g., developmental hormones) to inhibit the developmental cycles of specific parasites. These chemical are much less toxic to humans and the environment than, for instance, DDT; and they are specific for certain pest insects and do not affect environmentally necessary and benevolent insects, such as bees and other pollinators.

Another promising method used for the control of malaria is the substitution of wild-type vector insects with genetically modified strains that are resistant to parasites and

* E-mail: gerzinger@gmail.com.

cannot transfer malaria to humans [3, 4]. Individuals working in laboratories involving biological materials must receive training in biosafety, regardless of their level of research training. Coupled with the theoretical and practical knowledge acquired concerning biosafety, two elements are essential to ensure the quality of the system: prevention and a collective vision of security. If these requirements are met, the risk of accidents will be reduced substantially, and the quality of the work environment will be maintained.

INTRODUCTION

Malaria is one of the most important tropical infectious diseases in the world [5, 6]. At least 300 million cases of malaria occur each year, causing the deaths of more than 1 million people [4]. Most cases of malaria occur in Africa (approximately 2/3 of cases and 90 % of deaths), but certain tropical regions of Asia and Latin America are also severely affected [7].

Today, approximately 2.5 billion people, or 40% of the world's population, live in areas where there is a risk of dengue transmission (see WHO/Impact of Dengue). Dengue is endemic in at least 100 countries in Asia, the Pacific, the Americas, Africa, and the Caribbean. The World Health Organization (WHO) estimates that 50 to 100 million infections occur yearly, including 500,000 Dengue Hemorrhagic Fever (DHF) cases and 22,000 deaths, mostly among children [8].

Dengue fever is endemic over large areas of the tropics and subtropics. Outbreaks of dengue have repeatedly occurred in Brazil over the last 10 years. The etiological agent of dengue is an arbovirus, and the major vector is the *Aedes aegypti* mosquito, which is found in 3600 Brazilian municipalities [9]. Although most patients are asymptomatic, reinfection with different serotypes of dengue viruses might lead to hemorrhagic fever with high mortality [10].

Kurane [11] described that the effect of global warming on infectious disease vectors is indirect and might affect the geographic distribution and activity of vectors, especially mosquitoes.

Figure 1. Child suffering from malaria. Most deaths caused by malaria occur in sub-Saharan Africa in children younger than 5 years (photo: http://www.worldpress.org).

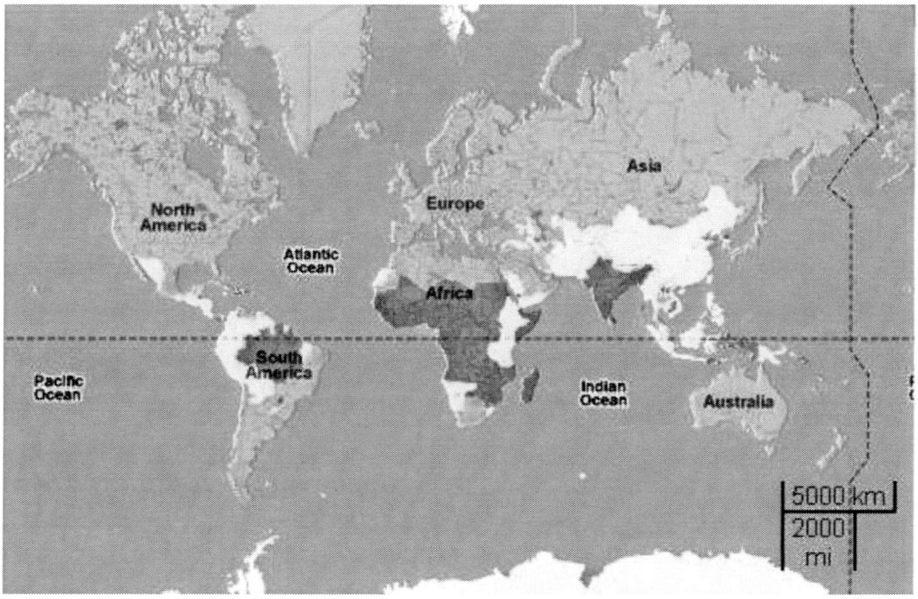

Figure 2. The malaria map used as a guide for the assessment of transmission risk throughout the world, as defined by the Centers for Disease Control and Prevention (CDC) (http://cdc-malaria.ncsa.uiuc.edu/). Red: widespread malaria; Yellow: at-risk area; Green: no known malaria.

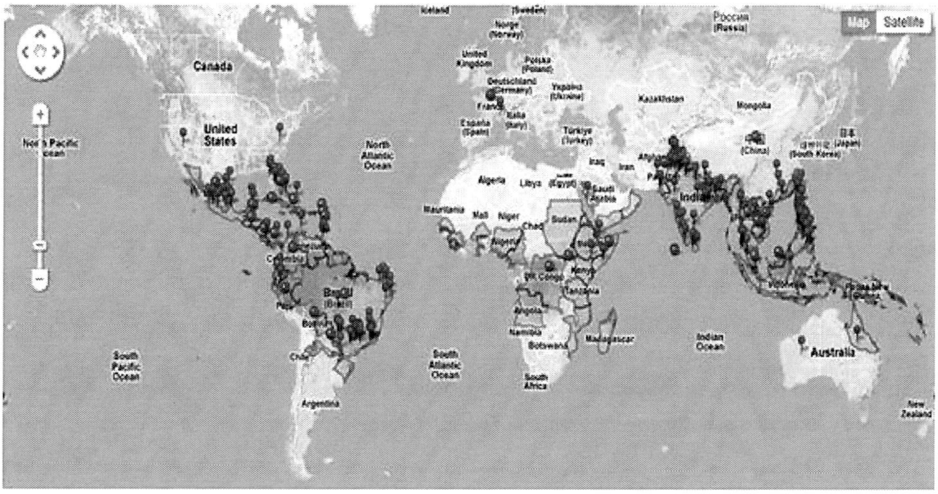

Figure 3. The world dengue map used as a guide for the assessment of dengue risk throughout the world, as defined by the Centers for Disease Control and Prevention (CDC) (http://www.healthmap.org/dengue/index.php).

The mosquito-borne diseases, such as malaria and dengue, have a strong relationship to global warming. Malaria has been considered as the most important vector-borne infectious disease in the world, and global warming alters its distribution, intensity of transmission, and seasonality in sub-Saharan Africa [12-13]. An association between the inter-annual variability in temperature and malaria transmission was detected in Africa. In Kenya, the number of malaria cases was associated with high rainfall and maximum temperatures in the previous 3 and 4 months [14]. In Ethiopia, malaria epidemics were associated with high minimum temperatures in recent months [15]. However, there are also reports suggesting that there is no clear association between climate change and malaria in South America [16] and the Russian Federation [17]. According to Kurane [11], global warming may affect malaria at the local level [16].

There are also reports suggesting the absence of an association between global warming and epidemics of dengue [17-20]. Several other factors in addition to climatic factors may affect the transmission of dengue, and the effects of global warming on dengue epidemics may vary by geographic area. The abundance estimated by the vector model was consistent with the distribution of dengue cases reported in some areas of the world [11, 21]. Thus, it is reasonable to expect that the positive effect of global warming on the abundance and distribution of mosquitoes is related to the abundance of water, which eventually leads to an increase in the number of dengue patients and the expansion of the endemic areas of dengue virus [11].

1. Malaria

Malaria is a protozoan parasitosis caused by several *Plasmodium* species (phylum Apicomplexa). The most important are *P. malariae*, *P. ovale*, and *P. vivax*. *P. falciparum* causes the most severe illness. *P. falciparum* causes the greatest incidence of illness and death (approximately 80 % of malaria infections and 90 % of deaths caused by malaria). Other species are reported to infect humans but do not play an important role. *Plasmodium* demonstrates the alternation of generations, with *Anopheles* being the final host and humans being the intermediate host.

1.2. The Malaria Parasite Life Cycle

In the female *Anopheles* mosquito (only females suck blood), gamonts taken up with the blood of an infected human transform into macro- and microgametes in the gut of the mosquito and subsequently fuse and form an ookinete. The ookinete penetrates the gut and encapsulates, forming an oocyst on the outer surface of the gut. The oocyst produces thousands of haploid sporozoites, which later infect the cells of the salivary glands of *Anopheles*. When the mosquito stings a human, sporozoites are transferred into the human's blood with the mosquito's saliva. The sporozoites infect liver cells, where they multiply one thousand fold via schizogonic division to merozoites (exoerythrocytic stage) in 2 weeks. The merozoites are released and infect erythrocytes, where they transform into trophozoites. The trophozoites take up amino acids (mostly from degraded hemoglobin) and subsequently divide into 32 schizonts each; the schizonts are released upon erythrocyte lysis to infect other erythrocytes. The release of the schizonts from all infected erythrocytes is synchronized and is accompanied by rhythmic fever attacks in the infected person. Tumor necrosis factor alpha (TNF-α) is secreted from macrophages, which remove the debris from disintegrated erythrocytes and most likely induce the fever. A fraction of the merozoites differentiate to gamonts, which infect *Anopheles* (see above), and the infection cycle starts again.

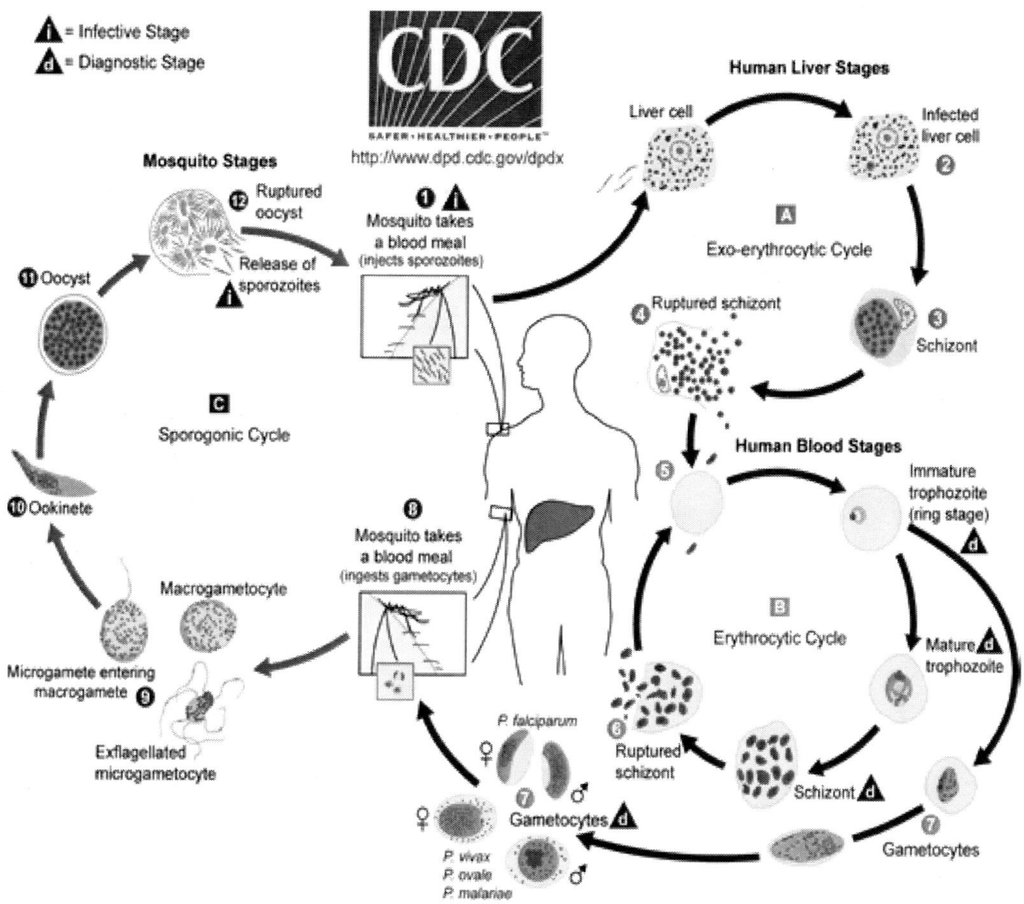

Figure 4. The malaria parasite life cycle involves two hosts. During a blood meal, a malaria-infected female *Anopheles* mosquito inoculates sporozoites into the human host 1. The sporozoites infect liver cells 2 and mature into schizonts 3. Subsequently, the liver cells rupture and release merozoites 4. (Notably, in *P. vivax* and *P. ovale,* a dormant stage [hypnozoites] can persist in the liver and cause relapses by invading the bloodstream weeks, or even years, later). After this initial replication in the liver (exo-erythrocytic schizogony A), the parasites undergo asexual multiplication in the erythrocytes (erythrocytic schizogony B). Merozoites infect red blood cells 5. The ring stage trophozoites mature into schizonts, and merozoites are released upon the rupture of red blood cells 6. Some parasites differentiate into sexual erythrocytic stages (gametocytes) 7. The blood stage parasites are responsible for the clinical manifestations of the disease. The *Anopheles* mosquito ingests male (microgametocytes) and female (macrogametocytes) gametocytes during a blood meal 8. The parasitic multiplication in the mosquito is known as the sporogonic cycle C. While in the mosquito's mid-gut, the microgametes penetrate each other, generating zygotes 9. The zygotes, in turn, become motile and elongated (ookinetes) 10 and invade the midgut of the mosquito, where they develop into oocysts 11. The oocysts grow, rupture, and release sporozoites 12, which migrate to the mosquito's salivary glands. The inoculation of the sporozoites into a new human host perpetuates the malaria life cycle 1. Biological cycle of *Plasmodium* (www.dpd.cdc.gov/dpdx).

All human pathogenic *Plasmodia* cause severe diseases. *P. falciparum* often causes fatal plague and is one of the factors inhibiting the development of many poor countries in Africa [9]. The high severity of this strain is a consequence of physiological properties, such as protein synthesis, that lead to the aggregation and agglutination of erythrocytes that block the capillary systems of the kidneys, brain, and other organs. Notably, humans can acquire a partial, but not complete, immune defense against this parasite. *P. falciparum* has promoted a gene defect in human hemoglobin that leads to sickle cell anemia (i.e., abnormal erythrocytes that are not infected by the *schizonts*). In Brazil, approximately 99% of malaria cases are concentrated in the Amazon region. Brazil's endemic area for malaria comprises Acre, Amapá, Amazonas, Para, Rondonia, Roraima and parts of Maranhão, Mato Grosso and Tocantins. Tocantins is currently considered a non-endemic area with risk for malaria transmission. In extra-Amazonian regions, 64% of the cases reported are imported: approximately 55% of the cases are from states belonging to Legal Amazonia, and 9% of the cases originate from other South American countries (French Guiana, Paraguay and Suriname) or from Africa. These cases are of great importance for transmission in municipalities located on the shores of the lake of the Itaipu hydroelectric plant and in the areas covered by the Atlantic Forest in the states of Espirito Santo, Minas Gerais, Rio de Janeiro, Sao Paulo and Bahia in the center-west region (states of Goias and Mato Grosso do Sul). The remaining 36% are sporadic cases occurring in restricted focal areas [22].

2. DENGUE

2.1. Impact of Dengue

During the 19th century, dengue was considered a sporadic disease that caused epidemics at long intervals, which reflected the slow pace of transport and limited travel at that time.

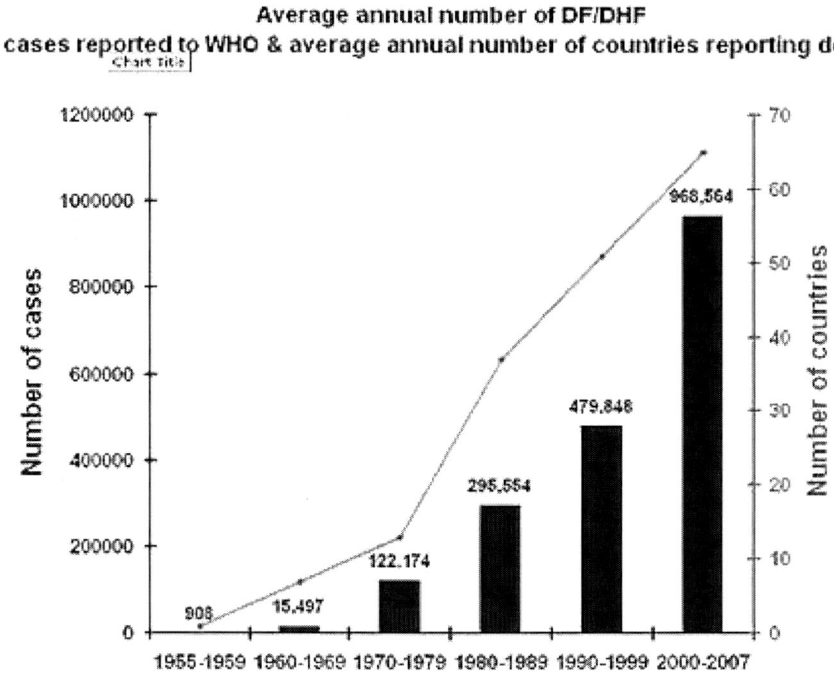

Figure 5. Average annual number of DF/DHF cases reports to WHO and the average annual number of countries reporting dengue [23].

Today, dengue ranks as the most important mosquito-borne viral disease in the world. In the last 50 years, the incidence of dengue has increased 30-fold. There are an estimated 2.5 billion people living in over 100 endemic countries and areas where dengue viruses can be transmitted. Up to 50 million infections occur annually, with 500,000 cases of dengue hemorrhagic fever and 22,000 deaths among children. Prior to 1970, only 9 countries had experienced cases of dengue hemorrhagic fever (DHF); since that year, the number has increased more than 4-fold and continues to rise[23].

The pandemic of 1998, in which 1.2 million cases of dengue fever and DHF were reported from 56 countries worldwide, was unprecedented. Data for 2001-2002 indicate a situation of comparable magnitude. In 2001, the Americas alone reported over 652,212 cases of dengue, of which 15,500 were DHF, nearly twice the number of cases reported for the same region in 1995. The challenge for national and international health agencies is to reverse the trend of increasing epidemic dengue activity and incidence of DHF [23].

2.2. Epidemiology

Dengue fever (DF) is caused by any of four closely related viruses, or serotypes, known as dengue 1-4. Infection with one serotype does not protect against the others, and sequential infections increase the risk for dengue hemorrhagic fever (DHF) and dengue shock syndrome (DSS).

In 2000, Schatmayer [24] reported that the Department of Virology of the Instituto Oswaldo Cruz (IOC) detected dengue virus type 1 in a large epidemic close to the city of Rio de Janeiro in 1986 [24]. In the same year, the virus quickly spread to other cities along the northeast coast, where the vector *Aedes aegypti* had been reintroduced in the late 1970s. This vector was eliminated from most American countries, including Brazil, in a continental effort by the Pan American Health Organization to eradicate yellow fever in the 1960s. However, the vector was not eradicated in Suriname, the Guyanas, Venezuela, the Caribbean Islands and the United States. Due to a lack of funds to maintain the campaign and the emergence of other priorities as the years passed, the mosquito was reintroduced to the region from infested countries, and in 1995, the distribution of *Ae. aegypti* reached the same level as before the eradication campaign [25]. In Brazil, the vector was detected in Bahia in 1975 and in Rio de Janeiro by 1977. The expansion of *Ae. aegypti* in the 1970s and 1980s increased the movement of dengue viruses in the American region. Dengue viruses of type 2 and type 3 were present in the Americas prior to 1977 and caused epidemics especially in the Caribbean region. Type 3 apparently disappeared from the Americas in the mid-1970s but returned in 1994, causing a major epidemic in Nicaragua and becoming established in Mexico and Central America. In 1977, type 1 was reintroduced with epidemics in Jamaica, Cuba, Puerto Rico and Venezuela. This serotype spread quickly throughout the Caribbean, Mexico, Central America and northern South America and subsequently reached Brazil and almost all countries in South America. In 1981, a new strain of dengue type 2, which was probably of Asian origin, was introduced into Cuba and caused the first epidemic of severe dengue hemorrhagic fever (DHF) and dengue shock syndrome (DSS) in the Americas, with 158 deaths. The same strain reached other countries, including Brazil, and was associated with an increased severity of clinical cases. In 1981, dengue type 4 was detected in the eastern Caribbean Islands and, similar to type 1, spread to many countries in the following years [26,

27]. Although only types 1 and 2 are currently present in Brazil, the data emphasize the high risk of the introduction of types 3 and 4.

As in many other countries of the region, dengue soon became a national public health problem in Brazil. More than 1.8 million cases in total, with an annual peak of 570,148 cases in 1998, have been reported since June 2000 [28]. In 1990, dengue type 2 was again detected in the State of Rio de Janeiro, and a clear increase in the number of severe cases was observed [29, 30].

During the first type 1 epidemic, which was referred to as a "virgin soil" epidemic, studies conducted in the area showed a 40% rate of silent infections in the affected populations, confirming data obtained in other countries [31]. The dengue cases reported in Brazil in recent years represent 80% of the total number reported in the Americas, emphasizing the magnitude of Brazil's problem. Currently, only dengue types 1 and 2 are present in Brazil; however, as previously noted, there is a high risk for the introduction of types 3 and 4, which are present in Central America, the Caribbean and northern South America.

Dengue epidemics have been described in 24 of the 26 Brazilian states and in the Federal District/ Brasília. An increase in the severity of the clinical course of dengue infections was observed after the introduction of type 2, although one fatal case due to type 1 was observed before dengue type 2 was introduced in the country. As of June 2000, 893 dengue hemorrhagic fever/dengue shock syndrome cases had been reported in Brazil, with 45 deaths [28].

The efforts to reduce and control the vectors have been based largely on campaigns to control the adult mosquito, which is often associated with epidemics, and house-by-house larvae control has been discontinued throughout Brazil, especially in underdeveloped urban areas. The participation of the population in vector control is also quite limited, although sufficient to control in-house breeding sites. *Ae. albopictus*, a species known to transmit dengue in Asia, particularly in endemic situations, was introduced in Brazil during the 1980s by Asian ships transporting iron ore. This vector's spread could potentially increase dengue virus transmission in the Southeast region, which is the most populated, and could present another challenge for dengue control.

With IOC support, a network of laboratories was established in Brazil for diagnosis of dengue following confirmation of the first cases in Rio de Janeiro. Isolation of the virus clone C6/36 of the *Aedes albopictus* cell line, dengue IgM antibody detection and immunoperoxidase and RT-PCR analysis of sera and vectors are the typical diagnostic methods used in the laboratory [32, 33, 34].

In addition to the classical symptoms of dengue fever and the seldom-observed severe shock syndrome, which is primarily associated with secondary infections, involvement of the central nervous system has been described, including encephalopathy and encephalitis [35]; however, more studies on this topic are needed. Arterial hypertension has been confirmed as an important risk factor for the development of more severe dengue clinical presentations [36].

During outbreaks, public health authorities in Brazil have standardized the use of aerosolized pyrethroid insecticides that can cause allergies. However, this measure only partially controls the mosquito population because it eliminates adult flying insects but not their breeding places. The larvicide used at these breeding sites is usually the organophosphorate Temephos, which might be slightly toxic, causing headaches, loss of memory, and irritability [37].

The Secretary of Health Surveillance in the Ministry of Health (SVS/MS) recorded 438,949 cases of dengue fever from January to July 2007, 926 cases of dengue hemorrhagic fever, and 98 dengue-related deaths. As compared with 2006, these numbers indicate an increase of 136,488 cases of dengue in Brazil, with 102,011 of these cases being reported in the month of March. Importantly, this increase in the absolute number of cases is related to the occurrence of epidemics with high incidence rates in particular states: Mato Grosso do Sul, Parana and Rio de Janeiro reported surpluses of 59,370, 39,391 and 18,181 cases, respectively.

In 2008, Rio de Janeiro experienced an epidemic of dengue caused by dengue virus types 2 (most cases) and 3. According to the Secretary of State for Health and Civil Defense, in 2008 (until March 26) 43,523 cases of dengue were recorded. Most of the cases occurred in Angra dos Reis (3141), Campos dos Goytacazes (2144), Nova Iguaçu (2643), Duque de Caxias (1075) and Rio de Janeiro (28,233). There were 54 confirmed deaths (60 are still being investigated), of which 27 occurred in children between 2 and 13 years of age [9].

The report from the Ministry of Health predicted that 16 Brazilian states might be affected by an outbreak of dengue in 2009 (Santa Catarina, Rio Grande do Sul, Sao Paulo, Rio de Janeiro, Minas Gerais, Goias, Bahia, Alagoas, Pernambuco, Rio Grande do Norte, Piaui, Tocantins, Maranhao, Para, Amazonas and Roraima). The infectologist and Superintendent of the Health Surveillance of Rio de Janeiro, Vitor Berbara, identified the state of Santa Catarina as one of the most susceptible to the three types of dengue detected in Brazil because no type of dengue has been observed there in the last 5 years [37].

2.4. Transmission of the Dengue Virus

Dengue is transmitted between people through the mosquitoes *Aedes aegypti* and *Aedes albopictus*, which are found throughout the world. Insects that transmit disease are known as vectors. Symptoms of infection usually begin 4 to 7 days after the mosquito bite and typically last 3 to 10 days. For transmission to occur, the mosquito must feed on a person during a 5-day period when large amounts of virus are in the person's blood; this period usually begins shortly before the person becomes symptomatic. Some people never display significant symptoms but can still infect mosquitoes. After entering the mosquito in the blood meal, the virus requires an additional 8 to 12 days of incubation before it can be transmitted to another human. The mosquito remains infected for the remainder of its life, which might be days or a few weeks.

In rare cases, dengue can be transmitted in organ transplants or blood transfusions from infected donors, and there is evidence of transmission from an infected pregnant woman to her fetus. However, in the vast majority of infections, a mosquito bite is responsible.

In many parts of the tropics and subtropics, dengue is endemic; that is, it occurs every year, usually during a season when *Aedes* mosquito populations are high, often when rainfall is optimal for breeding. These areas are additionally at periodic risk for epidemic dengue, when large numbers of people become infected during a short period. Dengue epidemics require a coincidence of large numbers of vector mosquitoes, large numbers of people with no immunity to one of the four virus types (DENV 1, DENV 2, DENV 3, and DENV 4), and the opportunity for contact between the two. Although *Aedes* species are common in the southern U.S., dengue is endemic in northern Mexico, and the U.S. population has no immunity, dengue transmission is not observed in the continental U.S., primarily because contact between people and vectors is too infrequent to sustain transmission.

3. MOSQUITO LIFE CYCLE

Aedes aegypti and other mosquitoes have a complex life cycle with dramatic changes in shape, function, and habitat. Female mosquitoes lay their eggs on the wet inner walls of vessels containing water. Larvae hatch (picture 1, inset) when water inundates the eggs as a result of rain or the addition of water. In the following days, the larvae (picture 2) will feed on microorganisms and particulate organic matter, shedding their skins three times to grow from the first to fourth instars. When the larva has acquired enough energy and size and is in the fourth instar, metamorphosis is triggered, changing the larva into a pupa (picture 3). Pupae do not feed; they merely change in form until the body of the adult, flying mosquito is formed. Then, the newly formed adult emerges from the water after breaking the pupal skin (picture 4, inset). The entire life cycle lasts 8-10 days at room temperature, depending on the level of feeding. Thus, there is an aquatic phase (larvae, pupae) and a terrestrial phase (eggs, adults) in the *Ae. aegypti* life cycle.

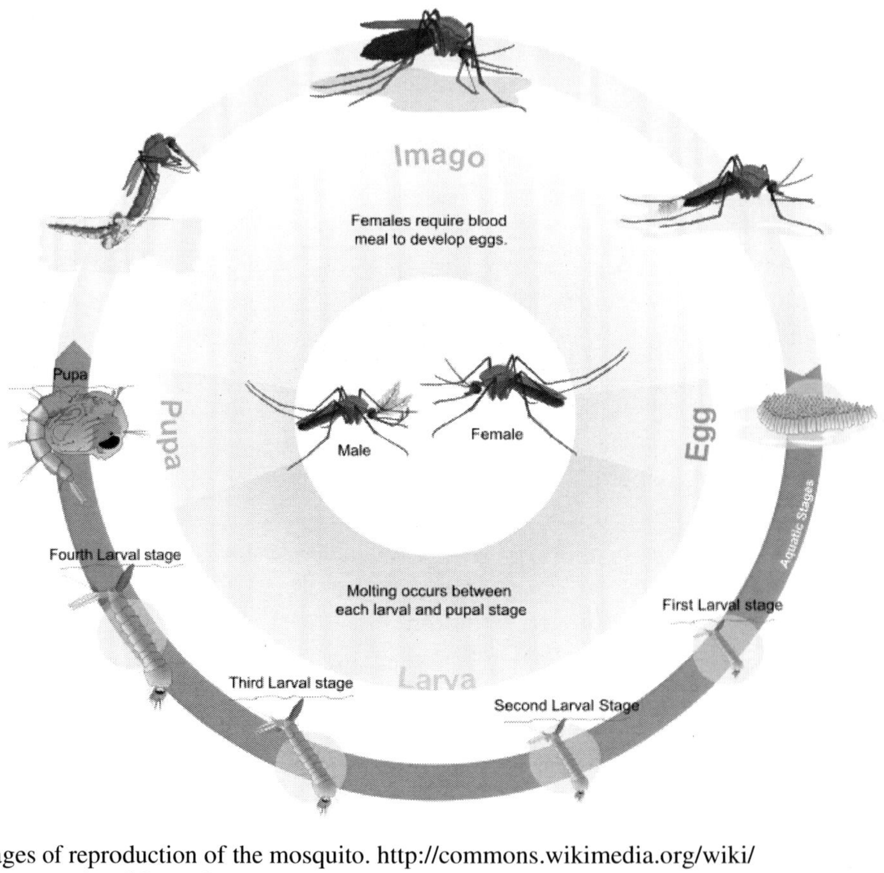

Figure 6. Stages of reproduction of the mosquito. http://commons.wikimedia.org/wiki/Media:Culex_mosquito_life_cycle_en.svg.

It is this life-cycle complexity that makes it difficult to understand the origin of mosquitoes. Similar complex life cycles with aquatic and terrestrial forms are observed in amphibians. For educational and training purposes, it is rather useful to make life cycle kits to observe how the aquatic stages turn into terrestrial ones.

4. Vector Control of Malaria and Dengue

The WHO has stated that the principal objective of vector control is the reduction of malaria morbidity and mortality by reducing levels of transmission. Vector control methods vary considerably in their applicability and cost and in the sustainability of their results.

Interestingly, malaria was close to extinction in the 20th century due to control of the *Anopheles* vector by DDT. Later, *Anopheles* became resistant, and malaria spread again with the same virulence as before. At least in Europe (Italy and the Netherlands), the DDT campaigns were successful for some time. Both *Anopheles* and *Plasmodia* developed resistance against anti-malaria pharmaceuticals, such as chloroquine.

As Massad and Bezerra Coutinho [38] described, dengue is thought to be the most important vector-borne disease, with approximately 2.5 billion people (two-fifths of the world's population) living in regions affected by dengue. There are 50 to 100 million new infections annually. According to the WHO, the incidence of dengue is increasing because of human population growth and the wider spread of mosquitoes due to climate change [38-40]. The total yearly cost of treatment in dengue-endemic areas can reach U.S. $2 billion.

Dengue control follows three main strategies [41]: source reduction (location and destruction of mosquitoes' breeding sites); use of larvicides; and use of ultralow-volume aerosolized adulticides. The first two strategies have been used with varying success [42].

However, the efficacy of current methods to control adult mosquitoes is controversial. *Ae. aegypti* was highly susceptible to DDT [42], with DDT-based control strategies resulting in the eradication of *Ae. aegypti* from 22 countries in the Americas in 1962 and from all countries in the Mediterranean region in 1972. The fate of DDT, however, is well known—this insecticide was abandoned because of mosquitoes' development of resistance and because of its environmental effects. Thereafter, control of dengue shifted to source reduction and use of larvicides and adulticides from other chemical classes.

In 1952, MacDonald [43] proposed that killing adult mosquitoes would be the most effective control strategy against vector-borne infections. Studying the dynamics of dengue [44], we showed that transmission models describing infections transmitted by haematophagous insects are indeed sensitive to mosquitoes' mortality rate. Our model showed a relative efficacy of 1 for source reduction, 15.8×10^3 for larvicides, and 28×10^3 for adulticides. Therefore, theoretically, dengue models are most sensitive to adulticides.

According to the WHO, vector control is based on three approaches. The first is integrated vector management (IVM), a rational decision-making process for the optimal use of resources in the management of vector populations to reduce or interrupt the transmission of vector-borne diseases. This approach seeks to improve the efficacy, cost-effectiveness, ecological soundness and sustainability of disease-vector control. This approach's characteristic features include the selection of proven vector control methods based on knowledge of local vector biology and ecology, disease transmission and morbidity;

utilization of a range of interventions, separately or in combination, and often synergistically; collaboration within the health sector and with other public and private sectors that influence vector breeding; engagement with local communities and other stakeholders; a public health regulatory and legislative framework; rational use of insecticides; and good management practices [45]. An IVM approach takes into account the available health infrastructure and resources and integrates all available effective measures, whether chemical, biological, or environmental. IVM also encourages an integrated approach to disease control [45].

The second strategy is indoor residual spraying (IRS). The main purpose of IRS is to reduce transmission by reducing the survival of malaria vectors entering houses or sleeping units.

According to the WHO, IRS remains a valuable intervention for malaria control when the following conditions are met: a high percentage of the structures in an operational area have adequate surfaces and can be expected to be well-sprayed; the majority of the vector population is endophilic, i.e., rests indoors; and the vector is susceptible to the insecticide being used [47,49].

The third strategy involves insecticide-treated materials (ITMs). In community-wide trials in several African settings, insecticide-treated mosquito nets have been shown to reduce all-cause mortality by approximately 20%. The use of nets has repeatedly been shown to reduce severe disease and mortality due to malaria in endemic regions.

Insecticide-treated materials include ITMs and other materials. ITMs are an effective form of vector control when coverage rates are high and a large proportion of human biting by local vectors takes place while people sleep. ITMs can also be used for personal protection. Their use has repeatedly been shown to reduce severe disease and mortality due to malaria in endemic regions.

4.1. Chemical Insecticides

Dichloro-diphenyl-trichloroethane (DDT), an organochlorine developed during the Second World War, which, when applied to the walls and ceilings of houses, remained active against insects for several months [49].

The other organic compounds belong mainly to groups of organophosphate.

Chemical control with organic or inorganic pesticides is one of the methodologies adopted as part of sustainable management and integrated vector control in public health [48]. The development of insecticides that remain active for long periods was a major advance in insect control in the 20th century. The first insecticide with a prolonged effect, or residual property, was s, carbamates and pyrethroids. All of these groups act on the central nervous system of insects and have been used in programs to control diseases transmitted by vectors [50-52].

4.2. Organochlorines

The organochlorines are insecticides that contain carbon, hydrogen and chlorine. These insecticides are classified into four groups: diphenyl aliphatic; hexachlorocyclohexanes; cyclodienes; and polychloroterpenes [53].

The group of organochlorine diphenyl aliphatics is the oldest. It includes DDT, probably the most notorious chemical of the last century. In 1948, the Swiss entomologist Paul Muller was awarded the Nobel Prize in Medicine for his discovery of the usefulness of DDT in vector control of malaria, yellow fever and many other diseases [53]. Although the mode of action of the insecticide has never been clearly established, it is known that it acts on the sodium channel, probably by keeping it open and destroying the balance of sodium and potassium ions in the axons, thereby preventing normal transmission of nerve impulses in insects and mammals. The effect of DDT is inversely proportional to temperature, with more toxic effects being observed at lower temperatures [53].

Benzene hexachloride (BHC), in the group of hexachlorocyclohexanes (HCH), also marketed under the name of Lindane [53], has an action similar to that of DDT.

The cyclodiene insecticides, such as chlordane, aldrin and dieldrin, emerged after World War II. Most of them are persistent and stable in the soil and relatively stable when exposed to sunlight or ultraviolet light. Given these characteristics, they were used primarily as insecticides for control of termites and other insects whose larval stages feed on plant roots [53]. They inhibit gamma-aminobutyric acid (GABA) receptors, which, after binding of the neurotransmitter, increase the permeability of the neuron to chloride ions. The cyclodienes block entry of chloride ions into neurons, acting as antagonists of the effects of the GABA receptor. Unlike DDT and HCH, the cyclodienes show a positive correlation between temperature and toxicity.

Only two polychloroterpenes, with modes of action that are equivalent to the cyclodienes, were developed: toxaphene (in 1947) and estrobane (in 1951). These compounds came to be used most intensively in agriculture [53]. Despite being widely used by programs aiming to control malaria, organochlorines were discontinued and even banned in several countries, due to their persistence in the environment and accumulation in the body tissues of animals and humans [52, 54-56]. These prohibitions and restrictions referred to the agricultural use of DDT, although it remains indicated by the WHO for use in controlling vectors [57, 58].

The use of DDT in public health campaigns was discontinued, mainly due to enduring national and international pressure combined with aggressive marketing tactics by producers of other, more expensive insecticides [59]. The publication of research reporting the presence of DDT in breast milk and its association with the occurrence of cancer in humans [60-62] led the WHO to order a committee of experts to review the literature on the insecticide. The conclusions of this committee, presented in 1993, were to the effect that "due to lack of sufficient and convincing evidence of adverse effects from exposure to DDT for residual applications for vector control, there is no toxicological or epidemiological reasons to change the policy to use the product in the control of malaria and leishmaniasis"[57]. The document, however, suggests that countries that have replaced DDT with other insecticides consider using it again in the future.

Currently, many researchers working in malaria control support the use of DDT in countries with high transmission and insufficient resources to purchase another product with a similar effect [59, 63]. During a meeting of the Inter-governmental Committee on Persistent Organic Pollutants (CPOP) in South Africa at the end of 2000, which had the purpose of establishing an international agreement that would allow implementing actions relating to persistent organic compounds (POCs), DDT was not included on the list of restricted POCs. These recommendations should survive until the establishment of a replacement policy in the countries that continue to use DDT. To that end, deadlines and financial support are necessary because, generally, countries that continue to use DDT do not have the resources to replace it with a different compound.

4.3. Organophosphates

The generic term "organophosphate" (OP) includes all insecticides containing phosphorus. These insecticides were later discovered to be superior to organochlorines [64]. The group of organophosphorus insecticides is classified into three subgroups: the aliphatics (e.g., malathion, dichlorvos, bidrin); phenyl derivatives (e.g., ethyl and methyl parathion, fenitrothion); and heterocyclics (e.g., chlorpyrifos, chlorpyrifos-methyl). These insecticides are widely used in public health because of their many advantages over organochlorines, such as biodegradability and lack of bioaccumulation. However, these insecticides' main disadvantage is their chemical instability. Moreover, these insecticides are more toxic to vertebrates than organochlorines, even at relatively low doses [52, 53].

The organophosphate temephos was registered in the U.S. in 1965 for use in agriculture and mosquito control and is the only larvicide in the group with widespread use in the control of mosquito larvae. It is recommended by the WHO for use in potable water [58].

Organophosphates act by inhibiting acetylcholinesterase (AChE), an important enzyme in the central nervous system. This enzyme is phosphorylated by the insecticide, thereby becoming irreversibly inactivated. Inhibition of AChE results in accumulation of acetylcholine between neurons (or in the synapses), which prevents the interruption of propagation of the electrical impulse. Consequently, the central nervous system is continuously stimulated, triggering a process of paralysis that can lead to the death of the insect.

4.4. Carbamates

Carbamates are insecticides derived from carbamic acid that were first marketed in the 1960s. Among the most used carbamates is carbaryl. Similar to organophosphates, carbamates have rapid lethal action on insects but short residual power. They are systemic to plants and are relatively water-soluble. Organophosphorous compounds also inhibit acetylcholinesterase; though, in this case, the reaction involved is carbamylation. Although both act in a similar manner in biological systems, carbamates differ from organophosphates in two ways. First, some carbamates are potent inhibitors of aliesterase (an aliphatic whose exact function is unknown) and have pronounced selectivity against certain species of AChE. The second difference is that the inhibition of AChE by carbamates is reversible [53].

4.5. Pyrethroids

The synthetic pyrethroids are stable compounds that are produced in the laboratory, starting from a natural substance, pyrethrum, extracted from the chrysanthemum. These compounds are biodegradable, non-bioaccumulative and rarely cause acute poisoning in birds and mammals, although they may cause sensitization and irritation of mucous membranes in these animals. For aquatic animals, however, they are extremely toxic. Pyrethroids also have the advantage of being active in small doses; their only drawback is high cost [53].

Pyrethroids developed in an interesting way and are divided into four generations. The first consists of a single product, allethrin, launched in 1949, the synthesis of which was complex and involved 22 chemical reactions to obtain the final product. The second generation includes tetramethrin (1965), resmethrin (1967), bioresmethrin (1967), bioallethrin (1969) and phenothrin (1973). The third generation includes fenvalerate and permethrin, which appeared in 1972-1973 and were the first pyrethroids used in agriculture, due to their high insecticidal activity and photostability. The fourth and current generation is highly effective at low doses and includes bifenthrin, lambda-cyhalothrin, cypermethrin, cyfluthrin, deltamethrin, esfenvalerate, fenpropathrin, flucythrinate, fluvalinate, prallethrin, tau-fluvalinate, tefluthrin, tralomethrine and zeta-cypermethrin, all of which are photostable [53].

Pyrethroids present a mode of action similar to DDT by maintaining open sodium channels in neuronal membranes. There are two types of pyrethroids: those that have a negative temperature coefficient, resembling DDT (type 1), and those with a positive temperature coefficient (type 2), i.e., the mortality of insects exposed to them varies directly with temperature. Pyrethroids affect the central and peripheral nervous systems of the insect, encourage nerve cells to produce repetitive discharges and eventually cause paralysis. The stimulating effect of pyrethroids is much more pronounced than that of DDT [53].

4.6. Resistance to Insecticides

Insecticides have been used extensively both in agriculture and in the area of public health. The continued use of insecticides has caused the emergence of resistant populations and caused problems for vector control. Resistance has been detected for all classes of insecticides with direct and profound effects on the re-emergence of vector-borne diseases [65]. Despite the substantial progress made in developing alternative methods of vector control, chemical insecticides remain an important tool of integrated control programs [48]. In this context, monitoring and resistance management and the use of substances with different modes of action from conventional chemical insecticides are very important elements in any vector control program [65].

Resistance is defined as the ability of a population of insects to tolerate a dose of insecticide that, under normal conditions, would cause death. Physiological resistance is a genetic trait, similar eye color in humans. Thus, populations of insects, mites and other arthropods can include a proportion of individuals who have alleles that confer resistance to a particular chemical. Resistant strains can arise as a result of the persistent use of pesticides that kill individuals with susceptible alleles and do not kill those who have resistant alleles.

Resistance to insecticides can considered as a process of accelerated evolution in a population that responds to intense selective pressure, resulting in the survival of individuals with alleles that confer resistance. Pre-adaptive resistance is the result of random mutations. Thus, a small number of individuals possess characteristics that allow their survival under normally lethal doses of insecticides. The insecticide itself does not produce a genetic change, although its continued use can select resistant individuals. Despite the many documented studies on insecticide resistance, the number of mechanisms involved is quite small and includes reducing the rate of penetration through the cuticle, increased metabolic detoxification and decreased sensitivity of the target site. All of these mechanisms are nonspecific and often confer cross-resistance to structurally related insecticides [66].

4.7. Reduction of the Rate of Penetration of the Insecticide

This mechanism of resistance is observed in various insects and has been studied mainly in the housefly. Compared with other mechanisms, it is considered to be a minorsz cause of resistance [67]. However, in combination with other mechanisms, it can result in an increase in resistance. The biochemical means of reducing the rate of penetration of an insecticide is not clear, although it has been proposed that the protein composition of the integument plays an important role.

4.8. Metabolic Resistance

There are many reported cases of resistance resulting from the increased capacity to metabolize insecticides into less toxic products. The increase in metabolism may be the result of modifications of the enzyme that breaks down insecticides to enhance the production of enzyme molecules that are already available in susceptible insects, though in small quantities [65, 66]. Generally, the enzymes involved in the metabolism of xenobiotics are divided into Phases 1 and 2. Phase 1 enzymes recognize and act directly on the exogenous compound, introducing a reactive polar group into the molecule, which gives rise to the substrate for Phase 2 enzymes. The detoxification contributes to the insecticide resistance achieved by Phase 1 enzymes, such as monooxygenases and esterases, or Phase 2 enzymes, such as glutathione S-transferases.

4.9. Monooxygenases

Increased oxidative metabolism has been implicated as a mechanism of resistance to all classes of insecticides except cyclodienes and is one of the most common mechanisms of resistance. The enzymes associated with oxidative metabolism increases are monooxygenases that are dependent on the cytochrome P450 [68-69] and are concentrated in the intestine, the fat body and the Malpighian tubules of insects. Initial evidence of the involvement of resistance to oxidative metabolism might be obtained by the addition of oxidase inhibitors (which act here as synergists), such as piperonyl butoxide, in bioassays with insecticides and by evaluating the ability of the synergist to reduce the magnitude of resistance. Although this technique is not conclusive, it is valuable as a first indication that the increased activity of monooxygenases contributes to resistance [66, 67].

4.10. Esterases

Several insecticides with chemical structures that contain ester groups (organophosphates, carbamates and pyrethroids) are susceptible to hydrolysis reactions. The enzymes that catalyze these reactions are collectively called hydrolases or esterases. Also known as carboxylesterases, these enzymes hydrolyze compounds containing carboxylic esters in their structure, such as most organophosphates, which are esters of phosphoric acid [65, 70]. The involvement of esterases in resistance can be demonstrated by the addition of inhibitors to assays containing insecticides, such as S,S,S-tributyl phosphorothioate (DEF) [65, 70].

4.11. Glutathione S-Transferases (GST)

Glutathione S-transferases (GST) are multifunctional enzymes that can detoxify a number of xenobiotics. These enzymes catalyze the conjugation of the hydrophilic group SH of reduced glutathione (GSH) to the electrophilic center of lipophilic compounds [71]. Various forms of these enzymes have been identified in mosquitoes, houseflies, *Drosophila* and other insects [72-74].

4.12. Changing the Target Site

All of the chemical insecticides described here operate in the central nervous system at different target sites. An insect population might become resistant, due to the alteration of these sites to prevent or hinder interactions with the insecticides.

4.13. Acetylcholinesterase (AChE)

Acetylcholinesterase (AChE) is the target site of organophosphates and carbonates [71]. This enzyme is responsible for the degradation of acetylcholine, a neurotransmitter that is present in the synaptic cleft and promotes the propagation of nerve impulses and the opening of sodium channels in the cell upon synaptic excitation. Under normal circumstances, the cessation of the stimulus is marked by acetylcholine reuptake or enzymatic degradation through the enzyme acetylcholinesterase. Organophosphates and carbamates inhibit AChE; therefore, acetylcholine remains in the synaptic cleft and the momentum never stops, causing the insect to die. In these cases, resistance is associated with reduced affinity for the AChE insecticide, which allows the normal interruption of nerve stimulation [66, 71, 75].

4.14. Sodium Channels

Sodium channels are the primary target sites of pyrethroids and organochlorines, such as dichlorodiphenyltrichloroethane (DDT) and hexachlorocyclohexane (BHC). These insecticides keep sodium channels open and cause continuous nerve impulse propagation, which results in the death of the insect. Resistance is the result of reduced sensitivity of the sodium channel to these compounds [53, 71, 76]. Best known as a knockdown mechanism (or KDR), this type of resistance to DDT and pyrethroids has been recorded in various species [71] but is best characterized in houseflies, for which several variants of KDR are described, including the super-factor KDR [78]. Because of the cross-resistance between DDT and pyrethroids, biological assays that identify resistance to both classes of insecticides might be useful as an indicator of KDR resistance.

4.15. Gamma-Aminobutyric Acid (GABA) Receptors

The gamma-aminobutyric acid (GABA) receptors are the target sites of polychloroterpenes and cyclodienes (organochlorine). They belong to a superfamily of receptors present in synaptic junctions of the central nervous system and neuromuscular synapses of insects [71]. Insecticides act on GABA receptors, preventing the entry of chloride ions to the intracellular environment and thereby causing the emission of spontaneous impulses that lead to twitching, convulsions, paralysis and death. The resistance caused by GABA receptor insensitivity to insecticides has been associated with the mutation of a single base pair, resulting in an amino acid substitution (alanine for serine) [71].

5. Control of Mosquitoes by Insecticides and Bioinsecticides

In addition to chemical insecticides, other products have been used in vector control, such as biological insecticides and growth regulators. Pathogenic bacteria are an example of biological insecticides and have been used in agricultural pest control for nearly five decades. In 1970, bacteria were demonstrated to be effective against insects of medical importance, especially mosquitoes and black flies.

In 1964, a strain of *Bacillus sphaericus* (Bs) with effectiveness against mosquito was described [68]. Although its spectrum is restricted to certain types of mosquito larvae its effectiveness in polluted waters made Bs a particularly useful agent against species of *Culex* which are vectors for filariasis and viral encephalitis. The use of Bs in programs to control mosquitoes that thrive in polluted water in urban areas has increased over the last decade [88].

Another bacterium, *Bacillus thuringiensis israelensis* (Bti), was so effective that it has become a major component of the Onchocerciasis Control Program in West Africa. This bacterium is currently used as an alternative to synthetic chemical insecticides in many mosquito control programs [82].

The spores of the entomopathogenic bacteria Bti and Bs present crystals that contain pro-toxins. *Culicidae larvae* consume the crystals, which are dissolved in the alkaline gut. The digestive proteases cleave the pro-toxins present in the crystals and activate the insecticide component of the molecule. The resulting toxic peptides act on the intestinal epithelium of the immature forms of the vector, promoting the reduction of peristalsis and, consequently, the death of the larva [83, 84]. The WHO recommends Bti as a larvicide for use in drinking water to control *Aedes aegypti* larvae [58].

Bti-based products have been used in programs to control mosquitoes and black flies for over 20 years. Nevertheless, resistance has never been reported, probably because several proteins with insecticidal activity are present in these crystals. In the case of Bs, with a main active ingredient being a single toxin, resistance has already developed in some populations in

India, Brazil and France. Thus, the potential for developing resistance to more widespread use of Bs is high. Recent studies have shown, however, that this resistance can be controlled through a rotation scheme including both Bs- and Bti-based products or other larvicides [80-82].

6. Biosafety in Parasitology Labs

The manipulation of pathogenic organisms comprises one of numerous activities of professionals working in research laboratories or healthcare facilities. Every year, new diseases are discovered, new microbes are manipulated and more professionals come into contact with biological agents. The increasing need for rapid and accurate diagnosis must undoubtedly be accompanied by a structure that ensures safety in the handling of biological agents. Issues such as the proper use of equipment, control of accidents and inspection routines for risky activities are still not routinely addressed by the heads of laboratories and should be better supervised.

Individuals who work in clinical and research laboratories, or even professionals who care for patients, are at high risk for contact with known and unknown parasites through accidental exposure. Unfortunately, many professionals who handle parasites often do not know the clinical manifestations, monitoring protocols and treatment approaches after accidental exposure. Because of these uncertainties and the severity of some diseases caused by parasites, the first reactions after an accident are frequently anxiety and confusion [85].

It is important to remember that professionals who become contaminated in their work environment are capable of transmitting the disease to family and friends. When a professional immediately reports the occurrence of an accident at work, containment procedures can be implemented quickly to avoid contamination of the external environment. Moreover, a comprehensive record of accidents enables the laboratory or institution to identify the cause of an accident and develop programs to prevent the accident from happening again.

In recent years, scientific knowledge and practical control of biological hazards have evolved quickly. Accordingly, research has been undertaken to determine the extent of the problem, locate sources of contamination and develop standards that will ensure security and protection for the exposed individual. Since the development of biotechnology in the 1970s, the daily lives of people who come into contact with pathogenic organisms have changed in many ways. New products and techniques have been created, and a number of different activities have become part of the daily routines of researchers and health care professionals. To minimize the risks involved in laboratory activities and prevent accidents, compliance and biosafety procedures are essential and should be addressed in any laboratory handling pathogenic organisms.

7. Biosafety: Concept and Importance

Biosecurity is the application of knowledge, techniques and equipment to ensure the safety of a laboratory and the individuals who work there and to avoid environmental release of potentially infectious agents. Biosafety is defined as the containment of infectious agents that can be handled safely [86]. The main objectives of biosafety standards are to highlight the importance of security measures in day-to-day research and to avoid accidents when proper safety measures are taken.

In a parasitology laboratory, professionals handle potentially infectious biological agents; therefore, it is essential that the environment be controlled. Some of these microorganisms are pathogenic, which makes infection prevention a key competence of each professional. It is necessary to protect not only the professionals' subordinates and colleagues, but also the material used to avoid cross-contamination.

It is also necessary to pay attention to other risks associated with the work environment, such as handling of chemicals, physical factors, radioactivity, ergonomic hazards, fire and other factors that, if not taken seriously, can cause severe accidents. Safety precautions should be part of the routine work of a laboratory without hampering research activities [87].

8. Danger and Risk

Any laboratory involved in health-related research is exposed to danger and risk. Professionals in health-related fields should be focused on the basic principle of biosecurity: prevention.

When a certain activity could be dangerous for the person performing it, personal protective equipment (PPE), which is designed to provide security for workers, must be used. PPE is also necessary to meet biosafety standards and minimize risks. The use of PPE and biosafety are necessarily associated with appropriate training to avoid risky situations (Figure 1). It is not easy to identify risky situations when working under stressful conditions or when anxious or distracted. In these situations, the worker should stop the activity until he or she is able to focus again [88].

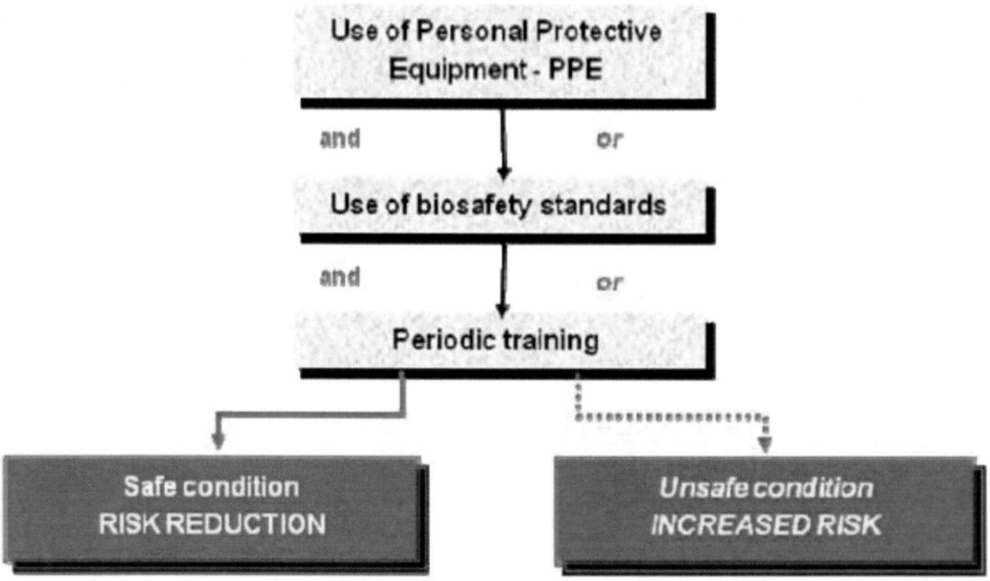

Figure 7. Factors that create safe or unsafe conditions in the workplace.

9. Clinical Laboratories

Clinical laboratories often receive contaminated materials for clinical diagnosis. Typically, the infectious nature of the clinical material is unknown, and samples are frequently subjected to a series of microbiological tests to identify multiple agents. The coordinator of the laboratory is responsible for compliance with standards and procedures to ensure sufficient security in handling these agents. Primary barriers, such as Biological Safety Cabinets (BSC) of classes I and II, should be used when working with dripping or spray dopant material [89].

10. Protective Equipment

Protective devices are designed to provide security during laboratory activities and minimize exposure to individual risks. Protective devices can be divided into Personal Protective Equipment (PPE) and Collective Protection Equipment (CPE). PPE includes a face shield, lab coat, gloves, respirators, footwear, eye protection, and goggles. The CPE includes BSC, emergency shower, and eye wash stations. Both PPE and PPCs are used as physical barriers to protect against pathogens and other risks associated with the work environment.

Protective equipment should not be required for routine work, but it is essential that professionals are given time to adapt to the use of the equipment until safety is ensured [90]. It is worth remembering that there is no PPE or PPC that can ensure full security against any agent by itself. These devices may create a false sense of security and encourage carelessness, thereby increasing risk. Therefore, safety practices should be designed, instituted, maintained and managed according to the manufacturer's recommendations.

11. Laboratory Accidents Involving Parasites

Although the risk of infection after exposure remains poorly defined, surveys of laboratory-acquired infections suggest that the *Brucella*, *Shigella*, and *Salmonella* species, *Mycobacterium tuberculosis*, and *Neisseria meningitidis* are the most common causes [91]. Laboratory-acquired parasitic infections are uncommon in diagnostic microbiology laboratories [92-94]. However, approximately 313 cases of laboratory-acquired infections involving a variety of blood and intestinal protozoa have been reported [85, 93].

The risk of contamination to individuals who work in parasitology laboratories depends on several factors, such as the stage of the parasite life cycle, incubation period, species and lab safety [85, 95]. Because accidents in laboratories are rare, there is no way to determine accurately the number of people involved. In 1979, Pike conducted a study that showed that the incidence of contamination increases with the discovery of new agents, with more people becoming infected through the handling of infectious agents [90]. Out of an estimated 4,079 accidents involving infection, 116 (3%) were caused by 17 different parasites. Of these accidents, only 64% were documented [90]. After accidental exposure to a parasite agent, the exposed person should be monitored through clinical and laboratory diagnosis. Although most parasite infections are usually diagnosed by conventional microbiological methods, more precise molecular biology techniques, such as polymerase chain reaction (PCR), exist [85]. The lack of a culture to establish prevention as the main tool to control the risks associated with work in research laboratories has been the main obstacle in reducing accidents. Many scholars and interns are not mandated to undergo training and therefore perform duties without being familiar with safety procedures, increasing their risk. The culture of not reporting accidents is another factor that complicates the establishment of biosafety policies for maintaining a controlled environment.

Tables 1 and 2 describe some parasites that cause diseases that are commonly acquired in laboratories [85].

Table 1. Parasites associated with accidents in parasitology laboratories (Adapted from HERWALDT [86])

Parasite	Routes of exposure	Stage of infection	Protective measures	Clinical manifestations[a]
Blood and tissue protozoa				
Acanthamoeba spp.	Sore, eyes	Trophozoite, Cyst	Gloves, mask, lab coat, BSC, precautions against percutaneous injuries	Headache, abscess lesions, conjunctivitis, keratitis
Babesia spp.	Sore, needle, vector	Sporocyst	Gloves, precautions against percutaneous injuries	Fever, anemia, fatigue
Leishmania spp.	Sore, needle, vector	Amastigote, promastigote	Gloves, precautions against percutaneous injuries	Nodules, ulcers, hepatosplenomegaly, fever
Naegleria fowleri	Mucosa, aerosol	Trofozoíto	Gloves, mask, lab coat, BSC, precautions against percutaneous injuries	Headache, neck tightening, coma
Plasmodium spp.	Sore, needle, vector	Erythrocyte stages, sporocyst	Gloves, precautions against percutaneous injuries and wounds	Fever, fatigue, anemia, chills
Sarcocystis spp.	Oral	Oocyst or sporocyst	Gloves and hand washing	Fever, fatigue, anemia, chills
Toxoplasma gondii	Oral, sore, needle, mucosa	Oocyst, bradyzoites	Gloves, hand washing precautions against percutaneous injuries	Adenopathy, fever
Trypanosoma cruzi	Sore, needle, mucosa, vector	Trypomastigote	Gloves, hand washing precautions against percutaneous injuries	Fever, rash, adenopathy, redness and/or swelling at the infection site
Protozoa of the intestine[b]				
Cryptosporidium parvum	Oral, mucosa	Oocyst	Gloves, hand washing precautions with the mucosa	Symptoms of gastroenteritis
Entamoeba histolytica	Oral	Cyst	Gloves, mask and hand washing	Symptoms of gastroenteritis
Giardia lamblia	Oral	Cyst	Gloves, mask and hand washing	Symptoms of gastroenteritis

[a] Clinical manifestations may be variable and will depend on such factors as the species of parasite, the amount of inoculum and the stage of infection.

[b] Possibility of contamination by aerosol inhalation or by droplets of contaminated material.

Table 2. Parasites involved in accidents in parasitology laboratories (Adapted from HERWALDT [86])

Parasite	Routes of exposure	Stage of infection	Protective measures	Clinical manifestations[a]
Isospora belli	Oral	Oocyst	Gloves, mask and hand washing	Symptoms of gastroenteritis
Other protozoa				
Microsporidian spp.	Oral, eyes, mucosa[c]	Spore	Gloves, mask, gown, hand washing, precautions against percutaneous injuries	Skin ulceration, cystitis, diarrhea, pneumonitis
Helminths				
Ascaris lumbricoides	Oral	Eggs	Gloves, mask and hand washing	Cough, fever, abdominal cramps, diarrhea, constipation
Enterobius vermicularis	Oral	Eggs	Gloves, mask, hand washing, cleaning of the nails	Perianal itching
Fasciola hepatica	Oral	Metacercaria	Gloves, mask and hand washing	Jaundice, increased transaminase levels, biliary colic
Hymenolepis nana	Oral	Eggs	Gloves, mask and hand washing	Diarrhea and abdominal pain
Schistosoma spp.	Percutaneous[d]	Cercaria	Gloves, lab coat and hand washing	Dermatitis, fever, cough, adenopathy, hepatosplenomegaly
Strongyloides stercoralis	Percutaneous[d]	Larva	Gloves, lab coat and hand washing	Chest pain followed by abdominal pain and cramping
Taenia solium	Oral	Eggs, cysticercus	Gloves and hand washing	Cysticercosis: neurological symptoms Worm: usually asymptomatic, but may cause mild abdominal pain
Trichinella spiralis	Oral	Larva	Gloves, mask and hand washing	Abdominal pain and muscle
Trichuris trichiura	Oral	Eggs	Gloves, mask and hand washing	Abdominal pain

[c] Some species of *Microsporidia* have been recognized as pathogenic to immunocompromised individuals, especially AIDS patients, and have occasionally been responsible for causing disease in healthy people.

[d] Parasites that can penetrate intact skin.

Blood and Tissue Protozoa

Plasmodium spp., *Trypanosoma* spp. and *Leishmania* spp. are among the microorganisms handled in parasitology laboratories that are often associated with laboratory accidents. Although not commonly associated with laboratory infections, transmission through mosquitoes and malaria have also occurred [89]. Other direct sources of infection for people who work in health laboratories include contact with contaminated material extracted from rodents with cutaneous leishmaniasis and contact with the feces or blood of animals or insects experimentally or naturally infected with *T. cruzi* [95]. No cases of laboratory-associated infection with *Babesia* spp. have been described; transmission can occur only through contact with infected ticks or blood from infected animals or humans [85].

Infections can be transmitted through the blood, feces, cerebrospinal fluid, bone marrow or other tissue biopsies and infected arthropods. Depending on the parasite, the precautions include primary care to avoid ingestion, skin penetration through wounds, accidental parenteral inoculation and transmission by arthropod vectors. When cultures of *Leishmania* spp. are handled, precautions must be taken to avoid aerosol exposure to the mucous membrane of the eyes, nose and mouth. [89] Several authors have also reported laboratory-associated infections with *Toxoplasma gondii, Entamoeba histolytica, Isospora belli, Giardia lamblia, Sarcocystis* spp. and *Cryptosporidium* spp. [93,95]. In rodents inoculated with *Toxoplasma*, contact with the peritoneal fluid might result in exposure to infectious organisms. Infections are frequent in laboratories studying *Cryptosporidium*. Evidence suggests that oocysts can infect humans through the air [89]. The infectious stages can be found in the stool, body fluids and tissues. The transmission of trophozoites by aerosol to the mucosa of the eyes, nose or mouth can cause harm when manipulating cultures of amoebae, such as *Naegleria fowleri*, *Acanthamoeba* and *Balamuthia*, but the level of risk is still unknown. Women who may become pregnant should receive intensive training, due to the high risk of toxoplasmosis in the fetus. The manipulation of oocysts involves the potential risk of contamination. Infection with tachyzoites also presents a high risk of contamination through skin abrasions or mucosa [89].

Human Intestinal Protozoan Parasites

Potential protozoan parasites of the human intestine, such as *Entamoeba histolytica, Giardia lamblia, Cryptosporidium parvum, Isospora belli* and *Cyclospora cayetanensis*, are associated with infections caused by accidents in laboratories [85]. Infections associated with animal testing laboratories have been reported, and persons who come in contact with the feces of naturally or experimentally infected animals may become infected directly. In the case of rodents inoculated intraperitoneally with *Toxoplasma*, contact with the peritoneal fluid might result in exposure to infectious organisms [89].

Special attention should be directed to activities involving the handling of material contaminated with feces, including habitual hand washing after contact with such material. Even previously stored species must be handled carefully because some microorganisms might have been stored improperly, creating the potential for contamination [85].

Cryptosporidium infections occur regularly in almost all laboratories working with this agent [89]. If skin contamination occurs, thorough washing with water is necessary because there are no effective disinfectants against *Cryptosporidium, Isospora* and *Cyclospora* [85].

Nematode Parasites of Humans

There are a small number of reports of infections involving helminths in associated laboratories. The lack of information is related in part to the fact that helminth infections are less likely to occur compared with protozoan infections [85]. The most common infections associated with nematodes include laboratories working with *Ascaris lumbricoides, Strongyloides stercoralis* and *Enterobius vermicularis* [95, 96]. Allergic reactions to various antigenic components of nematodes might represent a risk to sensitive individuals. There have been no reported infections associated with laboratory animals. The eggs and larvae in fresh feces are not generally definitive host vectors because the development of infectious stages varies in duration from one day to several weeks. Special care should be taken if *Trichinella* is ingested, especially when present in tissues containing larvae [89].

Cestode Parasites of Humans

Although no incidents of infection have been reported in *Echinococcus granulosus* or *Taenia solium* laboratories, the consequences of infection through ingestion of the eggs are extremely harmful and deserve attention. *Hymenolepis nana* is a cosmopolitan parasite that does not require an intermediate host and is transmitted directly by ingestion of the feces of infected humans. Humans may become infected with *Echinococcus granulosus* eggs from the feces of infected dogs or other canids, which can act as definitive hosts of this parasite. Thus, people who work in research laboratories or veterinary diagnostic laboratories are more susceptible to exposure to these agents [85]. Special attention should be paid to personal hygiene practices, particularly hand-washing [89].

Trematode Parasites of Humans

Several laboratories have reported *Schistosoma* spp. and *Fasciola* spp. infections that were not associated with laboratory animals [96]. People who work in research laboratories that handle snails should take special precautions because snails are the intermediate hosts of these parasites [89]. The infectious stages of *Schistosoma* spp. (Cercaria) and *Fasciola* spp. (metacercariae) can be found in water and aquatic plants present in the aquariums of laboratories that maintain these intermediate hosts. The metacercariae can be transferred from hand to mouth through contact with fingers following contact with contaminated aquatic vegetation in an aquarium. Most incidents of *Schistosoma* spp. contamination occur in laboratories after contact with cercariae that have the potential to cause disease. The snails and cercariae contained in the aquarium water and vegetation should be decontaminated using chemicals (hypochlorite) or boiled before being discarded [89]. Gloves should be used at all times to prevent skin contact with the parasites [85].

13. Good Laboratory Practices

Various infections acquired through accidental exposure in laboratories are associated with inadequate safety practices. Surely, prevention is always the best approach, and preventive measures should be implemented when designing an effective treatment after an accident involving contaminated material. To minimize the risk of accidents, all professionals working in parasitology laboratories should adopt the standards of biosecurity described for BSL-2 laboratories, which are based on microbiological practices using appropriate PPE and EPC for each type of activity [85]. Good laboratory practices also extend to health services and refer to technical standards and work procedures that aim to minimize and control workers' exposure to the risks inherent in their activities. The application of best practices is essential to worker, production and environmental safety and should be part of routine work. The use of good practices should be collective and not only individual. All workers must receive continual updates on safety techniques; failure to implement these techniques will surely create a risk not only to the individual but to all who work in that environment [97]. The following are examples of technical standards and procedures that provide safety in research laboratories and health services and facilitate the work routine [89, 97].

- Keep your hands clean, and always wash them at the end of the work day.
- Never pipette by mouth. Always use automatic pipettors, rubber pipette bulbs and cotton filters.
- Never prepare food in the laboratory.
- Do not store food or personal items in the laboratory.
- Always wear closed-toed shoes that protect your feet completely.
- Try to avoid the formation of aerosols and splashes. When necessary, use the BSC.
- Do not open the door or answer the phone while wearing gloves.
- Always wear a coat in the laboratory. When you need to leave the lab, remove the coat and keep it in the laboratory.
- The bench tops should be disinfected before and after routine work.
- Focus when handling chemicals or pathogens and avoid talking to coworkers.
- Try to avoid the use of sharps.
- When it is not possible to avoid the use of needles, dispose of them immediately after application. Never try recapping a needle.
- Avoid working alone in the lab, especially at night.
- When you become sleepy, stop your activity and try to rest and return to work at another time.

Conclusion

Individuals working in laboratories involving biological materials must receive training in biosafety, regardless of their level of research training. Coupled with the theoretical and practical knowledge acquired concerning biosafety, two elements are essential to ensure the quality of the system: prevention and a collective vision of security. If these requirements are met, the risk of accidents will be reduced substantially, and the quality of the work environment will be maintained.

References

[1] Cox, F. E. G. (2002) History of human parasitology. *Clin. Microbiol. Rev.* 15, 595-612.
[2] Goklany, I. M. (2004) Climate change and malaria. *Science* 306, 55-56.

[3] Ito, J., A. Ghosh, L. A. Moreira, E. A. Wimmer, and M. Jacobs-Lorena (2002) Transgenic anopheline mosquitoes impaired in transmission of a malaria parasite. *Nature* 417, 387-388.

[4] Anonymous (a). Malaria drug offers new hope. 22-3-2007. *BBC News*. Ref. Type: Electronic Citation.

[5] Trampuz, A., M. Jereb, I. Muzlovic, and R. M. Prabhu (2003) Clinical review: Severe malaria. *Crit. Care* 7, 315-323.

[6] Greenwood, B.M., K. Bojang, C.J.M. Whitty, and G.A.T. Targett (2005) Malaria. *Lancet* 365, 1487-1498.

[7] Anonymous (b). Malaria. Geographic Distribution. 2007. Ref Type: Electronic Citation.

[8] WHO Dengue Bulletin – Volume 34, 2010.

[9] Cives - Centro de Informação em Saúde para Viajantes 2002. Dengue. Available at: www.cives.ufrj.br/informacao/dengue/den-rio2008.html. Accessed on 03 abril.

[10] Cavalcanti, E.S.B.; Morais, S.M.; Lima, M.A.A.; Santana, E.P.W. 2004 Larvicidal Activity of essential oils from Brazilian plants against *Aedes aegypti* L. Mem. Inst. Oswaldo Cruz v.99 n.5 Rio de Janeiro.

[11] Kurane, I. The Effect of Global Warming on Infectious Diseases. Public Health Res Perspect 2010 1 (1),4 e 9 doi:10.1016/j.phrp.2010. 12.004.

[12] Hales S, de Wet N, Maindonald J, Woodward A. Potential effect of population and climate changes on global distribution of dengue fever: an empirical model. *Lancet* 2002;360:830e4.

[13] Hopp MJ, Foley JA. Worldwide fluctuations in dengue fever cases related to climate variability. *Clim. Res.* 2003;25:85e94.

[14] Githeko AK, Ndegwa W. Predicting malaria epidemics in the Kenyan highlands using climate data: a tool for decision makers. *Glob. Change Hum. Health* 2001;2:54 e 63.

[15] Abeku T, van Oortmarssen G, Borsboom G, et al. Spatial and temporal variations of malaria epidemic risk in Ethiopia: factors involved and implications. *Acta Trop.* 2003;87:331 e 40.

[16] Khasnis AA, Nettlelman MD. Global warming and infectious disease. *Arch. Med. Res.* 2005; 36: 689–96.

[17] Semenov SM, Gelver ES, Yasyukevich VV. Temperature conditions for development of two species of malaria pathogens in Russia in 20th century. *Dokl. Akad. Nauk.* 2002;387:131e6.

[18] Hay SI, Rogers DJ, Randolph SE, et al. Hot topic or hot air? Climate change and malaria resurgence in East African highlands. *Trends Parasitol.* 2002;18:530e4.

[19] Craig MH, Kleinschmidt I, Nawn JB, et al. Exploring 30 years of malaria case data in KwaZulu-Natal, South Africa. Part I. The impact of climatic factors. *Trop. Med. Int. Health* 2004;9:1247e57.

[20] Confalonieri U, Menne B, Akhtar P, et al. Human health. Climate change 2007: impacts, adaptation and vulnerability. In: Parry ML, Canziani OF, Palutikof JP, editors. *Contribution of Working Group II to the Fourth Assessment Report of the Intergovernmental Panel on Climate Change*. Cambridge, UK: Cambridge University Press; 2007. p. 391 e 431.

[21] Burattini MN, Chen M, Chow A, et al. Modelling the control strategies against dengue in Singapore. *Epidemiol. Infect.* 2008; 136: 309–19.

[22] Portal de saúde - 20005 www.saude.gov.br- /arquivos/pdf/be_malaria_ 01_2005.pdf.

[23] http://www.who.int/csr/disease/dengue/impact/en/[cited 2012 dec 14].

[24] Schatmayer, HG. Dengue Situation in Brazil by Year 2000. *Mem. Inst. Oswaldo Cruz.* v.95. s.1. 2000.

[25] Schatzmayr HG, Nogueira RMR, Travassos da Rosa APA 1986. An outbreak of dengue virus at Rio de Janeiro. *Mem. Instituto Oswaldo Cruz. 81*: 245-246.

[26] Gubler DJ 1997. Dengue and dengue hemorrahgic fever: its history and resurgence as a global public health problem. In DJ Gubler, G Kuno (eds), *Dengue and Dengue Hemorrhagic Fever*, Cab International, New York, p. 1-22.

[27] Pinheiro FP, Corber SJ 1997. Global situation of dengue and dengue hemorrhagic fever and its emergence in the Americas. *Wld. Health Statist. Quart. 50*: 161-169.

[28] Ministério da Saúde 2000. National Epidemiology Center, National Health Foundation.

[29] Nogueira RMR, Miagostovich MP, Lampe E, Schatzmayr HG 1990. Isolation of dengue virus type 2 in Rio de Janeiro. *Mem. Inst. Oswaldo Cruz. 85:* 253.

[30] Miagostovich MP, Nogueira RMR, Schatzmayr HG, Lanciotti RS 1998. Molecular epidemiology of den-2 in Brazil. *Mem. Inst. Oswaldo Cruz. 93*: 625-626.

[31] Dietz VJ, Gubler DJ, Rigau-Pérez JG, Pinheiro F, Schatzmayr HG, Bailey R, Gunn RA 1990. Epidemic dengue 1 in Brazil, 1986: evaluation of clinically based dengue surveillance system. *Am. J. Epidemiol. 131*: 693-701.

[32] Nogueira RMR, Miagostovich MP, Cavalcanti SMB, Marzochi KBF, Schatzmayr HG 1992. Levels of IgM antibodies against dengue virus in Rio de Janeiro, Brazil. *Res. Virol. Institut Pasteur. 143*: 423-427.

[33] Miagostovich MP, Nogueira RMR, Cavalcanti SMB, Marzochi KBF, Schatzmayr HG 1993. Dengue epidemic in the state of Rio de Janeiro, Brazil: virological and epidemiological aspects. *Rev. Inst. Med. Trop. São Paulo 35*: 149-154.

[34] Miagostovich MP, Santos FB, Araujo ESM, Dias J, Schatzmayr HG, Nogueira RMR 1997. Diagnosis of dengue by using reverse transcriptase-polymerase chain reaction. *Mem. Inst. Oswaldo Cruz. 92*: 595-600.

[35] Chimelli L, Dumas H, Barretto-Netto M, Ramos RG, Dias M, Gray F 1990. Dengue neuropathological findings in 5 fatal cases from Brazil. *Clin. Neuropathol. 9*: 157-162.

[36] Cunha RV, Zagne SMO, Miagostovich MP, Schatzmayr HG, Nogueira RMR 1998. Arterial hypertension as a possible risk factor for dengue hemorrhagic fever/dengue shock syndrome in Niterói, Rio de Janeiro. *Rev. Inst. Med. Trop. São Paulo 31*: 246-247.

[37] LIRAa. 2008-portal.saúde.gov/portal-arquivos/pdf/LIRAs_relatorio. Accessed on 12 abril.

[38] Massad E, Bezerra Coutinho FA. 2011. The cost of dengue control. *The Lancet,* Volume 377, Issue 9778, Pages 1630 – 1631.

[39] Khasnis AA, Nettlelman MD. Global warming and infectious disease. *Arch. Med. Res.* 2005; 36: 689–96.

[40] Zaim M, Guillet P. Alternative insecticides: an urgent need. *Trends Parasitol.* 2002; 18: 161–63.

[41] Luz PM, Vanni T, Medlock J, Paltiel AD, Galvani AP. Dengue vector control strategies in an urban setting: an economic modelling assessment. *Lancet* 2011; published online May 3. DOI:10.1016/S0140-6736(11)60246-8.

[42] Reiter P, Gubler DJ. Surveillance and control of urban dengue vectors. In: Gubler DJ, Kuno G, eds. *Dengue and dengue hemorrhagic fever.* Wallingford, UK: CABI Publishing, 2001: 425–62.
[43] MacDonald G. The analysis of equilibrium in malaria. *Trop. Dis. Bull.* 1952;49:813–28.
[44] Burattini MN, Chen M, Chow A, et al. Modelling the control strategies against dengue in Singapore. *Epidemiol. Infect.* 2008; 136: 309–19.
[45] WHO Position statement on integrated vector management. WHO/HTM/NTD/VEM/ 2008.2.
[46] WHO The technical basis for coordinated action against insecticide resistance: preserving the effectiveness of modern malaria vector control. WHO Geneva, 4–6 May 2010:*Meeting report.*
[47] Luz PM, Vanni T, Medlock J, Paltiel AD, Galvani AP. Dengue vector control strategies in an urban setting: an economic modelling assessment. *Lancet* 2011; 377: 1673–80.
[48] Rose RI. Pesticides and public health: integrated methods of mosquito management. *Emerging Infectious Diseases* 2001;7(1):17-23.
[49] Rozendaal JA. *Vector control methods for use by individuals and comnunities.* Geneve: World Health Organization; 1997.
[50] Mellon RB, Georghiou GP. Rotational use of insecticides in mosquito control programs.
[51] *Proceedings and papers of the fifty-second annual Conference of the California and Vector Control Association* 1984;65-67.
[52] Rathburn Jr CB. Insecticide formulations – types and uses: a review. *Journal of American Mosquito Control Association* 1985; 1 (1): 80-84.
[53] Palchick S. Chemical Control of Vectors In: *The Biology of Disease Vectors.* Colorado: University Press of Colorado; 1996.
[54] Crinnion WJ. Environmental Medicine, Part 4: Pesticides – Biologically Persistent and Ubiquitous Toxins. *Alternative Medicine Review* 2000;5(5):432- 447.
[55] Chen A, Rogan WJ. Nonmalarial infant deaths and DDT use for malaria control. *Emerging Infectious Diseases* 2003;9(8):960-964.
[56] Ferrer A. Pesticide poisoning. *Anales del Sistema Sanitario de Navarra* 2003;26(1):155-171.
[57] Waliszewski SM, Gomez-Arroyo S, Infanzon RM, Villalobos-Pietrini R, Hart MM. Comparison of organochlorine pesticide levels between abdominal and breast adipose tissue. *Bulletin of Environmental Contamination and Toxicology* 2003;71(1): 156-162.
[58] Chavasse DC, Yap HH. *Chemical methods for the control of vectors and pests of public health importance.* Geneve: WHO; 1997. WHO/CTD/ WHOPES/97.2.
[59] World Health Organization. *Vector Control for Malaria and other Mosquito-Borne Diseases Report of a WHO Study Group.* Geneva: WHO; 1995. WHO TRS no 857.
[60] Roberts DR, Laughlin LL, Hsheih P, Legters LJ. DDT, global strategies, and a malaria control crisis in South America. *Emerging Infectious Diseases* 1997;3(3):295-302.
[61] Bouwman H, Cooppan RM, Reinecke AJ, Becker PJ. Levels of DDT and metabolites in breast milk from Kwa-Zulu mothers after DDT application for malaria.
[62] Garabrant DH, Held J, Langholz B, Peters JM, Mack TM. DDT and related compounds and risk of pancreatic cancer. *Journal of the National Cancer Institute* 1992;84: 764-771.

[63] Wolff MS, Toniolo PG, Lee EW, Rivera M, Dubin N. Blood levels of organochlorine residues and risk of breast cancer. *Journal of the National Cancer Institutem.* 1993;85(8):648-652.

[64] Roberts DR, Manguin S, Mouchet J. DDT house spraying and re-emerging malaria. *Lancet* 2000; 356:330-332.

[65] Crinnion WJ. Environmental Medicine, Part 4: Pesticides – Biologically Persistent and Ubiquitous Toxins. *Alternative Medicine Review* 2000;5(5):432- 447.

[66] Brogdon WG, McAllister JC. Insecticide resistance and vector control. *Emerging Infectious Diseases* 1998;4(4):605-613.

[67] Ferrari JA. Insecticide resistance In: *The Biology of Disease Vectors.* Colorado: University Press of Colorado; 1996.

[68] Oppenoorth FJ. Biochemical and genetic in insecticide resistance. In: Kerkut GA, Gilbert LI, editors. *Comprehensive Insect Physiology Biochemistry and Pharmacology* 1985;12:731-773.

[69] Bergé J, Feyerensen R, Amichot M. Cytochrome P450 monoxygenases and insecticide resistance in insects. Philosophical Transactions Royal Society London Series B: *Biological Sciences* 1998;353:1701-1705.

[70] Scott JG. Cytochromes P450 and insecticide resistance. *Insect Biochemistry and Molecular Biology* 1999;29(9):757-777.

[71] Hemingway J, Karunaratne SH. Mosquito carboxylesterases: a review of the molecular biology and biochemistry of a major insecticide resistance mechanism. *Medical and Veterinary Entomology* 1998;12 (1):1-12.

[72] Hemingway J, Ranson H. Insecticide resistance in insect vectors of human disease. *Annual Review Entomology* 2000;45:371-391.

[73] Clark AG, Shamaan NA, Dauterman WC, Hayaoka T. Characterization of multiple glutathione transferases from the house fly, Musca domestica (L.). *Pesticide Biochemistry and Physiology* 1984;22:51-59.

[74] Clark AG, Dick GL, Martindale SM, Smith JN. Glutathione S-transferases from the New Zealand gras grub, Costelytra zealandica. *Insect Biochemistry* 1985;15:35-44.

[75] Toung YP, Hsieh TS, Tu CP. Drosophila glutathione S-transferase 1-1 shares a region of sequence homology with the maize glutathione S-transferase III. *Proceedings of the National Academy of Sciences USA* 1990;87(1):31-35.

[76] Ffrench-Constant RH, Pittendrigh B, Vaughan A, Anthony N. Why there so few resistance-associated mutation target genes? *Philosophical Transactions of the Royal Society London* 1998;353:1685-1693.

[77] Bloomquist JR. Toxicology, mode of action and target site-mediated resistance to insecticides acting on chloride channels. *Comparative Biochemistry and Physiology* 1993;106(2):301-314.

[78] Vais H, Williamson MS, Goodson SJ, Devonshire AL, Warmke JW, Usherwood PN, Cohen CJ. Activation of Drosophila sodium channels promotes modification by deltamethrin. Reductions in affinity caused by knock-down resistance mutations. *Journal of General Physiology* 2000;115(3):305-318.

[79] Farnham AW, Murray AWA, Sawicki RM, Denholm I, White JC. Characterization of the structure activity relationship of kdr and two variants of super-kdr to pyrethroids in the housefly (Musca domestica). *Pesticide Science* 1987;19:209–220.

[80] Kellen WR, Meyers CM. Bacillus sphaericus Neide as a pathogen of mosquitoes. *Journal of Invertebrate Pathology* 1964;7:442-448.

[81] Chevillon, Christine; Raymond, Michel; Guillemaud, Thomas; Lenormand, Thomas and Pasteur, Nicole (1999): Genetics da população da resistência do insecticide no mosquito Pipiens do mosquito. *Biol. J. Linn. Soc.* 68(1-2): 147–157. Pdf fulltext

[82] Conselho Coordinating de Florida no mosquito Controle (1998): Controle do mosquito de Florida: O estado da missão como definido por Mosquito Controlador, reguladores, e gerentes ambientais. Universidade de Florida.

[83] Mullen, G. and Durden, L. (2002): *Entomology médico e veterinário.* Ciência de Elsevier.

[84] Serviço, M. W. (1993): *Ecology do mosquito: Métodos de amostragem do campo.* Ciência aplicada de Elsevier.

[85] Ware, G. W. (1994) *O livro do Pesticide*. Universidade do Arizona.

[86] Herwaldt BL. Laboratory-Acquired Parasitic Infections from Accidental Exposures *Clinical Microbiology Reviews*. 14(4): 659-88, 2001.

[87] ORCBS. Biosafety Manual, *Office of Radiation, Chemical and Biological Safety.* Michigan: Michigan State University Press, 1998.

[88] World Health Organization (WHO). *Laboratory Biosafety Manual*, 3rd ed. Geneva, 2004.

[89] Mastroeni MF. Biossegurança em laboratórios de parasitologia. In: De Carli GA. (org.). *Parasitologia clínica: Seleção de métodos e técnicas de laboratório para o diagnóstico das parasitoses humanas*. 2a ed. São Paulo: Atheneu. 906 p., 2007.

[90] Centers for Disease Control and Prevention and National Institute of Health. Biosafety in Microbiological and Biomedical Laboratories. 4th ed. Washington: Center for Disease Control and Prevention National Institutes of Health, 1999.

[91] Pike RM. Laboratory-associated infections: Incidence, Fatalities, Causes, and Prevention. University of Texas. Health Science Center, Dallas, Texas. *Ann. Rev. Microbiol.* 33:41-66, 1979.

[92] Kamaljit S. Laboratory-acquired infections. *Clin. Infect. Dis.* 49:142-7, 2009.

[93] Sewell DL. Laboratory associated infections and biosafety. *Clin. Microbiol. Rev.* 1995; 8:389–405.

[94] Pike RM. Laboratory-associated infections: Summary and analysis of 3.921 cases. *Health Lab. Sci.* 13:105-114, 1976.

[95] Pike RM, Sulkin SE, Schulze ML. Continuing importance of laboratory-acquired infections. *Am. J. Public Health Nations Health* 1965; 55:190–9.

[96] Herwaldt BL, Juranek D.D. Protozoa and Helminthes. In: Fleming DO, Richardson JH, Tulis JI et al. *Laboratory Safety. Principles and Practices.* 2nd ed. Washington (DC): ASN Press, 1995.

[97] Lettau LA. Nosocomial transmission and infection control aspects of parasites and ectoparasitic diseases. Part I. Introduction/enteric parasites. *Infect. Control Hosp. Epidemiol.* 12: 59-65, 1991.

In: Parasites
Editor: Gilmar S. Erzinger
ISBN: 978-1-62257-692-0

Chapter 2

PHOTODYNAMIC DESTRUCTION OF MOSQUITO LARVAE USING CHLOROPHYLL DERIVATES AS PHOTOSENSITIZERS

Peter Richter[1,*], Gilmar Sidnei Erzinger[2], Suellen Carolina Souza[2], Fernanda Vollrath[2], Stephanie Wohllebe[3] and Donat-Peter Häder[#]

[1]Department of Biology, Cell Biology, Friedrich-Alexander-University Erlangen-Nuremberg, Germany
[2]Department of Pharmacy, Medicine and Masters Sciences Program in Health and Environment of University of Joinville Region, Brazil
[3]Bavarian State Research Centre for Agriculture, Institute for Fisheries, Starnberg, Germany

ABSTRACT

Chlorophyll derivates such as chlorophyllin and pheophorbid can be utilized to photodynamically destroy mosquito larvae. These substances were produced from natural chlorophyll extracted from green plant material. By simple chemical modifications hydrophobic chlorophyll can be transformed to water soluble chlorophyllin, which in turn can be modified to pheophorbid by means of acidification. Due to their water solubility these substances can be applied in aquatic environments. Mosquito larvae and other larvae from biting insects in the water accumulate these substances mainly in their intestine. Upon illumination photodynamic reactions start leading to apoptosis and necrosis of the cells in the gut resulting in the death of the larvae. This method employs considerably low concentrations of chlorophyll derivates. The production of chlorophyll derivates is inexpensive and easy and therefore a promising approach to control water-born diseases in developing countries. This chapter gives an overview over the basics of the photodynamic reactions and about the results obtained so far using photodynamic properties of chlorophyll derivates in order to control the vectors of parasitic diseases.

* E.mail: prichter@biologie.uni-erlangen.de.
# Neue Str. 9, Möhrendorf, Germany.

INTRODUCTION

Parasites and Their Impact on Humans and Livestock

Parasites have a large impact on the health situation and economic development of underdeveloped and developing countries. About 300 worm diseases and 70 protozoan-elicited diseases have a more or less severe impact on the human body [1]. Malaria alone costs between 1 million - 5 million lives every year mainly in Africa but also in tropical regions of Asia and Latin America [2]. It is estimated that about 300 million cases of malaria occur every year. Malaria is a protozoan disease (caused by different *Plasmodium* species) which is transmitted by *Anopheles* mosquitos, which is the final host, in whom sexual reproduction takes place. The most severe *Plasmodium* species is *P. falciparum*, which leads to the fatal Malaria tropica disease, which is responsible for about 90 % of *Plasmodium*-induced mortality [3]. Measures against malaria are medical treatment against *Plasmodium* protozoans (e.g. chloroquin) or control of the vector with pesticides such as DDT. In both cases resistances developed. Although DDT was extremely effective in controlling *Anopheles* flies and other pest insects it was banned because of environmental problems and possible health hazards. Today the use of DDT is highly restricted according to the Stockholm Convention [4]. After the ban of DDT the number of malaria cases increased ,and some experts claim that the ban of DDT was responsible for millions of cases of death [5-7]. Meanwhile the WHO states that DDT is still needed as long as no alternatives are available [8]. Malaria is not restricted to tropical countries. In Europe malaria was extinct in the last century by means of pesticides, killing the *Anopheles* vector. Some experts claim that due to global warming malaria may reappear in European countries [9]. Other important diseases elicited by protozoans are caused by different *Leishmania* species, which are transmitted by sandflies (*Phlebetomus*) [10]. In addition, different groups of dipters also transmit various viral infections such as the Dengue fever or the Japanese encephalitis ([11] and literature cited therein).

Worm diseases caused by nematodes, cestodes and trematodes pose considerable impact on public health. *Bilharziosis* (*schistosomiasis*), a worm disease caused by trematodes from the genus *Schistosoma,* is a major problem in Egypt and other countries worldwide. Estimated 200 million people are presently infected. Between 200,000 to 1 million people die every year from the severe complications caused by *Schistosoma* infestation [12]. *Schistosoma* has an interesting life cycle. The adult worms live in the venous system of liver and gut of infected humans. The filiform female worm lives in a dorsal fold of the male. In order to lay eggs it separates from the male and migrates into the capillary system of the gut or bladder (*S. haematobium*) to deposit the eggs. The eggs elicit inflammations leading to local tissue destruction allowing the eggs to break through into the gut or bladder, from where they are released with urine or feces. In case the eggs get into water, larvae (miracidia) are hatching infecting water snails (*S. mansoni*: *Biomphalaria,* S. *haematobium*: *Bilinus*, *S. japonicum: Tricula aperta*).

Inside the snail the miracidium develops to a sporocyst, a life stage which produces large amounts of so-called cercaria, which are released from the snail infecting humans again by penetration through the skin. As many parasitic diseases are transmitted by vectors such as flies or molluscs, control of these vectors is one of measures to control their impacts on humans or livestock (farm animals are also affected by parasitic diseases). Pesticides, wetland draining, spraying of oil on water surfaces (killing mosquito larvae), mosquito nets and natural predators are current approaches in order to mitigate the number of vectors. Very promising is the control of mosquito larvae by means of spores of the bacterium *Bacillus thuringensis israelensis* (Bti), which contains an insect-specific toxin. Production and application of Bti is relatively easy and there are almost no side effects on non-target organisms [13].

In recent experiments, Abdel Kader and coworkers from the NILES Institute in Cairo (Egypt) observed that mosquito larvae are killed after contact with the photodynamic substance hematoporphyrin [14]. As the chemical structure of hematoporphyrin is similar to chlorophyll, which can be found in almost all plants, it was hypothesized that chlorophyll may have the potential to be employed as a cheap and easy to produce measure against parasites and their vectors. Experiments were performed in which the potential of chlorophyll or chlorophyll derivates was investigated. This chapter summarizes the results of these investigations.

1. Photosensitizers and Photodynamic Reactions

In general photosensitizers are molecules which are excited by light [15]. The excited state can react with other molecules changing the chemical properties of the reaction partner. The reaction probability is higher when the photoreaction of the photosensitizer results in the relatively long-lived triplet stage. Photodynamic reactions with oxygen lead to the formation of the highly reactive singlet oxygen (formula 1), which can react with various biomolecules [16].

$$^{3}P + {^{3}O_2} \rightarrow P + {^{1}O_2} \qquad (1)$$

In addition, photosensitizers such as chlorophyll and chlorophyll derivates are capable to oxidize and reduce other molecules. In the excited state chlorophyll is a strong reductant, which can transfer electrons onto other molecules. But in the subsequent oxidized state chlorophyll is a strong oxidant, which may oxidize other biomolecules. The redox potential of chlorophyll ranges from +1.1 V (strong oxidant) in the not excited state to over 0.32 V (high reduction power) upon light excitation [17].

As a result, reactive oxygen species, such as superoxide or hydrogen peroxide are formed posing strong oxidative stress to the cells. Excessive oxidative stress results in damage to cell membranes, proteins, DNA and other cell structures [18]. In photodynamic therapy (PDE) photodynamic reactions play a role in medicine [16]. Application of photodynamic substances, such as aminolaevulic acid or protoporphyrin onto diseased tissue (e.g. warts, ceratinosis or superficial tumors) often results in amelioration of the clinical pattern [19-20].

It is believed that the effects are mainly due to evolution of singlet oxygen. Also chlorophyll derivates were successfully tested on cell cultures [21-23]. Coupled to certain molecules, which accumulate in cancer tissue chlorophyll derivates may become promising drugs in cancer treatment. A pyropheophorbide-folate complex was shown to accumulate in cancer cells exhibiting a high expression of folate receptors [24-25].

2. Treatment of Pest Organisms by Means of Photodynamic Substances – Results and Concepts

Abdel Kader from the National Institute of Laser Enhanced Science (NILES) had the idea to treat pest organisms photodynamically by means of hematoporphyrin. In different experiments it could be shown that e.g. *Culex* larvae were killed in the presence of hematoporphyrin and sunlight. A concentration of about 0.07 µmol/ml in the water was reported to be sufficient to induce photodynamic mortality of the larvae [26]. The substance was also active against eggs of the snail *Lymnea natalensis*, which is a vector of *Fasciola hepatica*, as well as against *Musa domestica* larva.

As hematoporphyrin is too expensive for utilization on a large scale, chlorophyll was considered to be a cheap alternative. Like hematoporphyrin, chlorophyll consists of a porphyrin structure and is very effectively excited by light. It is a fundamental molecule in green plants from which it can easily be extracted by means of cheap solvents such as ethanol. Vegetable waste, such as e.g. banana leaves or weeds such as e.g. *Eichhornia* (a pest plant) can be used for extraction. More advantages of chlorophyll are the complete biodegradability and the fact that it is not toxic for humans (it is an allowed food additive - E 140).

3. Extraction and Modification of Chlorophyll

Standardized experiments were conducted with chlorophyll extracted from spinach. Wohllebe et. al. tested other sources of chlorophyll (a mixture of chlorophyll *a* and *b*) such as *Eichhornia*, grass from lawn mowing and other weeds [27]. Chlorophyll of all sources had the same effect on larvae. Due to the extended aerenchym and the relatively low pigment content, chlorophyll yield from *Eichhornia* was not very efficient. High water content of plant material dilutes the solvent making it more polar, which impairs extraction of chlorophyll. Reducing the water content of the plant material before extraction is recommended.

For small scale experiments plant material was extracted by means of 96 % EtOH at 65 °C in the presence of 0.1 M $CaCO_3$. Calcium carbonate buffers the acidic content of the plant vacuole, which would partially convert the chlorophyll to pheophytin. After filtration benzene oil or petroleum benzene was added to the extract. After shaking, chlorophyll accumulates in the lipophilic benzene phase. Lipophilic chlorophyll hardly mixes with water, which makes standardized experiments on a laboratory scale almost impossible, because it becomes attached to the walls of the flasks in which the experiments are performed. This inhibits exact dosage. For this reason it is transformed into the water-soluble chlorophyllin by saponification with methanolic KOH (1 M). Chlorophyllin accumulates at high concentration in the MeOH-KOH phase. After determination of the concentration by means of a spectrophotometer and using formulas from Lichtenthaler [28] it is stored in dark flasks at 4°C under nitrogen until further processing.

Pheophorbid (chlorophyllin without a central Mg^{2+}) was produced upon addition of ccHCl (10µl/ml). To test different modifications of chlorophyll different heavy-metal-chlorophyllins (hms-chl) were produced [29]. The chlorophyll extract was acidified (10µl/ml ccHCl) in order to eliminate the central Mg^{2+} ion.

The methanolic extract was saturated with $ZnSO_4$ or $CuSO_4$ (30 min at 30 °C), leading to heavy metal complexes of chlorophyll (Cu- or Zn-chlorophyll, respectively). In order to avoid allomerization the chlorophyll derivates were transferred to cyclohexane [30]. After evaporation under a stream of nitrogen the substances were dissolved in acetone before further processing. Upon saponification the hms-chl derivates were modified to the corresponding water soluble forms.

4. Effects of Chlorophyllin and Pheophorbid Treatment on Mosquito Larvae

Most experiments were performed with larvae of *Culex* spp. and *Chaoborus crystallinus*. The latter is not a biting insect, but due to its transparency an ideal model organism in order to monitor the in vivo uptake of chlorophyllin by light and fluorescence microscopy. In addition, it can be easily subjected to different methods using fluorescent markers and biochemical techniques [31]. Different other species of diptera larvae were successfully treated with chlorophyllin (unpublished results).

Before use, defined amounts of the chlorophyllin extracts were evaporated in order to remove the solvent. Different concentrations of chlorophyllin were adjusted in water, with water serving as controls with a KOH concentration (from the saponification process) of the highest chlorophyllin concentration. Then larvae were added. It was found that the larvae need some hours of incubation time in darkness in order to accumulate sufficient chlorophyllin in their intestine. Erzinger at al. determined that about 3-4 h incubation time is necessary before the accumulation of chlorophyllin reaches a plateau [32]. As long as the larvae are not exposed to light, almost no mortality is observed. After exposure to a standardized solar-simulator light source (Sol 1200 W, mercury lamp 0383, Dr. Hönle, Martinsried, Germany) for about 3 h a concentration-dependent increase of the mortality can be observed. Wohllebe et al. determined the EC50 in *Culex* larvae at a chlorophyllin concentration of 6.8 mg/l chlorophyllin or 8.2 mg/l pheophytin, respectively, in water. In *Chaoborus* the required concentration was higher: about 24 mg/l in the case of chlorophyllin and less than 2 mg/l for pheophorbid. Pheophorbid was also toxic for the larvae in darkness [33]. Toxicity of pheophorbid was also observed in mouse myeloma cell cultures [34].

Erzinger et al. determined the mortality of *Chaoborus* larvae in dependence of incubation time, temperature and light density. Larvae were incubated in water at a chlorophyllin concentration of 25 mg/l, which was determined to be the EC50 concentration for this species [32].

The mortality increased in dependency of the incubation time. The killing rate was closely related to the uptake kinetics observed by means of fluorescence microscopy. After 1 h incubation about 40 % of the larvae died after 3 h of irradiation, after 2 h incubation 70 % were killed and after 4 h about 90 % of the larvae died after subsequent irradiation. Microscopic analysis showed a pronounced increase of chlorophyllin in the intestine. Correlation of light intensity with mortality revealed that after sufficient preincubation time (3 h) even about 40 W/m^2 simulated solar radiation induced high mortality in the treated larvae. Temperature has no effect on mortality. Mortality was the same at 10°C, 20°C or 30 °C (Figure 1).

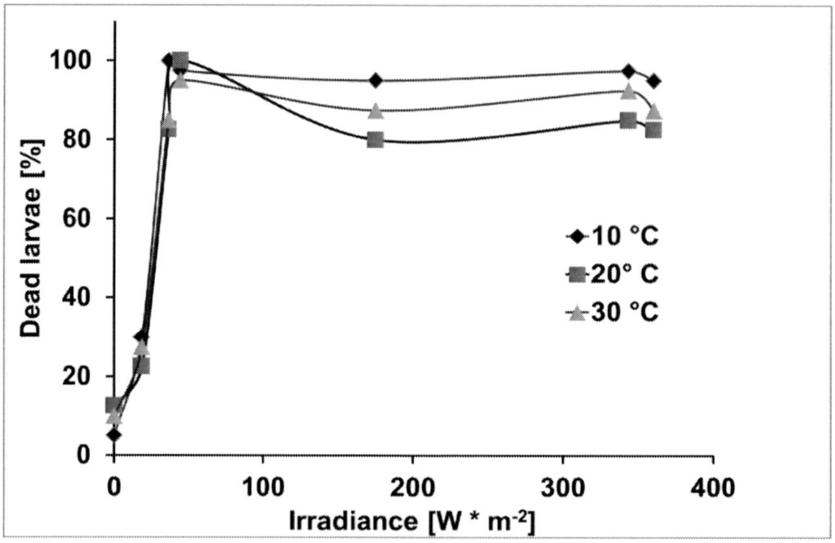

Figure 1. Mortality of *Chaoborus crystallinus* larvae at different irradiances and different temperatures (10 °C, 20°C or 30°C) after 6 h dark incubation in the presence of 25 mg/l Mg-chlorophyllin and subsequent 3 h illumination. (Data from [11]).

In order to determine the intraintestinal chlorophyllin concentration, defined numbers of larvae were extracted in MeOH and the extracted chlorophyllin was determined. Each larva contained about 8-15 ng chlorophyllin [31]. Among the hms-chl Zn-chlorophyll showed the highest activity. The EC50 was about 17 mg/l compared to Mg-chlorophyllin from the same lot (EC50 22.5 mg/l). Cu-chlorophyllin did not exhibit photodynamic properties, but induced high mortality probably due to copper toxicity. Although the efficiency of Zn-chl is higher than Mg-chl the use of the latter is probably more recommendable, because it is a permitted food aditive in contrast to Zn-chl. The ecotoxicological relevance of Zn-chl is not clarified yet. Erzinger could show that freeze drying immediately after extraction increases the activity of chlorophyllin. The EC50 of freeze dried chlorophyllin against *Chaoborus* larvae was about 14 mg/l. Allomerization is an oxidation of certain oxidizable groups in the chlorophyll molecule, which did not change the light absorption properties, but probably impairs its photodynamic activity [32].

Recently the InRad company (Heliopolis, Egypt) performed successful field experiments in Uganda [35]. It was shown that at suitable concentrations a formulation with chlorophyllin could kill mainly mosquito larvae in a pond, while other water organisms were not affected ([36]. Tests with other aquatic organisms revealed different sensitivity [27]. Daphnids and fish larvae were highly sensitive against chlorophyllin, while carps and other fresh water fishes, water slater and Chironomidae were less sensitive against chlorophyllin treatment. While adult *Lymnea stagnalis* and *Bulinus* were not affected by chlorophyllin, initial experiments with *Biomphalaria* snails indicated high sensitivity against chlorophyllin treatment. This would be very interesting because this snail is one of the intermediate hosts of *Schistosoma*.

5. Determination of Cellular Damages of Chlorophyllin Treatment in Chaoborus Larvae

As mentioned above, *Chaoborus crystallinus* larvae are interesting test organisms because of their transparency. By means of staining with acridine orange, Hoechst 33342 and propidium iodide (PI) significant apoptotic and necrotic events could be observed after chlorophylline/light-treatment in the dissected intestines [27, 31]. Flow cytometry of isolated cells labelled with PI and annexine clearly demonstrated increased apoptosis and necrosis at a chlorophyllin concentration of 21 mg/l after irradiation). Protein analysis from irradiated and non-irradiated chlorophyllin-treated larvae showed clear differences in formation of heat shock proteins (HSP) [27]. HSP 70 and HSP 90 were increased in treated cells.

By means of specific antibodies different apoptotic proteins (caspases, and the anti-apoptotic protein BCl-2) were investigated. Upon irradiation cytochrome *c* and active caspase-9 were increased at high chlorophyllin concentrations and after irradiation. The data indicate a concentration-dependent shift of apoptotic pathways. The results clearly show that the uptake of chlorophyll leads to severe photodynamic destruction of cells in the intestine resulting in the death of the organisms.

6. Outlook

Because of the promising capacities of chlorophyll and chlorophyll derivates work in order to improve the extraction, production and application of chlorophyll is continuing [11, 35].

A recent project at the Landesanstalt für Fischerei und Teichwirtschaft, Außenstelle Karpfenteichwirtschaft in Höchstadt (Germany), in which the impact of chlorophyllin on fish parasites was investigated, yielded very promising results [37]. Chlorophyllin was effective against various fish parasites.

Conclusion

Photodynamic substances based on chlorophyll have been shown to be very effective against vectors of parasites. Chlorophyll can be extracted from all green plant material by means of inexpensive solvents. Further conversion to water-soluble chlorophyllin is technically not demanding. The applied chlorophyll derivates are biologically degradable in a short time. The results are of relevance to the field, because chlorophyll has the potential to become an important substance to control human pathogenic.

Acknowledgments

The authors gratefully acknowledge the technical support by Ulrike Trenz and Martin Schuster. Also the friendly and competent support of Dr. Martin Oberle and his team of the Institut für Fischerei und Teichwirtschaft, Außenstelle Karpfenteichwirtschaft, Höchstadt a. d. Aisch (Germany) is highly acknowledged. Thanks to the DAAD for financiation of a scientific visit of Prof. Erzinger to Germany.

REFERENCES

[1] F. E. G. Cox, *Clin. Microbiol. Rev.* 15, 595 (2002).

[2] J. Sachs and P. Malaney, *Nature* 415, 680 (2002).

[3] WHO, *Malaria. Fact sheet N°94.*, Geneva, Switzerland (2007).

[4] WHO, *10 things you need to know about DDT use under the Stockholm Convention. World Health Organization*, Geneva, Switzerland (2004).

[5] J. G. Edwards, J. *Am. Phys. Surg.* 9, 83 (2004).

[6] D. R. Roberts, *To control Malaria, we need DDT!,* 21st Century Fall (2002)

[7] M. Cone, *Sci. Am. May*, 4, 1 (2009).

[8] WHO, *The Use of DDT in Malaria Vector Control. WHO position statement (WHO/HTM/GMP/2007),* World Health Organization, Geneva (2007).

[9] M. Goklany, *Science*, 306, 55 (2004).

[10] R. Killick-Kendrick, *Ann. Parasitol. Hum. Comp.* 65, 37 (1990).

[11] G. S. Erzinger *Pharm. Anal. Acta* 2, 139 (2011).

[12] WHO, *fact sheet Fact sheet N°115,* World Health Organization, Geneva (2010).

[13] P. K. Mittal, *J. Vect. Borne Dis.*, 40, 20 (2003).

[14]] M. H. Abdel-Kader, A. S. A. M. Al-Sherbini and T. A. A. El Tayeb, The 9th International Conference on ''Environmental Protection is a must,'' Alexandria University, May 4-6 (1999).

[15] B. C. Wilson and M. S. Patterson, Wilson, *Phys. Med. Biol.*, 31, 327 (1986).

[16] J. D. Spikes, *Light in Disease and Therapy*, Plenum Press, New York (1979).

[17] D. W. Lawlor, *Photosynthesis (3d* edition), Bios, Oxford (2001).

[18] R. Kohen and A. Nyska, *Toxicol. Pathol.*, 30, 620 (2002).

[19] B. W. Henderson and T. J. Dougherty, *Photochem. Photobiol.,* 55, 145 (1992).

[20] F., C. Fernandez-Marcos, N. Swamy, D. Dunn and R. Ray, *J. Cell. Biochem.*, 99, 665 (2006).

[21] P. M. K. Tang, P.-M. Hon, S. W. N. Au, S. K. W. Tsui, M. M. Y. Waye, S.-K. Kong, T. C. W. Mak, and K.-P. Fung, *Planta Medica*, 72, 28 (2006).

[22] W.-Y. Lee, D.-S. Lim, S.-H. Ko, Y.-J. Park, K.-S. Ryu, M.-Y. Ahn, Y.-R. Kim, D. W. Lee, and C.-W. Cho, *J. Photochem. Photobiol. B Biol.*, 75, 119 (2004).

[23] K. Stefflova, H. Li, J. Chen, and G. Zheng, *Bioconjugate Chem.*, 18, 379 (2007).

[24] X. Sun and W. N. Leung, *Photochem. Photobiol.*, 75, 644 (2002)

[25] F. Gacio, , C. Fernandez-Marcos, N. Swamy, D. Dunn, and R. Ray, *J. Cell. Biochem.*, 99, 665 (2006).

[26] T. A. A. Tayeb, *Laser scanning microscopy for determination of the efficiency of hematoporphyrin in control of Culex pipiens larvae and the snail vector of Fasciola gigantica*, PhD thesis. National Institute of Laser Enhanced Sciences (NILES), Cairo University (2003).

[27] S. Wohllebe, *Bekämpfung humanpathologischer Parasiten mittels natürlicher Photosensitizer*, diploma thesis, Friedrich-Alexander-Universität Erlangen-Nürnberg. (in German) (2008).

[28] H. K. Lichtenthaler and A. R. Wellburn Lichtenthaler, *Biochem. Soc. Transact.*, 11, 591 (1983).

[29] H. Küpper, M. Spiller, *Analytical Biochemistry*, 286, 247 (2000).

[30] P. S. Woolley, A. J. Moir, R. E. Hester, B. J. Keely, *J. Chem. Soc. Perkin Trans.*, 2, 1833 (1998).

[31] S. Wohllebe, C. Ulbrich, D. Grimm, J. Pietsch, G. Erzinger R, , R. Richter, M. Lebert, P. R. Richter and D.-P. Häder, *Photochem. Photobiol.*, 87, 1113 (2011).

[32] G. S. Erzinger, S. Wohllebe, F. Vollrath, S. C. Costa, P, Richter, M. Lebert, and D.-P. Häder, *Parasitol. Res.*, 109, 781 (2011).

[33] S. Wohllebe, R. Richter, P. Richter, D.-P. Häder, *Parasitol. Res.*, 104, 593 (2009).

[34] S. Chernomorsky, A. Segelman and R. D. Poretz, *Teratogen. Carcinogen. Mutagen*, 19, 313 (1999).

[35] http://inradcorp.org/page.php?section=Our-Projects.

[36] http://inradcorp.org/photos/letter.pdf.

[37] S. Wohllebe, R. Richter, P. Richter und D.-P. Häder, *Bekämpfung von Parasiten in aquatischen Ökosystemen durch Einsatz der photodynamischen Substanzen Chlorophyll und Phaeophytin*, XII. Gemeinschaftstagung der Deutschen, Österreichischen und Schweizer Sektionen der European Association of Fish Pathologists (EAFP), 148 (2009).

In: Parasites
Editor: Gilmar S. Erzinger

ISBN: 978-1-62257-692-0

Chapter 3

TRANSPORT OF ESSENTIAL METABOLITES IN TRYPANOSOMATIDS

Claudio A. Pereira* and Carolina Carrillo
National Council of Scientific and Technical Research (CONICET)
and University of Buenos Aires (UBA), Argentina

ABSTRACT

Infections caused by the protozoan parasites from the genus *Trypanosoma* and *Leishmania* are among the most relevant public health problems in the developing countries. These parasites present several metabolic characteristics which are completely absent in their insects and mammalian hosts.

Some of the differences were acquired after millions of years of adaptation to parasitism, in which this protozoan replaced, along its evolution, many biosynthetic routes for transport systems. In the proposed chapter we will describe the advances in the knowledge of transport systems and the molecules involved in such processes from the first biochemical description to the post-genomic era including the future perspectives in this area. In particular, we will focus the attention on amino acids and polyamines transporters from the AAAP family (Amino Acid/Auxin Permeases), because they seem to be unique to Trypanosomatids. Considering that these permeases are completely absent in mammals they could be put forward as a potential target against Trypanosomatid parasites.

INTRODUCTION

Trypanosomatids are parasites belonging to the taxonomical group Kinetoplastidae. They are distinguished by having a single flagellum and a peculiar mitochondrial DNA structure called kinetoplast. All Trypanosomatids are parasites of insects, and some genera have life-cycles involving a secondary host, which may be a vertebrate, an invertebrate or a plant. It is remarkable that some kind of Trypanosomatids cause severe human diseases. In particular,

* E-mail address: cpereira@retina.ar.

two species of the genus *Trypanosoma*, *T. brucei* and *T. cruzi*, are the causative agents for the sleeping sickness, or African trypanosomiasis, and the Chagas´ disease, or American trypanosomiasis, respectively. Several species of the genus *Leishmania* from the subgenus *Leishmania* and *Viannia* are the causative agents of a group of diseases known together as leishmaniasis. Taken together, these parasites affect approximately to 25 million people in the endemic areas all over the world, with an estimated population in risk of acquiring the infection of more than 350 million people. Unfortunately, the therapeutic alternatives to treat the human infection by these parasites are unsatisfactory: their low therapeutic efficiency, high toxicity, and the appearance of resistant strains constitute serious drawbacks that should be overcome [1].

Life cycles of Trypanosomatid parasites are complex, involving morphological cell changes and a wide variety of environments with different composition, mainly in the insect vector gut, mammalian blood and mammalian host cell cytoplasm. In consequence, the parasite survival mainly depends on their adaptive ability and metabolic plasticity that in many cases involve nutrient permeases. The life cycles of *Leishmania* and *T. cruzi* have an obligatory intracellular stage in mammals, in contrast to the exclusively extracellular parasites of the *T. brucei* group. Both *Leishmania* and *T. cruzi* invade host cells; *Leishmania* lives inside parasitophorous vacuoles while *T. cruzi* escapes from the vacuole and lives in the cytoplasm of the host cell. In both cases, the parasite has to adhere to the host cell surface in order to invade the cell and survive under harsh conditions of the host cytoplasm to establish an effective infection.

The host cell is now the source of nutrients for parasite metabolism. The transport of those nutrients from the host cell cytoplasm to the parasite (*T. cruzi*) or from the host cytoplasm and through the membrane of the parasitophorous vacuole before reaching the membrane of the parasite (*Leishmania*) are essential steps that can be targets for drug design, in addition to metabolic routes that are specific for the parasites and absent in the host [2].

Each month we can find many papers reporting new synthetic or natural compounds with trypanocidal activity, however at present it is still using drugs that were discovered almost 40 years ago.

The probable reason is that, as it occurs with normal or transformed cell of any organism, when parasites are isolated in culture conditions or in experimental biological model, are much more vulnerable than in wild-type media, where they are adapted to survive and have the resources of the mammalian or insect host. Many of these resources are substrates of transport molecules.

1. Parasites Nutrition: The Role of Transport Mechanisms

Since the parasite survival mainly depends on their adaptive ability and metabolic plasticity, these parasites have changed, through the evolution process, biochemical routes – useful just in a cycle phase and with a high "burden cost" for the other stages- by transport systems that function as extracellular sensors and adaptors to the environmental changes [3]. A critical aspect to achieve these adaptations is the parasite potential to take advantage of the available metabolites in the different extracellular milieus using a large and adaptive repertoire of nutrient permeases. The ability of these cells to deal with these environmental changes is reflected on their flexible metabolism, being these cells able to use glucose or amino acids as the main carbon and energy source [4, 5].

Usually, these organisms consume amino acids when they are submitted to glucose deprivation conditions [6]. Transporters are considered essential proteins in most of cells, and are among the first molecules involved in the detection of their substrates in the extracellular medium [7]. Despite the relevance of amino acids in Trypanosomatids, the way by which they are internalized to become available for metabolism remains relatively unexplored.

T. cruzi has no intracellular glucose reserve such as glycogen or starch. Thus, glucose has to be imported from the environment, as the mammalian blood, through a glucose transport system. The amino acids can be obtained from three main sources: biosynthesis from metabolic precursors, active transport from the medium and protein degradation. Proteins could be incorporated from the surrounding medium by pynocitosis [8] and kept in reservosomes.

A metabolic characteristic that is unique for kinetoplastids is the fact that the glycolysis is compartmentalized between cytoplasm, as usual, and specialized organelles called glycosomes [9]. Glucose is taken up by hexose transporters, which are present in a number varying from one to three depending on the species: for example, only one transporter was described for *T. cruzi*, while two were described for *T. brucei* and three for *Leishmania* spp. [10].

It is assumed that once in the cytoplasm, glucose is transported into the glycosomes, to be oxidized. A peculiar characteristic of glycolysis in kinetoplastids is the fact that the hexokinase and phosphofructokinase, as opposed to mammals, are not allosterically regulated [11-14] and the glycolytic flux it is not influenced negatively by the presence of oxygen (Pasteur Effect) [15]. It was thus proposed that compartmentalization of glycolysis may constitute an evolutionary alternative to regulate the glycolytic flux in trypanosomes at the level of glucose transport from the cytoplasm to the glycosome [16]. However, the mechanisms and kinetics of the latter are unknown and no proteins were identified that qualify for this function.

The relevance of amino acids along the *T. cruzi* life cycle is well established. The early works of Zeledón [17] and Sylvester and Krassner [18] have shown the ability of epimastigotes to oxidize certain amino acids. Besides glucose, the addition of three amino acids, aspartate, glutamate and proline, to defined media promotes differentiation from epimastigote to metacyclic trypomastigote (metacyclogenesis) in a variety of *T. cruzi* strains [19].

Along the chronic infection, *T. cruzi* lives mostly restricted to some specific tissues, such as cardiac or striated muscle, and in particularly restricted tissue regions (called "amastigote nests" by most of pathologists) of the digestive tube epithelium or kidney epithelium, maintaining its infection cycle without causing major symptoms.

During infection, *T. cruzi* spends a considerable time replicating as the intracellular forms (mostly amastigote, but also the transient intracellular epimastigote) and it appears that these stages have low, if any, glucose transport activity. This fact allows the supposition that the amastigote form would have a low sensitivity to therapeutic drugs targeted to glycolytic enzymes. It has proposed a proline-based metabolism for the replicative amastigote forms and intracellular epimastigotes that could be a promissory target [20, 21].

2. Biochemical Studies of Transport Systems: The First Steps

Amino acid transporters are the first cell proteins that become in contact with solutes in the surrounding medium and in several cases they function not only as permeases to carry the solutes into the cytoplasm, but also as environmental sensors. Metabolites transport studies in Trypanosomatids have begun at the end of the 60', in *Crithidia*, a parasite of arthropods, when it was made the initial carbohydrate transport study [22], followed by other reports in *Trypanosoma* [23, 24]. The first work specifically about amino acids transport, was made in *T. cruzi* [25] subsequently followed by phosphate and nucleosides transport analyzes [26, 27]. Polyamines uptake was first investigated twenty years after, when putrescine transport was compared between different Trypanosomatids in relation to their polyamine metabolism [28]. During the last decades, biochemical amino acids transport studies in trypanosomes were focused mainly in arginine and proline.

Following it is described the first biochemical evidence of the existence of transporters of amino acids and derivatives, mainly in *T. cruzi*:

2.1. Arginine Transport

In *T. cruzi*, arginine is an essential amino acid for the *in vitro* cultivation of epimastigote cells. An arginine transport system related to the uptake of pentamidine, an antiparasitic drug used to treat Leishmaniasis, was described by Kandpal et al [29]. In *T. cruzi* was characterized a highly specific and ATP-dependent arginine transport system [30]. Interestingly, this high-affinity transport system showed to be modulated by the parasite replication ability, and by the nutrient availability in the media, showing an adaptive regulation by substrate [31]. A continuous decrease in arginine uptake accompanying the reduction in parasite replication rates, together with an increase in cell density and consequent nutrient depletion of the medium, was observed. Furthermore, L-arginine uptake variations were observed treating the cells with conditioned medium or fresh medium. In addition, omission of any carbon source in medium increases the L-arginine uptake at a lower rate when compared to the complete medium. It is postulated that changes in medium conditions and/or the appearance of a density-derived factor that affects cell replication ability, modulate L-arginine uptake [31]. More recently, a complementary low-affinity arginine transport system was also described in *T. cruzi* [32]. This low-affinity arginine transporter presented a strong variation in terms of activity between 25 °C and 40 °C. This characteristic could be related with an adaptive advantage as *T. cruzi* is subjected to a wide variation in environmental temperatures around 28 °C in the midgut of the insect vector and 37 °C in the mammalian host. Thus, the increase in arginine uptake capacity may have physiological significance when temperature is raised from 28 °C, the optimal growth temperature of insect stage forms, to 37 °C, the temperature at which the infection in the mammalian host is carried out. The existence in *T. cruzi* of two arginine transport systems showing a wide variation of activation energies confers on the parasite the ability of having functionally stable arginine transport systems in a broad range of thermodynamic conditions. The arginine transport through the low-affinity system is developmentally regulated since it is higher in epimastigotes than in trypomastigotes. However, it is worth to note that arginine is actively transported in trypomastigotes by this system instead of the previously reported high-affinity one [32].

Our group has described an arginine kinase which is present in *T. cruzi* and *T. brucei*, but not in *Leishmania*, that converts arginine into phosphoarginine, a phosphagen with a role as spatio-temporal energy buffer [33-35]. Phosphoarginine phosphagen system is equivalent to the mammalian phosphocreatine, but is completely absent in higher-eukaryotes. Arginine kinase increases continuously during the parasites' exponential growth, reaching maximum values at the late stationary phase [36]. When the parasites were subjected to an oxidative media, arginine kinase was also up-regulated suggesting a regulatory function under starvation and oxidative stress conditions [37, 38]. This hypothesis was confirmed using a transgenic *T. cruzi* model that over-expresses the arginine kinase gene. The relationship between arginine metabolism and transport is addressed in later paragraphs.

2.2. Proline Transport

L-proline is the main energy source in insect vector stages of most Trypanosomatids, including *Trypanosoma cruzi* epimastigotes. It is an important metabolite that is also involved in the differentiation of intracellular epimastigote to trypomastigote forms and, which is required for the establishment of infection in the mammalian host [20]. Previous work showed that L-proline is taken up from the extracellular environment through two active transporters and converted into five intermediates of the Krebs cycle (e.g., citrate, isocitrate, malate, succinate and oxaloacetate), pyruvate and the amino acids glutamate and aspartate, which are rapidly metabolized [18], this seemed to comprise a conventional proline oxidation pathway. This idea is supported by the presence of two putative genes encoding proline oxidases and one putative gene encoding a pyrroline-5-carboxylate dehydrogenase annotated in the *T. cruzi* genome database [4]. Uptake of proline occurred in *T. cruzi* by two components with different substrate affinity, called system B (low affinity) and system A (high affinity). System B consistently appeared more specific than System A when an excess of competing amino acids were used in transport inhibition assays. Furthermore, the high affinity system is inhibited by L-tryptophan, but the low affinity system is not. Both systems were found to be insensitive to the intracellular proline concentration and D-proline did not inhibit L-proline uptake showing that both systems are stereospecific. Both systems were Na+ and K+ independent but dependant on energy since ATP depletion impairs L-proline uptake. The combined action of carbonyl cyanide p-trifluoromethoxyphenyl hydrazone (FCCP) and oligomycin, and the dependence of activity on pH, further differentiated between the two systems leading to the conclusion that the high affinity system is a H+ gradient-dependant transporter whereas the low affinity system depends directly on ATP [21].

2.3. Aspartate and Glutamate Transport

A high-affinity L-aspartate transport system was characterized in *T. cruzi*. The estimated *Km* value for aspartate is similar to those reported for other eukaryotic organisms, i.e. glutamate/aspartate transporters from the "Excitatory Amino Acid Transporters" family

(EAAT) [39]. Several amino acids were used to estimate substrate recognition by measuring the percent of inhibition of L-aspartate uptake. Aspartate transport was only competed by L-glutamate suggesting that a glutamate/aspartate transporter mediates the transport activity. Some adaptive changes in the aspartate transport process were also studied. When compared parasites from different days of culture, an increase in aspartate uptake was observed accompanying the reduction in parasite replication rates, the increase in cell density, and nutrient depletion of the medium. The uptake of aspartate in trypomastigotes was measured at the same rate of mid-log epimastigotes indicating that aspartate is transported by the parasite not only in the insect vector, but also in the mammalian host. These regulation mechanisms are quite different to those previously reported for cationic amino acid transport in *T. cruzi* [21, 30, 32]. In addition, aspartate transport seems to be a substrate-regulated process, since the amino acid starvation induces an increase in the transport rate. This might indicates the presence of a rapid mechanism to restore the intracellular aspartate concentration. On the other hand, kinetic data show a single saturable system with a medium-low affinity for glutamate. Transport was not affected by parasite nutrient starvation but aspartate, alanine, glutamine, asparagine, methionine, oxaloacetate and a-ketoglutarate competed with the substrate uptake. Glutamate transport was strongly dependent on pH, but not on Na+ or K+ concentrations in the extracellular medium. These data were consistent with the sensitivity of the system to H+ ionophores suggesting that transport is driven by H+ concentration gradient across the cytoplasmic membrane [40].

2.4. Cysteine Transport

L-cysteine plays an important role in stability, structure, catalytic activity, and regulation of numerous proteins due to the unique properties of sulfur and thiol. In addition, an oxidized derivative of L-cysteine, cystine form disulfide bonds which stabilize the tertiary structure of proteins [41]. L-cysteine is also utilized for the production of ubiquitous iron–sulfur clusters, which participate in electron transfer, redox regulation, nitrogen fixation, and sensing for regulatory processes, and are probably present in all living organisms [42]. In the specific case of Trypanosomatids, L-cysteine is particularly relevant as a substrate in the synthesis of trypanothione. Despite it could be synthesized de novo, L-cysteine is actively transported in *Trypanosoma cruzi* epimastigote cells. L-cysteine uptake is highly specific; none of the amino acids assayed produce significative differences on transport rates. This transport is finely regulated by amino acids starvation, extracellular pH and between the parasite growth phases. In addition, L-cysteine is incorporated post-translationally to proteins, suggesting its role in iron-sulfur cores formation.

3. A Multigene Family of Permases Unique to Trypanosomatids? The AAAP Family

As described in the previous point, similar biochemical properties and regulation mechanism have been observed for *T. cruzi* amino acid systems, such as arginine, proline and aspartate/glutamate and cysteine suggesting that all permeases involved in these transport systems may belong to the same family of amino acid transporters. For this reason different bioinformatic approaches were performed using all known members of each of the families of amino acids permeases previously described, and the preliminary data from the *T. cruzi* genome project , in pre -genomic era [7].

One of the major amino acid transporter families is the “Amino Acid/Auxin Permease Family” (AAAP; TC 2.A.18) which is largely represented in plants [43]. In *T. cruzi*, many members of this family were first identified by our group [7] and then confirmed by the TriTryps genome project [44]. This *T. cruzi* family, called TcAAAP, has more than 30 genes coding for proteins with lengths of 400-500 amino acids and 10-12 predicted transmembrane α-helical spanners (Figure 1 and 2).

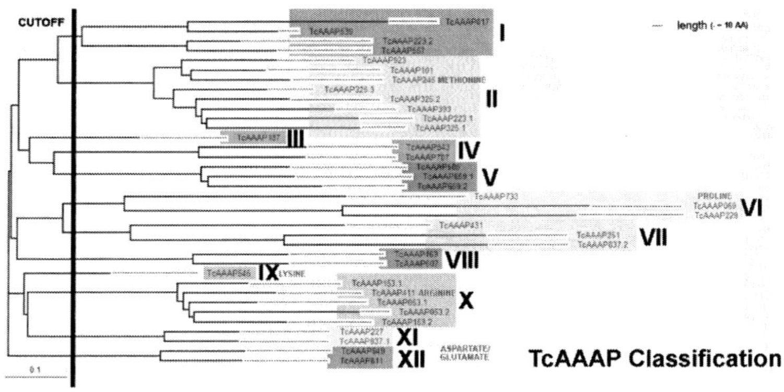

Figure 1. *Phenogram representation of AAAP sequences from Trypanosoma cruzi.* The phenogram was constructed with all 36 members of the Amino Acid / Auxin Permeases Family from *T. cruzi* genome, using Clustal W and the Neighbor-Joining programs; and applying a PAM250 score matrix. Related sequences were clustered in 12 major groups which are indicated with numbers. Amino Acid transporters with a known substrate are also indicated. Genes have been named according to the organism *T. cruzi* (Tc), the transporter gene family (AAAP, TCDB - 2.A.18) and the three last numbers of the systematic ID from the TriTrypDB.

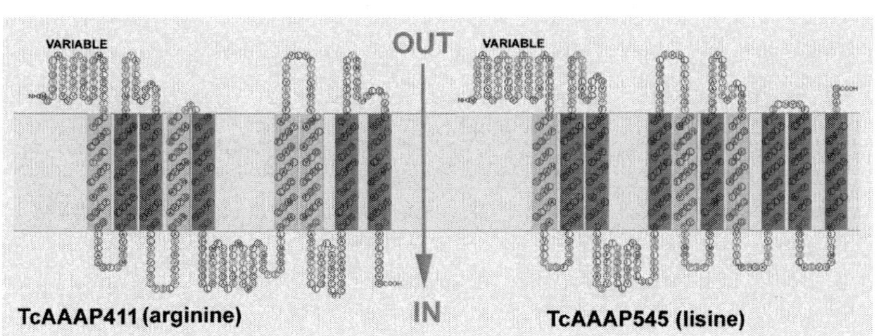

Figure 2. *Schematic representation of the arginine permease (TcAAAP411) and the lysine permease (TcAAAP545).* N-terminal variable region of the TcAAAP permeases are indicated. Vertical bars represent the transmembrane spans, and the arrow the direction of the transport flux.

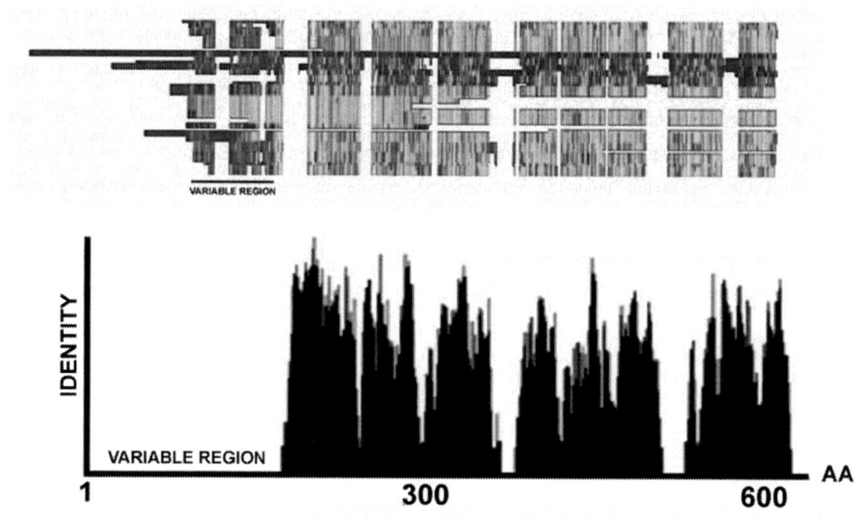

Figure 3. *Sequence analysis of TcAAAP family.* Sequence identity grayscale-based scheme performed with 36 TcAAAP genes using ClustalW alignment. Areas shaded in light gray denote conserved amino acids, whereas areas shaded in dark gray denote lack of identity, green are intermediate values. The same result is represented using a graphic where the identity values are plotted as a function of the amino acid position.

It was also demonstrated by RT-PCR that all the members identified corresponded to expressed sequences. As figure 3 shows, it is evident that the N-terminal domain (about 90-aa) represents a highly variable region (5% of consensus) in contrast with the central and C-terminal domains which have more than 70% of consensus amino acid positions. Finally, the presence of orthologous amino acid transporter genes has been shown in *Trypanosoma brucei, Leishmania major, Leishmania donovani, Plasmodium falciparum* and *Plasmodium yoelii*. One interesting feature of this permease family is the absence of similar sequences in mammalian genomes; however, the presence of unidentified orthologs could not be discarded [45].

4. Polyamine Transporters, the Unique Way to Obtain These Essential Molecules in *Trypanosoma Cruzi*

Polyamines are polycationic compounds that play a critical role as regulators of cell growth and differentiation [46]. In trypanosomes polyamines are involved in crucial cellular processes including the synthesis of the antioxidant compound trypanothione (N1,N8-bis(glutathionyl)spermidine) which has been found exclusively in Trypanosomatid protozoa [47]. Polyamines could be obtained by synthesis de novo from ornithine and in some cases from arginine (via agmatine), or transported from the extracellular medium. In contrast with other protozoa, *Trypanosoma cruzi* is auxotrophic for polyamines because of their inability to synthesize putrescine due to the lack of both, arginine decarboxylase (ADC) and ornithine decarboxylase (ODC) (Figure 4) [48, 49]. Therefore, the intracellular availability of polyamines in *T. cruzi* depends exclusively on transport processes. Polyamines are essential compounds for parasites growth, replication and stage differentiation [50]. In the specific case of *Trypanosoma cruzi*, polyamine transport is critical because it is the only Trypanosomatid organism lacking the ability to synthesize putrescine [49, 51]. Consequently, polyamines must be obtained from the extracellular medium exclusively through transport processes. Because the essentiality and differences in metabolic pathway respect to the hosts, polyamine transport and metabolism constitutes an appropriate target for chemotherapeutic strategies against parasitic diseases [52].

It was postulated since Bacchi and coworkers reported that blocking ornithine decarboxylase enzymatic activity using α-difluoromethylornithine (DFMO) was able to cure acute infections of mice with *Trypanosoma brucei* brucei [53]. The same drug has also been successfully used in humans against sleeping sickness caused by *Trypanosoma brucei* gambiense [54].

Respect to *T. cruzi*, a high-affinity diamine transport system has been biochemically characterized in epimastigotes [28, 55]. This transporter showed a high specificity for the diamines putrescine and cadaverine and a low specificity for the polyamines spermidine and spermine. Recently, a high-affinity polyamine permease from *Leishmania* major was cloned and functionally characterized [56]. This was the first report of a eukaryotic polyamine permease studied at a molecular level. Once the *L. major* LmPOT1 report in *Leishmania* was made we were able to identify a member of a previously described *T. cruzi* putative amino acid transporter (TcPAT) family [7] as a candidate for a polyamine permease. When TcPAT12 was expressed in Xenopus laevis oocytes a high-affinity spermidine uptake was detected with similar kinetic parameters to those measured in *T. cruzi* epimastigotes and in other polyamine transporters previously reported.

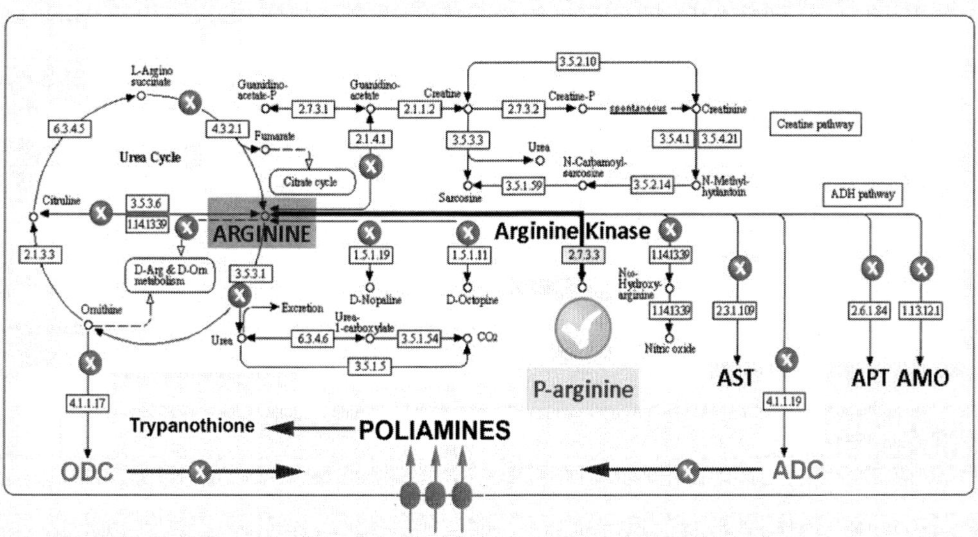

Figure 4. *Arginine and Polyamine metabolism in T. cruzi.* Using the KEGG arginine and proline metabolic pathways as a reference, genes coding for putative arginine and polyamine metabolism enzymes in *T. cruzi* were indicated. All enzyme EC numbers are indicated into the boxes, routes marked with a cross were not identified in *T. cruzi*. AST: arginine N-succinyltransferase; ADC: arginine decarboxylase; ODC: ornithine decarboxylase; APT: arginine-pyruvate transaminase; and AMO: arginine 2-monooxygenase. Gray circles indicate the polyamine permeases.

Both Km values corresponding to spermidine uptake in oocytes or epimastigotes parasites, were calculated around the same order as for *L. major* LmPOT1 [57] or putrescine uptake biochemically characterized by different groups in *Trypanosoma cruzi*, *Leishmania mexicana* and *Crithidia fasciculata* [28, 58].

A permease from *Saccharomyces cerevisiae* named AGP2 was first described as an amino acid (L-carnitine) transporter, and recently was redefined as a high-affinity polyamine transporter [59]. Similarly, TcPAT12 in the oocyte context also showed a reduced transport activity for putrescine and L-arginine. Subsequently, it was demonstrated that this gene, renamed TcPOT1.2 and one alternative allele, TcPOT1.1, function as polyamine transporters, and when they are overexpressed in *T. cruzi* epimastigotes revealed that TcPOT1.1 and TcPOT1.2 were high-affinity transporters for both putrescine and cadaverine but not for spermidine or spermine. This last work revalidated these permeases as key components of the *T. cruzi* polyamine transport pathway, amenable to therapeutic manipulation [56].

Studies carried out with different trypanosomes have shown that polyamines are indispensable for many life processes including macromolecular synthesis and progression through the cell cycle in this phyla, as is also the case in other eukaryotic cells [46]. Polyamines are natural and ubiquitous polycationic compounds, being the most abundant putrescine (in rigor a diamine), spermidine (an aminorpopil putrescine) and spermine (an aminopropil spermidine).

In the particular case of Trypanosomatids, it has been seen absence or extremely low concentration of spermine but significant concentration of other less common polyamines such as cadaverine (a diamimopentane), a diamine previously found in prokaryotes [51]. In addition to common and essential roles of polyamines in cells, in trypanosomes polyamines are involved in other crucial cellular processes including infectivity (Carrillo et al; unpublished data) and metacyclogenesis [50] and the synthesis of the antioxidant compound trypanothione (bis-glutathionyl spermidine) which has been found exclusively in Trypanosomatid protozoa [60, 61]. This molecule participates in the redox intracellular equilibrium of Trypanosomatid parasites [47, 62].

The biosynthesis of polyamines starts with the conversion of ornithine into putrescine, catalysed by the enzyme ornithine decarboxylase (ODC). This key enzyme in polyamine biosynthesis has been characterised at biochemical and molecular level in numerous Trypanosomatids such as *Trypanosoma brucei* [63, 64], *Crithidia fasciculata* [65], *Phytomonas* Jma [66] and different species of *Leishmania* [67, 68], and it has been proposed that the differences in the stability of the ODC enzyme in Trypanosomatids could be associated with the degree of complexity and type of life cycles [69]. Polyamines could be obtained by synthesis de novo from ornithine and in some cases from arginine (for convertion to ornitine or, alternatively, via agmatine), or transported from the extracellular medium. In contrast with other protozoa, *T. cruzi* is auxotrophic for polyamines because of their inability to synthesize putrescine due to the lack of both ADC and ODC [48, 49, 70], as it was mentioned previously. Therefore, the intracellular availability of polyamines in *T. cruzi* depends exclusively on transport processes.

Polyamine transport and metabolism constitutes an appropriate target for chemotherapeutic strategies against parasitic diseases. An example to illustrates this is that the newest drug to be approved by the U.S. Food and Drug Administration (FDA) for use against African trypanosomiasis is the ODC irreversible inhibitor DFMO.

Athough this compound was originally designed as an antitumor agent, it has proven to be highly effective against *Trypanosoma brucei* gambiense infections.

5. Amino Acids Transporters from the AAAP Family

As it was said above, Trypanosomatids' permeases are key proteins in parasite metabolism since they participate in the adaptation of parasites to different environments, but, despite its importance, recently a few years ago the molecular features of these transporters began to be studied. In this point the molecular aspects of Trypanosomatids' amino acid transporters of the Amino Acid/Auxin Permease (AAAP) family, are described.

As it was mentioned, *T. cruzi* arginine transport systems have been largely studied during the last decade [30]; even so, the first evidence of the molecular determinants of this process was recently reported. TcAAAP411, a member of the TcAAAP family, has an arginine transport activity in a yeast model [71]. A similar arginine transporter (LdAAP3) was also identified in *Leishmania donovani* [72]. LdAAP3 regulation mainly depends on the availability of the extracellular substrate since amino acid starvation produces an increase in arginine transport and LdAAP3 protein abundance [73]. Interestingly, a similar mechanism of regulation was described for arginine and other *T. cruzi* amino acids transport systems [74], supporting the hypothesis that most of the previously characterized transport systems in Trypanosomatids involve members of the AAAP family. On the other hand, lysine transporters from this family were also identified and characterized in *Trypanosoma* spp. and *Leishmania* spp. [75]. In higher eukaryotes most of cationic amino acid transporters incorporate arginine and lysine by a single permease [76-78]. On the contrary, parasite permeases translocate cationic amino acids by different mono-specific transporters [30, 75]. *T. cruzi* amino acid transporters have to translocate their substrates against a concentration gradient that can reach difference of about two orders of magnitude between the mammalian host plasma and the parasite cytosol. These features suggest that amino acid transporters could be used as a gateway for traditional anti-parasitic drugs by the addition of suitable functional groups recognized by these permeases [79]. Most recently, to test if the arginine permease activity regulates downstream processes of the arginine metabolism (Figure 5), the expression of the single *T. cruzi* enzyme that uses arginine as substrate, arginine kinase, was evaluated in TcAAAP411-overexpressing epimastigotes. In this parasite model, intracellular arginine concentration significantly increases and ATP level remains constant until cultures reach the stationary phase of growth, when dramatically decrease respect to the controls. Interestingly, Western Blot analysis demonstrated that arginine kinase is significantly down-regulated during the stationary phase of growth in TcAAAP411 epimastigotes. This decrease could represent a compensatory mechanism for the increase in ATP consumption as a consequence of the displacement of the reaction equilibrium of arginine kinase, when the intracellular arginine concentration augments and the glucose from the medium is exhausted [Miranda, et al. PLoS One, in press].

Unlike arginine transporters in higher eukaryotes that transport also lysine, *Trypanosoma* and *Leishmania* parasites transporters translocate only arginine. This phenomenon prompted us to identify and characterize parasite lysine transporters. Working in collaboration with Dr. Zilberstein group, we demonstrated that LdAAP7 and TcAAAP545 encode lysine-specific permeases in *L. donovani* and *T. cruzi*, respectively. These two lysine permeases are both members of the large AAAP family and share certain biochemical properties, such as specificity and Km. However, we evidence that LdAAP7 and TcAAAP545 differ in their regulation and localization; such differences are likely a reflection of the dissimilar *L. donovani* and *T. cruzi* life cycles. Failed attempts to delete both alleles of LdAAP7 or TcAAAP545 support the premise that these are essential genes that encode the only lysine permeases expressed in *L. donovani* promastigotes and *T. cruzi* epimastigotes, respectively. Importantly, the likely essential nature of these parasite lysine permeases together with their lysine specificity raise the possibility that LdAAP7 or TcAAAP545 could serve as vehicles for selective drug delivery or drug targets [80].

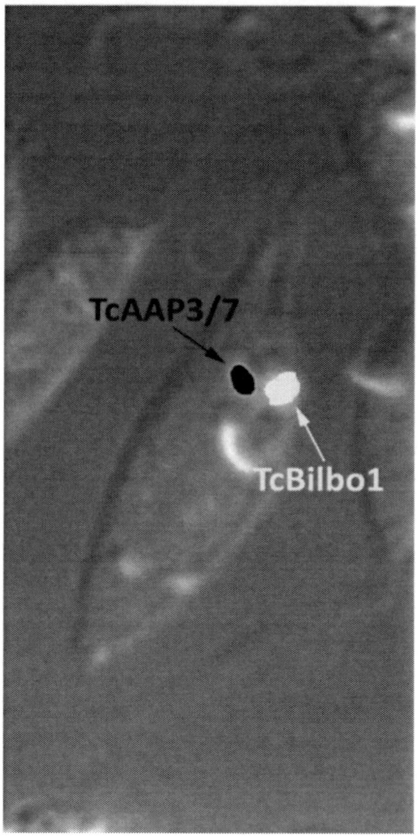

Figure 5. *Subcellular localization of T. cruzi amino acid permeases. T. cruzi* epimastigotes transfected with expression plasmids containing a flagellar pocket marker (TcBilbo1, white) and the permeases TcAAAP411 and TcAAAP545 (black) were analyzed by fluorescence microscopy.

Using immunofluorescence techniques we also determined that the *T. cruzi* arginine permease TcAAAP411 and the specific lysine transporter TcAAAP545 co-localize in a specialized region of the plasma membrane named flagellar pocket, staining a single locus close to the flagellar pocket collar (Figure 5). The flagellar pocket is an invagination in the Trypanosomatids plasma membrane that constitutes a multiorganelle complex that is involved in the exchange with the extracellular medium and the cell polarity and division [81].

TcAAAP411 also co-localized with TcAAAP545 suggesting that the TcAAAP family could be concentrated in that unique region of interchange with the extracellular medium, the flagellar pocket.

6. Nutrient Permeases as Future Therapeutic Targets

Transporters can be considered as cellular environmental sensors, responding to the presence of substrate molecules in extracellular media, and making them available for physiological processes. This is particularly relevant in organisms such as trypanosomes that endure broad range variations in their environment. However, Trypanosomatid transporters have been poorly studied as possible drug targets in spite of their potential already demonstrated for *Trypanosoma* spp. and *Leishmania* spp.

In fact, the uptake of trypanocidal drugs used for *T. brucei* therapy such as the melamine phenyl arsenical melarsoprol and diamidines occurs through adenosine transporters [82, 83]. Other important feature of Trypanosomatids' permeases is the possibility of being used for the entry of drugs mimicking natural substrates. For example, *Trypanosoma brucei*, have some unusual plasma membrane transporters, in particular the P2 aminopurine transporter and related permeases, which have been used for the selective targeting of trypanocidal compounds to the organism. The addition of melamine-based P2-targeting motifs to three different classes of compound to try and improve activity through increased selective uptake was recently reported [79].

Taking these evidences into account, the permeases in trypanosomatid parasites seem to be valuable drug targets, but further research on their characterization would be needed to validate their use.

References

[1] Barrett, M.P., et al., The trypanosomiases. *Lancet,* 2003. 362(9394): p. 1469.

[2] Barrett, M.P. and I.H. Gilbert, Targeting of toxic compounds to the trypanosome's interior. *Adv. Parasitol.*, 2006. 63: p. 125.

[3] Fairlamb, A.H., Metabolic pathway analysis in trypanosomes and malaria parasites. *Philos. Trans. R. Soc. Lond. B Biol. Sci.,* 2002. 357(1417): p. 101.

[4] Silber, A.M., et al., Amino acid metabolic routes in Trypanosoma cruzi: possible therapeutic targets against Chagas' disease. *Curr. Drug Targets Infect. Disord.,* 2005. 5(1): p. 53.

[5] Cazzulo, J.J., Aerobic fermentation of glucose by trypanosomatids. *FASEB J.,* 1992. 6(13): p. 3153.

[6] Silber, A.M., et al., Glucose uptake in the mammalian stages of Trypanosoma cruzi. *Mol. Biochem. Parasitol.,* 2009. 168(1): p. 102.

[7] Bouvier, L.A., et al., Post genomic analysis of permeases from the amino acid/auxin family in protozoan parasites. *Biochem. Biophys. Res. Commun.,* 2004. 321(3): p. 547.

[8] Soares, M.J. and W. De Souza, Cytoplasmic organelles of trypanosomatids: a cytochemical and stereological study. *J. Submicrosc. Cytol. Pathol.,* 1988. 20(2): p. 349.
[9] Opperdoes, F.R. and P. Borst, Localization of nine glycolytic enzymes in a microbody-like organelle in Trypanosoma brucei: the glycosome. FEBS Lett., 1977. 80(2): p. 360.
[10] Barrett, M.P., et al., Trypanosome glucose transporters. *Mol. Biochem. Parasitol.,* 1998. 91(1): p. 195.
[11] Racagni, G.E. and E.E. Machado de Domenech, Characterization of Trypanosoma cruzi hexokinase. *Mol. Biochem. Parasitol.,* 1983. 9(2): p. 181.
[12] Urbina, J.A. and V. Azavache, Regulation of energy metabolism in Trypanosoma (Schizotrypanum) cruzi epimastigotes. II. NAD+-dependent glutamate dehydrogenase. *Mol. Biochem. Parasitol.,* 1984. 11: p. 241.
[13] Aguilar, Z. and J.A. Urbina, The phosphofructokinase of Trypanosoma (Schizotrypanum) cruzi: purification and kinetic mechanism. *Mol. Biochem. Parasitol.,* 1986. 21(2): p. 103.
[14] Taylor, M. and W.E. Gutteridge, The regulation of phosphofructokinase in epimastigote Trypanosoma cruzi. *FEBS Lett.,* 1986. 201(2): p. 262.
[15] Cazzulo, J.J., M.C. Cazzulo Franke, and B.M. Franke de Cazzulo, On the regulatory properties of the pyruvate kinase from Trypanosoma cruzi epimastigotes. *FEMS Microbiol. Lett.,* 1989. 50(3): p. 259.
[16] Bakker, B.M., et al., Contribution of glucose transport to the control of the glycolytic flux in Trypanosoma brucei. *Proc. Natl. Acad. Sci. U S A,* 1999. 96(18): p. 10098.
[17] Zeledon, R., Comparative physiological studies on four species of hemoflagellates in culture. II. Effect of carbohydrates and related substances and some amino compounds on the respiration. *J. Parasitol.,* 1960. 46: p. 541.
[18] Sylvester, D. and S.M. Krassner, Proline metabolism in Trypanosoma cruzi epimastigotes. *Comp. Biochem. Physiol. B,* 1976. 55(3B): p. 443.
[19] Contreras, V.T., et al., In vitro differentiation of Trypanosoma cruzi under chemically defined conditions. *Mol. Biochem. Parasitol.,* 1985. 16(3): p. 315.
[20] Tonelli, R.R., et al., L-proline is essential for the intracellular differentiation of Trypanosoma cruzi. *Cell Microbiol.,* 2004. 6(8): p.733.
[21] Silber, A.M., et al., Active transport of L-proline in Trypanosoma cruzi. *J. Eukaryot. Microbiol.,* 2002. 49(6): p. 441.
[22] Min, H.S., Evidence of dual mechanisms for transport of carbohydrate in cells of Crithidia sp. from Arilus cristatus. *Life Sci.,* 1968. 7(20): p. 1105.
[23] Sanchez, G. and C.P. Read, Carbohydrate transport in Trypanosoma lewisi. *Comp. Biochem. Physiol.,* 1969. 28(2): p. 931.
[24] Southworth, G.G. and C.P. Read, Carbohydrate transport in Trypanosoma gambiense. *J. Protozool.,* 1969. 16(4): p. 720.
[25] Hampton, J.R., Lysine uptake in cultured Trypanosoma cruzi: interactions of competitive inhibitors. *J. Protozool.,* 1970. 17(4): p. 597.
[26] De Boiso, J.F. and A.O. Stoppani, Phosphate transport in Trypanosoma cruzi. *Experientia,* 1972. 28(10): p. 1162.
[27] Manjra, R. and D.G. Dusanic, Transport of nucleosides in Trypanosoma lewisi. *Comp. Biochem. Physiol. B,* 1973. 44(2): p. 587.

[28] Gonzalez, N.S., C. Ceriani, and I.D. Algranati, Differential regulation of putrescine uptake in Trypanosoma cruzi and other trypanosomatids. *Biochem. Biophys. Res. Commun.,* 1992. 188(1): p. 120.

[29] Kandpal, M., et al., Kinetics and molecular characteristics of arginine transport by Leishmania donovani promastigotes. *Mol. Biochem. Parasitol.,* 1995. 71(2): p. 193.

[30] Pereira, C.A., et al., L-arginine uptake and L-phosphoarginine synthesis in Trypanosoma cruzi. *J. Eukaryot. Microbiol.*, 1999. 46(6): p. 566.

[31] Pereira, C.A., et al., Arginine metabolism in Trypanosoma cruzi is coupled to parasite stage and replication. *FEBS Lett.,* 2002. 526(1-3): p. 111.

[32] Canepa, G.E., et al., Biochemical characterization of a low-affinity arginine permease from the parasite Trypanosoma cruzi. *FEMS Microbiol. Lett.,* 2004. 236(1): p. 79.

[33] Pereira, C.A., et al., Trypanosoma cruzi arginine kinase characterization and cloning. A novel energetic pathway in protozoan parasites. *J. Biol. Chem.,* 2000. 275(2): p. 1495.

[34] Pereira, C.A., et al., Arginine kinase: a common feature for management of energy reserves in African and American flagellated trypanosomatids. *J. Eukaryot. Microbiol.,* 2002. 49(1): p. 82.

[35] Canepa, G.E., et al., Arginine kinase in Phytomonas, a trypanosomatid parasite of plants. *Comp. Biochem. Physiol. B Biochem. Mol. Biol.,* 2011. 160(1): p. 40.

[36] Alonso, G.D., et al., Arginine kinase of the flagellate protozoa Trypanosoma cruzi. Regulation of its expression and catalytic activity. *FEBS Lett.,* 2001. 498(1): p. 22.

[37] Pereira, C.A., et al., Arginine kinase overexpression improves Trypanosoma cruzi survival capability. *FEBS Lett.,* 2003. 554(1-2): p. 201.

[38] Miranda, M.R., et al., Trypanosoma cruzi: Oxidative stress induces arginine kinase expression. *Exp. Parasitol.,* 2006. 114(4): p. 341.

[39] Umesh, A., et al., Functional characterization of a glutamate/aspartate transporter from the mosquito Aedes aegypti. *J. Exp. Biol.,* 2003. 206 (Pt 13): p. 2241.

[40] Silber, A.M., et al., Biochemical characterization of the glutamate transport in Trypanosoma cruzi. *Int. J. Parasitol.,* 2006. 36(2): p. 157.

[41] Nozaki, T., V. Ali, and M. Tokoro, Sulfur-containing amino acid metabolism in parasitic protozoa. *Adv. Parasitol.,* 2005. 60: p. 1.

[42] Johnson, D.C., et al., Structure, function, and formation of biological iron-sulfur clusters. *Annu. Rev. Biochem.,* 2005. 74: p. 247.

[43] Young, G.B., et al., The amino acid/auxin:proton symport permease family. *Biochim. Biophys. Acta,* 1999. 1415(2): p. 306.

[44] Berriman, M., et al., The genome of the African trypanosome Trypanosoma brucei. *Science,* 2005. 309(5733): p. 416.

[45] Akerman, M., et al., Novel motifs in amino acid permease genes from Leishmania. *Biochem. Biophys. Res. Commun.*, 2004. 325(1): p. 353.

[46] Igarashi, K. and K. Kashiwagi, Polyamines: mysterious modulators of cellular functions. *Biochem. Biophys. Res. Commun.,* 2000. 271(3): p. 559.

[47] Fairlamb, A.H., et al., Trypanothione: a novel bis (glutathionyl) spermidine cofactor for glutathione reductase in trypanosomatids. *Science,* 1985. 227(4693): p. 1485.

[48] Carrillo, C., et al., Lack of arginine decarboxylase in Trypanosoma cruzi epimastigotes. *J. Eukaryot. Microbiol.,* 2003. 50(5): p. 312.

[49] Carrillo, C., et al., Trypanosoma cruzi epimastigotes lack ornithine decarboxylase but can express a foreign gene encoding this enzyme. *FEBS Lett.,* 1999. 454(3): p. 192.

[50] Barclay, J.J., et al., Trypanosoma cruzi Coexpressing Ornithine Decarboxylase and Green Fluorescence Proteins as a Tool to Study the Role of Polyamines in Chagas Disease Pathology. *Enzyme Res.*, 2011. 2011: p. 657460.

[51] Ariyanayagam, M.R. and A.H. Fairlamb, Diamine auxotrophy may be a universal feature of Trypanosoma cruzi epimastigotes. *Mol. Biochem. Parasitol.,* 1997. 84(1): p. 111.

[52] Algranati, I.D., Polyamine metabolism in Trypanosoma cruzi: studies on the expression and regulation of heterologous genes involved in polyamine biosynthesis. *Amino Acids,* 2010. 38(2): p. 645.

[53] Bitonti, A.J., et al., Catalytic irreversible inhibition of Trypanosoma brucei brucei ornithine decarboxylase by substrate and product analogs and their effects on murine trypanosomiasis. *Biochem. Pharmacol.,* 1985. 34(10): p. 1773.

[54] Van Nieuwenhove, S., et al., Treatment of gambiense sleeping sickness in the Sudan with oral DFMO (DL-alpha-difluoromethylornithine), an inhibitor of ornithine decarboxylase; first field trial. *Trans. R Soc. Trop. Med. Hyg.,* 1985. 79(5): p. 692.

[55] Le Quesne, S.A. and A.H. Fairlamb, Regulation of a high-affinity diamine transport system in Trypanosoma cruzi epimastigotes. *Biochem. J.,* 1996. 316 (Pt 2): p. 481.

[56] Hasne, M.P. and B. Ullman, Identification and characterization of a polyamine permease from the protozoan parasite Leishmania major. *J. Biol. Chem.,* 2005. 280(15): p. 15188.

[57] Carrillo, C., et al., Molecular and functional characterization of a spermidine transporter (TcPAT12) from Trypanosoma cruzi. *Biochem. Biophys. Res. Commun.*, 2006. 344(3): p. 936.

[58] Gonzalez, N.S. and I.D. Algranati, Regulation of putrescine uptake in Leishmania mexicana promastigotes. *Cell Mol. Biol.* (Noisy-le-grand), 1994. 40(7): p. 907.

[59] Aouida, M., et al., AGP2 encodes the major permease for high affinity polyamine import in Saccharomyces cerevisiae. *J. Biol. Chem.,* 2005. 280(25): p. 24267.

[60] Oza, S.L., et al., A single enzyme catalyses formation of Trypanothione from glutathione and spermidine in Trypanosoma cruzi. *J. Biol. Chem.,* 2002. 277(39): p. 35853.

[61] Hunter, K.J., S.A. Le Quesne, and A.H. Fairlamb, Identification and biosynthesis of N1,N9-bis(glutathionyl)aminopropylcadaverine (homotrypanothione) in Trypanosoma cruzi. *Eur. J. Biochem.*, 1994. 226(3): p. 1019.

[62] Birkholtz, L.M., et al., Polyamine homoeostasis as a drug target in pathogenic protozoa: peculiarities and possibilities. *Biochem. J.,* 2011. 438(2): p. 229.

[63] Bacchi, C.J., et al., Prevention by polyamines of the curative effect of amicarbalide and imidocarb for Trypanosoma brucei infections in mice. *Biochem. Pharmacol.,* 1981. 30(8): p. 883.

[64] Li, F., et al., Procyclic Trypanosoma brucei cell lines deficient in ornithine decarboxylase activity. *Mol. Biochem. Parasitol.,* 1996. 78(1-2): p. 227.

[65] Ceriani, C., N.S. Gonzalez, and I.D. Algranati, Ornithine decarboxylase from Crithidia fasciculata is metabolically unstable and resistant to polyamine down-regulation. *FEBS Lett.,* 1992. 301(3): p. 261.

[66] Marcora, M.S., et al., Polyamine biosynthesis in Phytomonas: biochemical characterisation of a very unstable ornithine decarboxylase. *Int. J. Parasitol.,* 2010. 40(12): p. 1389.

[67] Coons, T., et al., Alpha-difluoromethylornithine resistance in Leishmania donovani is associated with increased ornithine decarboxylase activity. *Mol. Biochem. Parasitol.,* 1990. 39(1): p. 77.
[68] Gonzalez, N.S., et al., Control of Leishmania mexicana proliferation by modulation of polyamine intracellular levels. *Biochem. Biophys. Res. Commun.,* 1991. 180(2): p. 797.
[69] Carrillo, C., et al., Sensitivity of trypanosomatid protozoa to DFMO and metabolic turnover of ornithine decarboxylase. *Biochem. Biophys. Res. Commun.,* 2000. 279(2): p. 663.
[70] Carrillo, C., et al., Heterologous expression of a plant arginine decarboxylase gene in Trypanosoma cruzi. Biochim Biophys Acta, 2004. 1674(3): p. 223.
[71] Carrillo, C., et al., Trypanosoma cruzi amino acid transporter TcAAAP411 mediates arginine uptake in yeasts. *FEMS Microbiol. Lett.,* 2010. 306(2): p. 97.
[72] Shaked-Mishan, P., et al., A novel high-affinity arginine transporter from the human parasitic protozoan Leishmania donovani. *Mol. Microbiol.,* 2006. 60(1): p. 30.
[73] Darlyuk, I., et al., Arginine homeostasis and transport in the human pathogen Leishmania donovani. *J. Biol. Chem.,* 2009. 284(30): p. 19800.
[74] Pereira, C.A., et al., [Trypanosoma cruzi: transport of essential metabolites acquired from the host]. *Medicina* (B Aires), 2008. 68(5): p. 398.
[75] Inbar, E., et al., Lysine transporters in human trypanosomatid pathogens. *Amino Acids,* 2010.
[76] Christensen, H.N., Role of amino acid transport and countertransport in nutrition and metabolism. *Physiol. Rev.,* 1990. 70(1): p. 43.
[77] Closs, E.I., et al., Identification of a low affinity, high capacity transporter of cationic amino acids in mouse liver. *J. Biol. Chem.,* 1993. 268(10): p. 7538.
[78] Kim, J.W., et al., Transport of cationic amino acids by the mouse ecotropic retrovirus receptor. *Nature,* 1991. 352(6337): p. 725.
[79] Chollet, C., et al., Targeted delivery of compounds to Trypanosoma brucei using the melamine motif. *Bioorg. Med. Chem.,* 2009. 17(6): p. 2512.
[80] Inbar, E., et al., Lysine transporters in human trypanosomatid pathogens. *Amino Acids,* 2012. 42(1): p. 347.
[81] Field, M.C. and M. Carrington, The trypanosome flagellar pocket. *Nat. Rev. Microbiol.,* 2009. 7(11): p. 775.
[82] Carter, N.S. and A.H. Fairlamb, Arsenical-resistant trypanosomes lack an unusual adenosine transporter. *Nature,* 1993. 361(6408): p. 173.
[83] Carter, N.S., B.J. Berger, and A.H. Fairlamb, Uptake of diamidine drugs by the P2 nucleoside transporter in melarsen-sensitive and -resistant Trypanosoma brucei brucei. *J. Biol. Chem.,* 1995. 270(47): p. 28153.

In: Parasites
Editor: Gilmar S. Erzinger
ISBN: 978-1-62257-692-0

Chapter 4

PEPTIDASES IN PARASITIC NEMATODES: A REVIEW

David Malagón[1,2], Rocío Benítez[2], Martin Kašný[3] and Francisco Javier Adroher[2,*]

[1]Parasite Cell Biology Laboratory, Queensland Institute of Medical Research, Herston, Brisbane, Australia
[2]Departamento de Parasitología, Facultad de Farmacia, Universidad de Granada, Granada, Spain
[3]Department of Parasitology, Faculty of Science, Charles University in Prague, Prague, Czech Republic

ABSTRACT

The nematodes are, after the insects, the group of organisms with the largest number of species identified. They include members of great medical, veterinary and agricultural significance, making this group one of the most important animal parasites. However there are many gaps in our knowledge of them. For example, there is still not a single nematode species for which we have detailed knowledge of feeding, digestion and nutritional requirements, showing that there are still many aspects to be learned about nutrition in nematodes [1]. Our understanding of the process of protein digestion, a very important function in the biology of any organism, is still poor since our knowledge is composed of fragmentary data for different groups of nematodes. It is believed that peptidases are essential during the development process and in the most critical moments of parasite-host interactions, and are thus directly involved with the growth and survival of the parasite. Their identification and characterization are important for basic understanding of the biology of the parasite, and their relevance to parasitic nematodes as virulence factors is clear. Consequently, peptidases are currently viewed as potential targets for vaccines, drugs and serodiagnosis. Despite this, in most cases, the precise physiological functions of peptidases in parasites are not known [2].

Peptidases comprise a large class of hydrolytic enzymes in parasitic nematodes, participating in nutrition through digestion of host proteins [3]. They also act in the

* E-mail address: fadroher@ugr.es.

moulting and resorption of the cuticle by processing and activating proenzymes or prohormones [4], degrading proteins that anchor the epidermis to the underlying cuticle (apolysis) [5], or by digesting the cuticle for resorption or facilitating its shedding (ecdysis) [6]. They are also active during embryonic development of the egg [7]. Peptidases are important in host-parasite relationships, being important virulence factors in some parasites [8]. The pathogenicity of several species of nematode has been significantly correlated with their peptidase activity. These include *Strongyloides stercoralis* [9], *Anisakis simplex* [10], *Onchocerca volvulus* [11], *Trichinella spiralis* [12], and *Ancylostoma caninum* [13].

All major types of peptidases have been described in nematodes. Aspartic peptidases have been described primarily in functions related to the digestion of nutrients. In invertebrates it is thought that, along with the cysteine peptidases, these have the same role as aspartic and serine peptidases in vertebrates [14]. In parasitic nematodes, the cysteine peptidases may be the class for which we have most information, since, owing to their great diversity, they cover virtually all functions in which peptidases are involved in parasitic nematodes. Cathepsins B and L are types of cysteine peptidases belonging to the papain family, and have been comprehensively studied in nematodes [15]. High variability has been found among the cathepsins B from different species of nematodes regarding optimum temperature and pH, and substrate affinity. It is thought that their main role is to digest nutrients and that the high interspecific variability observed is due to the nematode adapting to the ecological niche it occupies [16]. Cathepsins L also seem to be involved in the digestion of nutrients, as well as in processes of embryogenesis and moulting [2]. Many of these cathepsins L have counterparts in the free living nematode *Caenorhabditis elegans*, suggesting that they may be involved in conserved functions in different species of nematodes, but little is known about their precise functions [7]. The metallopeptidases are involved in the invasion of host tissues by the parasite, as they are able to degrade the extracellular matrix, and are also involved in the process of ecdysis and digestion of nutrients. The serine peptidases are also present in nematodes, and, along with the metallopeptidases, are believed to play the largest part in the invasion of host tissues by the parasites [10].

INTRODUCTION

The peptidases represent around 2% of the genes expressed in an organism [15]. The proteolytic enzymes of many parasites have been widely studied over the last two decades. In addition to their expected role in nutrition, at the expense of the host proteins, peptidases are considered essential during development and in the most critical moments of parasite-host interaction, and are thus directly involved in the growth and survival of the parasite. Their identification and characterization are important for our basic understanding of the biology of the parasite and it is easy to deduce their importance as virulence factors for parasitic nematodes. Consequently, peptidases are regarded as potential targets for vaccines, drugs and in serodiagnosis. Despite this, in most cases, the precise physiological functions of peptidases in these and other parasites are still unknown [2]. So, the main functions known for peptidases in parasitic nematodes are the following:

1) To facilitate penetration in, and through, the host tissues by digesting their proteins until the site where the parasite will establish itself is reached. A large part of the morbidity and mortality associated with infection is the result of the extensive

migration through the host by some parasites [17]. This migration normally involves an external digestive process in which the products are not usually assimilated. During the progression of the infective larva of parasitic nematodes of animals from the site of penetration – the skin or the digestive tract – to the establishment site, it is often possible to observe the secretion of proteolytic enzymes, as reported in geohelminths [18]. It has also been shown that the penetration of parasitic nematode larvae through the skin of mammals can sometimes be halted by the presence of peptidase inhibitors [9].

2) To protect the parasite from the attack of the host immune system, since they are able to degrade immunoglobulins bound to surface antigens [19]. In some cases peptidases are the main source of parasite antigens, favouring the immune response of the host. Many are especially immunogenic [20], making them extremely useful in serodiagnosis and vaccine development, although they also favour allergy and hypersensitivity processes in the host.
3) To take part in the moulting and resorption of the cuticle. This has been shown in certain species by the fact that when peptidase inhibitors are present the nematode cannot rid itself of the old cuticle during moulting [3]. Peptidases can participate in several ways during moulting: activating prohormones or proenzymes [4], degrading the proteins which anchor the cuticle to the underlying epidermis (apolysis) [5], or digesting the cuticle to resorb part of the proteins of which it is composed to form the new one or to facilitate shedding of the intact old cuticle (ecdysis) [6]. They may also take part in processing propeptides before they are incorporated as proteins into the new cuticle [21]. In nematodes, as opposed to arthropods, the animal grows and the cuticle, being elastic, stretches until it can no longer stretch further, at which point moulting occurs. In many cases the adult continues to grow, as does the cuticle, from the centre of the animal towards the extremes. In the epidermis of the adult stage of *Ascaris suum* enzymes have been found which participate in synthesizing the collagens of the cuticle, in spite of there being no further moults [22].
4) To be actively involved during the embryonic development of the egg, in the remodelling of the tissues during larval development and in the degradation of the nematode cuticle [7], and, finally, during development after eclosion [23].
5) As might be expected, to process and activate relevant peptides in many other metabolic processes, such as during the extension of pseudopods in nematode sperm [24]. This is a process in which peptidases participate by processing peptides involved in signal transduction pathways. Knowledge is currently increasing with regard to the vital processes involving peptidases, such as protein turnover (proteasomes) and development [25], and apoptosis [26], amongst others. This knowledge is generally being acquired from study of the free-living nematode *Caenorhabditis elegans*, and, to a lesser extent, from parasites [42].
6) Finally, to digest nutrients of proteic origin. This function is easily deduced since it is widely found in metazoans and is highly important. In nematodes, as in other animals, it takes place to facilitate the assimilation and utilization of the component units of the peptide molecules and may be internal or external [3]. However, despite its importance in nematode biology, protein digestion is still not well understood and our knowledge is comprised of fragmented data from different nematode groups.

The Peptidases and their Functions in Nematodes

Even when considering the complete phylum of nematodes there is no species for which all the details of feeding, digestion and nutritional requirements are known [1]. In any case, wide knowledge of feeding in one particular species, such as the model species *C. elegans*, could not be used to generalize for all groups due to the great variety of their feeding habits, related to the habitats they occupy. This is reflected in a series of anatomical characteristics and associated behaviours in both the digestive and absorptive processes. Sometimes in nematodes the occupation of a particular alimentary niche has phylogenetic implications, as in the ascarids. Depending on species, food generally consists of bacteria, algae, diatoms, protozoans, fungi, other nematodes, other invertebrates, vegetable matter, sap, haemolymph, blood, cell and fluids of animal tissues and gut contents. In some parasitic nematodes the larval stages may have completely different feeding habits to those of the adult, as in *Haemonchus contortus*, in which the larva develops free and exhibits survival stages until it is able to infect a host [1]. On the other hand, in mermitids, the larvae develop while parasitizing other invertebrates, usually insects, whereas the adults are free-living and feed on stored reserves [27]. The peptidases involved in feeding are found in the principal structures for this, these being the excretory/secretory system (E/S) and the digestive system (Figure 1).

The E/S system of nematodes may participate in feeding by secreting digestive enzymes. When this system was first described its function was thought to be to secrete manufactured substances and to excrete waste into the external medium. However, its main functions are currently thought to be secretion and osmoregulation, although functions can vary with species and developmental stage. Some evidence for this is provided by the observation that, in some nematodes, such as *Anisakis* sp., the enzymatic activity of structures associated with the E/S system is high and varied with many types of enzyme described (dehydrogenases, aminopeptidases, oxidoreductases, esterases and phosphorylases) [226]. It is also supported by ultrastructural studies of E/S tissue [28] and other evidence, such as the fact that excretion is mainly carried out in the intestine. However, excretion also takes place through the E/S system as it is known that CO_2 and different organic acids, amongst other products, are excreted through this system, which is of two basic types: glandular and tubular. The former is typical of free-living forms while the latter is more common in the parasitic forms, exhibiting a variety of designs but basically consisting of two lateral canals connected by a transverse canal. This system opens to the exterior via an excretory pore whose location is a taxonomic feature (Figure 1). This is the case of the genus *Hysterothylacium*, in which the excretory pore is located level with the nerve ring on the side of the front third of the parasite, unlike in the genus *Anisakis* (although both are anisakids) where it is found next to the mouth, through which peptidases can be released. Together with those from the oesophageal glands these can take part in invasion and migration by the infective third larval stage (L3) through the host tissues, as in the other ascaridoids. It has been shown that the secretions release antigenic substances, digestive enzymes and enzymes for moulting of the cuticle. In another anisakid, *Pseudoterranova decipiens*, a leucinoaminopeptidase is released during ecdysis of the final moult following stimulation of the neurosecretory cells of the E/S system [29]. In the trichostrongylids both cells of the E/S system and the oesophageal glands are involved in the release of moulting fluid by the larva during the second moult [30]. In *Angiostrongylus cantonensis* the main E/S antigens have been studied, revealing a significant quantity of peptidases and peptidase inhibitors (aldolase; CBR-PYP-1 protein; beta-amylase; heat shock protein 70; proteosome subunit beta type-1; actin A3; peroxyredoxin; serine carboxy-peptidase; protein disulphide isomerase 1; fructose-bisphosphate aldolase 2; aspartyl protease inhibitor; lectin-5; hypothetical protein F01F1.12; cathepsin B-like cysteine proteinase 1; hemoglobinase-type cysteine proteinase; putative ferritin protein 2; and a hypothetical protein) [31].

The majority of nematodes have a complete digestive system (Figure 1), consisting of the buccal cavity and oesophagus, intestine and rectum. At the anterior end of the animal there are sensitive structures which are usually involved in the localization of food. This is sucked up by the oesophagus (circular, muscular and triradiate) and propelled up to and through the intestine by rapid contractions. The intestine is occluded by the pressure exerted on it by the hydrostatic skeleton of the pseudocele since it does not have associated musculature. When the muscles of the oesophagus are relaxed it occludes, thus preventing the food from returning towards the mouth. Among the oesophageal muscles there are several glands, usually three, one between each interradial zone (the dorsal being larger than the two ventrolateral), and which open into the oesophageal lumen. These glands generally secrete digestive substances such as proteolytic enzymes and amylases. In *Necator* spp. and *Ancylostoma* spp. the glandular secretions contain peptidases and hyaluronidases which degrade the tissue and capillaries engulfed by the parasite when it anchors itself with the buccal capsule to the intestine wall of the host [1], as well as anticoagulant proteins. The oesophagus can take several forms, depending on the order and species, and is thus an important taxonomic feature. In fact, some authors, such as Hsu [32], have related the level of development of the oesophageal glands of some nematodes to their feeding habits. Thus, in *Anisakis simplex*, the high level of development of the oesophageal glands has been related to the lesions which appear around the head of the parasite in the stomach wall of the cetacean definitive host [32-33]. It has also been observed in this species that proteolytic enzymes accumulate in granules within the oesophageal glands. It is thought that, due to the disposition of the glands, the secretions of the dorsal gland are involved in extracorporeal activities while those of the subventral glands are involved in intestinal digestion [34]. The oesophagus leads to the intestine, a simple tubular structure with an epithelium formed by a single layer of cells over the basal lamina, composed of collagens, with a brush border with numerous long folds supported by actin filaments in *Ascaris* sp. [35], increasing the surface area of the lumen by between 75 and 90 times [36]. These are very similar to the intestinal microvilli of mammals [37] and are covered by a glycocalyx. The front part of the intestine of *Ascaris* sp. is mainly for secretion and the rear part for absorption, with both apocrine and merocrine type secretions being described [36]. There are a large number of intestinal cells in the ascarids and transport processes take place in the intercellular spaces through the lateral membranes of the cells. In addition to absorbing nutrients, the epithelial cells probably excrete nitrogenated waste products, mainly ammonia. Excretion of this toxic substance is possible due to the availability of water. However, excretion is ureothelic in other nematodes which live in environments where water is a limiting factor [38]. Finally, the rectum and anus are situated at the end of the digestive tube. Some groups have rectal glands associated with the proctodeum, opening into the lumen of the rectum [1]. Waste products are expelled when the anus dilates, as there is a muscular sphincter, and hydrostatic pressure forces them out. Under experimental conditions, *Ascaris lumbricoides* (15-25 cm) renews the contents of its digestive tract in around 3 minutes while *C. elegans* (1 mm), which is much smaller, takes 45 seconds [39]. This gives an indication of the time required for digestion and/or absorption of nutrients.

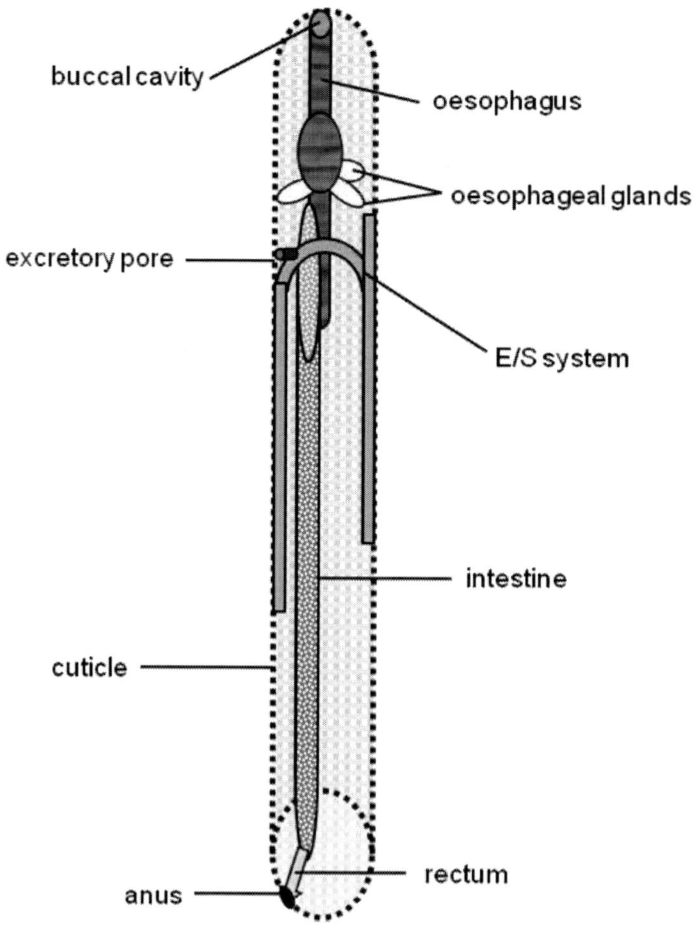

Figure 1. Diagram of digestive and Excretory/Secretory system of a nematode.

The digestive enzymes of nematodes comprise endopeptidases (of which the cysteine peptidases have been especially well studied in many organisms), exopeptidases, glycosidases, lipases, phospholipases and esterases. These may be secreted and released both externally and internally or are membrane proteins found on the luminal surface of the digestive tube. Van den Bossche and Borgers [37] reported that *A. suum* exhibited a low level of endocytosis in the intestinal epithelium suggesting that the acid hydrolases present in the intestinal microvilli might perform extracellular digestion of macromolecules in order to absorb the nutrients later. Several types of enzymes have been reported to be involved in the digestion of carbohydrates. In nematodes digestive enzymes have been described for all the principal nutrient groups. For example, in the case of the saccharides, the following were found in the intestine of *Ascaris* sp.: α-amylase, maltase, invertase, trehalase and isomaltase, but not lactase [36]. In *Nippostrongylus brasiliensis* and in other nematodes different glycosidases have been found and it is believed that these may be involved in the degradation of mucopolysaccharides. In parasitic nematodes of plants there are also cellulases, pectinases and chitinases [36]. Lipases have also been reported in nematodes. They have been found in different locations such as the intestine (*Ascaris* sp., *Leidynema* sp. and *Strongylus* sp.) and the oesophagus (*Ancylostoma caninum*; in this hematophage they express themselves in the intestine where they combine with lipids, the saposins, which are believed to be active in the lysis of red blood cells [40]. As might be expected, the nutrient transport proteins are found in the luminal and basal surfaces of the digestive tube. In *Ascaris* sp., the absorption of hexoses in the intestine has been studied in detail [41]. The transport of cholesterol (which nematodes cannot synthesize) has been observed in the digestive tube of *A. lumbricoides* [36]. Although nutrient absorption takes place primarily in the intestine some nematodes use other structures as well. Nutrients may be absorbed through the cuticle (composed of modified keratinized cells), as in *Ascaridia galli* or *Ascaris* sp., and, especially, in certain parasites of insects, which lodge in the haemocoel, such as *Bradynema* sp. or *Mermis nigrescens*. The entomophile larva of *M. nigrescens* develops in the haemocoel of an orthopteran, absorbing glucose through the cuticle, mediated by a specific transport system [43] which is probably related to the ecological niche of this parasite. Furthermore, in some filarians absorption of glucose through the cuticle, together with leucine and adenine, has also been reported [36]. The great importance of peptidases for the assimilation of nutrients as well as in many other processes should be emphasized. They include a huge class of hydrolytic enzymes which are essential for all animals in the processes of digestion, development, cell dynamics, blood clotting, inflammation, scar tissue formation, processing of prohormones and many others.

In the nematodes, as in other organisms, the peptidases are distributed in different locations throughout the body of the animal, according to their function. For example, gelatinolytic activity in *A. simplex* [34] is mainly found in the intestinal lumen but is also present in the pseudocelom, the cells of the body wall and the muscles of the oesophagus. Accumulations of digested gelatin granules are found in the oesophageal glands. However, the excretory cells exhibit very little gelatinolytic activity, as occurs in *A. suum* with a specific aminopeptidase, which, furthermore, has been reported in the female reproductive tissues [44]. The glandular oesophagus of the filarians *Brugia pahangi* and *Brugia malayi* only appears in L3 with a granular distribution of cystein peptidases in which E/S products are released during the third moult [45]. Thus, as well as being spatial, the distribution of peptidases can also be temporal, depending on the developmental stage of the parasite. In electrophoresis of extracts of adult *Onchocerca volvulus* no gelatinolytic activity is observed, whereas it is present in larval extracts [46].

The main absorption route of amino acids is probably intestinal and it appears that some species are able to synthesize different amino acids from acetate. However, the majority of amino acids essential for mammals are probably also essential for nematodes. Thus, *C. elegans* and *Caenorhabdithis briggsae* require histidine, leucine, isoleucine, lysine, methionine (or its precursor homocysteine), phenylalanine, threonine, tryptophan, valine and arginine in their diet [1]. In the larva of *M. nigrescens* amino acid transport through the cuticle has been reported, and, in *Ascaris* sp., in the digestive tube, muscle and reproductive tissue. In the intestine of *Ascaris* sp. the assimilation of amino acids is a non-passive mediated process although it is not known whether this can occur against the gradient or whether it is sensitive to ions, as in the case of hexoses. However, in the muscle and reproductive tissue of *Ascaris* sp. the assimilation of a non-metabolizable amino acid, cycloleucine, through active transport in the reproductive tissue and diffusion in the muscle, has been observed. In the larva of *M. nigrescens* amino acid assimilation through the cuticle is a mediated process [47]. *A. simplex* ingests no food during L3 [48-49,224], demonstrated by Sommerville and Davey [50] on cultivating L3 of *A. simplex*, which had recently been extracted from the fish

Leionura atun. Here, they are found encapsulated in the viscera and mesenteries. During the first few days they did not feed orally so it is probable that a type of absorption through the cuticle took place. This has also been proposed for *Hysterothylacium aduncum* [51-52].

In some haematophages, such as *H. contortus* and the hookworms, numerous peptidases of different types have been identified (aspartic and metallopeptidases and a cathepsin B type cysteine peptidase) which are involved in haemoglobin digestion [53]. At the time of this study there was no direct evidence for haemoglobin digestion in the intestinal lumen. However, many peptidases expressed in the intestine and absent from E/S products have been described, implying that they act here. Furthermore, the optimum pH for many of these peptidases is lower than that of the host tissues, suggesting that they act in the acidic environment of the parasite intestine. Following many studies with cDNA of a large number of peptidases of all types, biochemical characterizations, expression of recombinant peptidases and localization of peptidase expression in the main haematophage nematodes (*Ancylostoma* spp., *Necator americanus* and *H. contortus*), peptidases have been found which are homologous to those found in haemoglobin digestion [54] by the trematode *Schistosoma mansoni* (one of the helminths which best shows this) [14,55] and by the protozoans *Plasmodium* spp. In *A. caninum* the haemoglobinolytic enzymatic cascade has been described in vitro [56]; and in the other haematophage nematodes, the known intestinal peptidases, homologous to those of *A. caninum, S. mansoni* y *Plasmodium* spp., have been assigned a position in the digestive cascade. However, in nematodes the process has only been described in *A. caninum*. This position is still hypothetical and provisional, equivalent to that occupied by its peptidase homologue in the processes already described. In this cascade the first to act would be the aspartic peptidases, generating the first peptides derived from the haemoglobin tetramer, followed by cysteine peptidases, metallopeptidases, and, finally, the exopeptidases, which would degrade the smallest peptides, giving rise to free amino acids and small oligopeptides. These can be absorbed through the intestinal epithelium [54]. It is clear that lysis of red blood cells could also occur in the intestine. Pores are created in the cell membranes [57] by the action of saposins, molecules with an affinity for lipids and thus able to adhere to the cell membrane and disrupt the cells [40]. These peptidases are accessible for the host immune system and are consequently being considered as vaccine targets against the illnesses caused by these nematodes, which have serious health and economic implications [58]. Some peptidases from these nematodes have already been analyzed. They exhibited different degrees of protection and were sometimes orthologous antigens shared by these haematophage nematodes [59]. One of these antigens, H-gal-GP, is highly protective against *H. contortus*. It is an intestinal cell membrane glycoprotein complex, and contains predominantly digestive peptidases. A study showed that adding protective antibodies from H-gal-GP immunized sheep to the H-gal-GP catalysed the haemoglobin digestion reaction, reducing the rate by 70–90%, while this reduction was only 30% when nonprotective IgG from sheep immunized with denatured H-gal-GP was added. These results support the theory that the mechanism of protection is by specific antibodies blocking the activity of the proteolytic enzymes, and thus, the ability of the parasite to feed [60].

In the free-living nematode *C. elegans* it appears that there is also a synergic process within the peptidases during the digestion of proteic nutrients. Thus, although the cysteinic cathepsins are able, in vitro, to digest some easily degradable proteins like casein by themselves, they require the prior collaboration of aspartic cathepsins to degrade other proteins. Cathepsin D (aspartic) degrades other proteins by itself, and, additionally, is able to digest and inactivate itself when there is a lack of substrate. In effect, this is what takes place in vivo during fasting: after 4-8 hours without ingestion of food, levels of cathepsin D decrease first to 65%, while the cysteinic cathepsins do not decrease for at least 24 hours. The data show that, apart from the animal's age, its nutritional state is important for modulating lysosomal peptidase levels [61-63].

According to Delcroix et al. [14], the papain-type aspartic and cysteine peptidases assume the same role in invertebrate digestion as trypsins in that of vertebrates. In addition, this process occurs in an acidic environment. It was proposed that luminal or cellular acidic micro-environments may exist [64-65], a notion subsequently supported histochemically by the researchers [14]. This fits with proposals for other types of gastrointestinal parasites other than hookworms, the ascaridoids, in which it is thought that nutrient digestion is carried out through an enzymatic cascade similar to those described earlier at acidic pH. In *A. suum* the main intestinal activity is a result of an aminometallopeptidase. It has been suggested that in the gastrointestinal tract of the final host the metallopeptidases of the parasite complete the digestion initiated by the pepsin and other enzymes from the host [66]. This may also occur in another ascaridoid parasite, *H. aduncum*, in which it has been demonstrated that its aspartic activity is dependent on the pepsin of the culture medium (that mimics the pepsin from the host) being modulated by the level of exogenous pepsin (Figure 2). When pepsin activity is not in the milieu, the parasite tries to cover this deficiency with its own aspartic activity. Without exogenous predigestion by the pepsin in the medium, the larvae would require greater aspartate-peptidase activity [52]. Furthermore, it has been observed that infective L3 of *A. simplex* stimulate pepsin expression in the stomach of guinea pigs, an experimental host of this anisakid [223]. It should also be noted that pepsin has been shown to be necessary in the in vitro development and moulting of *A. simplex*. For optimum in vitro development of both these anisakids, besides pepsin, an acidic pH (pH 4.0) is necessary to facilitate its proteolytic activity [49,51]. The use of other peptidases such as papain or trypsin or other pH values has a negative effect on these nematodes [49,51,67-68]. It is thus possible that the dependence of *H. aduncum* on aspartic activity is related to the digestion of nutrients, and the decrease in its expression in culture is directly related to the presence of pepsin, as suggested by Rhoads et al. [66] for proteolytic activity in *A. suum*. This was mainly related to aspartate peptidases, but also metallopeptidases, and, to a lesser extent, cysteine peptidases. This activity seems to depend on the exogenous availability of nutrients of proteic origin, originating from previous digestion by the host [52]. The many ascaridoids which live in an acidic environment are able to withstand it due to their capacity for osmoregulation. *Ascaris suum* can maintain the concentration of the hydrochloric acid and the osmotic pressure in its haemolymph at levels lower than those in the intestinal content of the pig host [76]. It has long been known that these are factors that trigger the activation and development of infective larvae. So, in addition to pH, there are other stimuli in the environment that affect the parasite culture, such as temperature, CO_2 and environmental redox potential [50,69-71]. It has been found that the addition of hydrochloric acid during experimental infections with *A. suum* in neonate pigs inoculated with fertilized eggs resulted in more parasites in the lungs than in the controls, which were not given hydrochloric acid [72]. There is also another nematode, *A. cantonensis*, in which a combination of these two factors, pepsin and hydrochloric acid, has been found. Curiously, this is not a gastrointestinal parasite like the ascaridoids. This is a parasite whose final host is the rat, which is infected on ingesting the intermediate host (thought to be many species of snails). Then, the parasite reaches the brain through the host tissues (the human is an accidental host in whom it can cause severe eosinophilic meningitis), and, finally, the adult migrates and settles in the pulmonary arteries. To facilitate isolation of infective larvae which are embedded in snail tissues, for experimental infection, specimens were prepared by digesting the infected snails with pepsin–HCl. It was shown that this method of collection contributed to more patent infections in rodents after inoculation of the larvae [73-74]. It was later found that during the pepsin preparation, transcripts of the infective larvae encode proteins that participate in a wide range of biological processes (among these, several peptidases). These findings suggest that treatment with pepsin–HCl not only digests the tissues of the snail host but also activates the infective larvae [75]. All these processes are likely to be carried out through the properties of the E/S system. Nevertheless, water and ion excretion are poorly understood in nematodes. The E/S systems seem to be involved in water excretion, but this has not yet been confirmed [76].

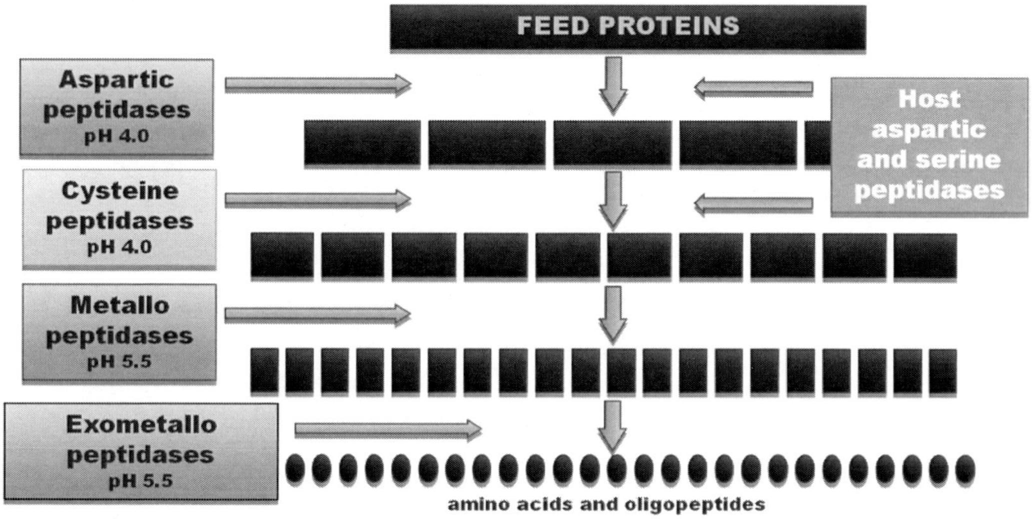

Figure 2. Proposed model for the digestion of feed proteins in the ascaridoid *Hysterothylacium aduncum* (gastrointestinal parasite of fishes). Based on Malagón et al. [52].

As mentioned previously, the peptidases have been considered potential targets for vaccines, drugs and serodiagnosis, and their importance is currently clear from the development of antiparasitic strategies initiated during the previous decade [77-79].This importance is largely due to their major role as virulence factors. In the nematodes in particular there are ample species which show a significant correlation between peptidase activity and pathogenicity. These include *Dirofilaria immitis* [80], *Nematospiroides dubius* [81], *Strongyloides stercoralis* [9], *A. simplex* [10], *N. americanus* [82-83], *Onchocerca* spp. [11,84], *Trichinella spiralis* [12], *Trichuris muris* [85], *Strongylus vulgaris* [219], *A. caninum* ([13] and *Teladorsagia* (*Ostertagia*) *circumcincta* [86]. A large part of the morbidity and mortality associated with infection is due to the extensive migration of the parasites through the host [17]. Migration of gastrointestinal parasites through the host tissues seems to be beneficial in that it favours more and faster growth than in those which do not migrate, leading to increased reproductive capacity of the parasite [87]. Balbuena et al. [88] found great variability in the location and routes followed by L3 of *H. aduncum* through the tissues of infected herring larvae and suggested that L3 migrate in search of an optimum environment for their development within the host tissues and that this migration depends on their ontogenic development. The decision to penetrate and move through the host tissues depends on such factors as the condition of the host and/or the development of the parasite. This decision is also conditioned by the differences in susceptibility to the parasite of different host species and also within the same host species [for example, the host responding by isolating the parasite in a thick capsule may reduce its chances of developing [222]. Furthermore, the ability to detect small variations in the signals emitted by the host and to modulate their response accordingly will determine the extent of penetration and movement of the parasites within the host tissues [89]. Evidence for the secretion of hyaluronidase by both a parasitic nematode (*A. suum*) [90] and a nonparasitic nematode (*C. elegans*) [91] allowed Rhoads et al. [90] to suggest that this enzyme not only acts by facilitating the migration of the ascarids through the host tissue by its histolytic action but also that hyaluronic acid must be a requirement or signal for development to L4, since this molecule is involved in cell development. The presence of several hyaluronidases which can take part in tissue invasion has been described in *A. simplex, A. caninum* [92] and *A. suum* [90], with optimum pH of 4.0, 6.0 and 5.0-6.0, respectively (Table 1). *Ascaris* penetrates through the cecum and ascending colon [93]; *Ancylostoma*, through the skin, blood vessels and lungs; and *Anisakis* follows a cycle very similar to that of *H. aduncum*, often using the same intermediate and paratenic fish hosts, which in both cases are infected principally by ingestion of small crustaceans. It is possible that its optimum pH of 4.0 for its hyaluronidase indicates that it penetrates through the stomach (as proposed for *H. aduncum* due to an optimum pH for metallopeptidase type activity expressed only in the infective larva [52]), whereas optimum pH values for the other two hyaluronidases are higher since the other two parasites penetrate zones with higher pH. It is clear that the site where the parasite initiates host tissue invasion and subsequent migration through it is very important, the zone of action being related to the physicochemical properties of the employed peptidase. To a great extent, this is what determines the ecology and life cycle of the parasite. Anderson y Bartlett [94] reported that *Hysterothylacium haze* appeared to have lost the ability to return to the gastrointestinal tract of the host to develop to adulthood, as practised by other worms of the genus *Hysterothylacium*, (*H. haze* attains the adult stage in the visceral cavity of the host fish). This nematode is an extreme example of precocious development by some parasites in the host as a result of diverse evolutive acquisitions in their life cycles and that of their hosts. *A. suum* is a monoxenous parasite, and, when the eggs hatch in the host intestine, after their development in an external medium, the larvae do not immediately penetrate but rapidly travel along the narrow intestine and penetrate through the mucus of the cecum and the ascending colon, and, instead of going directly to the lungs, they accumulate in the liver for 48 hours [93]. They then migrate to the lungs, via the blood. Here, they moult to L4 and then return to the digestive tube to develop to adulthood. A possible explanation for this is that the ascarids may originally have had a heteroxenous life cycle based on the predator-prey relationships of their hosts thereby having at least two hosts, one intermediate and one definitive. It is thought that the migration undertaken by monoxenous ascarids through the host tissues is a simulation of the migration they used to undertake through their ancestral intermediate host, resulting from "memories" of the behavior of their ancestors in order to remain in the tissues of their intermediate host (the prey) [95]. Many parasitic nematodes produce an infective L3 to infect the host, comparable to the resting-stage dauer larva found in free-living nematodes. This is known as developmental arrest or hypobiosis and serves as a resting stage that allows many species of nematode to withstand unfavourable conditions until a suitable host can be found [76]. It is clear that the invasion of host tissues has an important role in the development of parasitism, since the success of the parasite depends on it. Some species have attained a very wide distribution, such as the ascaridoid *H. aduncum*, which is regarded as one of the most abundant marine parasites [96-97]. This is also case of the insect parasite the *Steinernema carpocapsae*, which can parasitize a large number of different species of insects and has great potential for use in pest control. It has been demonstrated that the most virulent strains have more peptidases in their E/S products. But it seems that most of them come from *Xenorhabdus nematophila*, a bacterial species which lives in symbiosis with this insect parasitic nematode [98]. As a result, the peptidases have to evolve in this environment, as in the case of a metallopeptidase involved in invasion through the skin by the nematode *S. stercoralis*, which maintains its proteolytic activity after immunoprecipitation with IgG [99].

Table 1. Relationship between some hyaluronidase enzymes of nematodes and the tissues they invade

Nematode	Optimum pH for hyaluronidase	Place of invasion (pH)
Anisakis simplex	pH 4.0	Stomach (≈pH 2.0)
Ancylostoma caninum	pH 6.0	Skin (≈pH 5.5)
Ascaris suum	pH 5.0-6.0	Colon (≈pH 7.0)

In the plant-parasitic nematode, *Meloidogyne incognita*, the genome has recently been sequenced and the constitution of the degradome analyzed and compared with the peptidase genes of other nematodes, both free-living and parasites of animals [100]. The *M. incognita* degradome consists of at least 334 peptidases distributed in 43 families of the five known catalytic classes (i.e., 26 aspartic, 106 cysteine, 136 metallo, 52 serine and 14 threonine peptidases). A differential transcript level has been observed between eggs and infective juveniles and differences have also been found in the distribution of some peptidase families compared to those of five other nematodes (the plant-parasitic species *Meloidogyne hapla*, the model free-living species *C. elegans* and *C. briggsae,* the necromenic species *Pristionchus pacificus*, and the animal-parasitic species *B. malayi*). This could reflect specific aspects of the parasitic lifestyle of this organism; peptidases known to be used by other phytopathogenic microorganisms (bacteria), such as cysteine peptidases of the C48 sub-family, predicted to encode SUMO (small ubiquitin-like modifier) deconjugating enzymes and serine peptidases of the S16 family (Lon peptidases), were more abundant in *M. incognita* than *C. elegans*, which may reflect specific aspects of its interaction with host plant tissues. Together, these data reinforce the hypothesis that members of the nematode degradome may play a direct role in the host–parasite interaction. The comparative analysis of the *M. incognita* degradome with those from other nematodes has provided evidence that the global distribution of *M. incognita* peptidase classes closely resembles those of the other nematode genomes, which supports the current hypothesis that a core peptidase system is conserved throughout evolution [101-102]. The number of independent peptidase genes in parasitic species is very low and it is possible that genome compaction is an attribute of the parasitic lifestyle, as proposed for the root-knot nematode species *M. hapla*, considered the smallest metazoan genome yet completed [103], and the human parasite *B. malayi* [104]. These studies are beginning to determine whether nematode genes have similar roles in pathogenicity.

Peptidase Classification and Nomenclature

The peptidases are hydrolases specifically acting on peptide bonds, with a single molecule able to break up to 1,000,000 bonds per second when a peptide bond would take centuries to break spontaneously [15]. The enzymes able to hydrolyze peptide bonds are termed peptidases. The classification of peptidases presents certain problems due to their being enzymes that generally exhibit degenerate substrate specificity. That is, the same enzyme is often able to hydrolyze peptide bonds in different amino acid residues (although generally belonging to the same group, defined by their chemical properties).

Table 2. Classification of peptidases by EC system

Sub-subclass	Peptidase type	N° of entries
EC 3.4.11	Aminopeptidases	26
EC 3.4.13	Dipeptidases	12
EC 3.4.14	Dipeptidyl-peptidases and tripeptidyl-peptidases	9
EC 3.4.15	Peptidyl-dipeptidases	4
EC 3.4.16	Serine-type carboxypeptidases	4
EC 3.4.17	Metallocarboxypeptidases	20
EC 3.4.18	Cysteine-type carboxypeptidases	1
EC 3.4.19	Omega peptidases	9
EC 3.4.21	Serine endopeptidases	97
EC 3.4.22	Cysteine endopeptidases	58
EC 3.4.23	Aspartic endopeptidases	40
EC 3.4.24	Metalloendopeptidases	81
EC 3.4.25	Threonine endopeptidases	2
EC 3.4.99	Endopeptidases of unknown catalytic mechanism	0
	Total	363[a]

Notes: [a]Data from the section Enzyme Nomenclature (EC 3.4 (peptidases)) of the IUBMB website (Nomenclature Committee of the International Union of Biochemistry and Molecular Biology (NC-IUBMB)) in 2011[107].

A widely used classification system is that proposed by the EC (Enzyme Commission), an internationally recognized organization whose current successor is the IUBMB (International Union of Biochemistry and Molecular Biology) (Table 2). This is a mechanistic system which considers the type of catalysis employed by the peptidase during the breaking of the peptide bond. It depends on the main functional group responsible for the catalysis, which, in turn, depends on the amino acid residues that form the active centre of the enzyme. For example, in the cysteine peptidases the characteristic triad of catalytic residues in the active centre is distinguished by having a cysteine residue in the catalytic site, giving the family its name. This has implications in the conditions required for this family to be active, the cysteine peptidases mainly being active between pH 4.0 and 7.0.

On the other hand, the peptidases are also classified according to their site of activity. Thus, according to the IUBMB, the peptidases are placed in group EC 3.4, which is divided into two groups, exopeptidases (EC 3.4.11-19) and endopeptidases (EC 3.4.21-25). The former hydrolyze the peptide bonds of those amino acid residues situated at one of the two extremes of the polypeptide chain, the amino or carboxyl terminal, with the aminopeptidases starting hydrolysis at the amino terminal and the carboxypeptidases at the carboxyl terminal. The endopeptidases attack the internal bonds of the chain, that is, they perform well towards the middle of long chains, far from the ends, and badly on the resulting oligopeptides. They have traditionally been referred to as proteases but this is often a simplification and the IUBMB recommends they be called peptidases. The term oligopeptidase is used for the endopeptidases which perform well on the oligopeptides, the smallest substrates proceeding from the action of proteases on a proteine [105] (Table 3). In the classification of a peptidase by catalytic type, a simple initial approximation for the classification of a new peptidase in terms of its mechanistic class is usually based on its response to class-specific inhibitors. Further data will still be necessary since definitive classification requires knowledge of the complete sequence of amino acids [106]. Serine endopeptidases, cysteine endopeptidases, aspartic endopeptidases and metalloendopeptidases are the main peptidases.

Two main types of catalysis have been identified in the peptidases (both with an acid-base mechanism): a covalent type, with a covalent enzyme-substrate complex due to a nucleophilic attack carried out by a group present in the enzyme (cysteine sulfhydryl in the cysteine peptidases and serine hydroxyl in the serine peptidases); and another without the covalent complex, where the nucleophile is water, through a mechanism involving at least one aspartyl residue (aspartic peptidases) or a metal ion (metallopeptidases) [109] (Table 4). Peptidases from all catalysis types have been described in parasitic organisms [8]. Hydrolysis of the peptide bond takes place through the collaboration of an unstable tetrahedral intermediate, produced by the active centre of the proteolytic enzyme, and the contiguous residues at either side of the catalytic site help to bind the substrate. The properties of these contiguous residues confer specificity for one sequence or another. Each position of the residues is given a name, following the widely used nomenclature system introduced in 1967 by Schechter and Berger [110] to describe the interaction of proteolytic enzymes with their substrates. This names different binding subsites (points of the enzyme which interact with the polypeptide substrate residues) of the peptidase. The amino acid residues of the substrate which are going to bind with the enzyme and are on the aminoterminal side of the peptide bond to be hydrolysed are numbered P1, P2, P3, successively, starting from the bond to be hydrolysed. The residues on the carboxyterminal side of the peptide bond to be hydrolysed in the substrate are numbered P'1, P'2, P'3, successively, in the same manner. Up to 6 residues can be numbered on each side of the bond. The subsites are residues found at the active centre of the peptidase and are numbered S1, S2, S3, S'1, S'2, S'3. Each subsite interacts with its corresponding residue in the substrate. These interactions may be electrical, hydrophobic, steric, etc. These are the peptidase residues which will define their specificity for a particular sequence in the substrate and for the bond to be hydrolysed:

In the Substrate: N -...- P3 – P2 – P1 x P'1 – P'2 – P'3 -...- C
In the Enzyme: N -...- S3 – S2 – S1 ¤ S'1 – S'2 – S'3 -...- C

Where "¤" represents the catalytic site residue and "x" represents the peptide bond to be hydrolysed. Both class-specific inhibitors and peptidase specific substrates are designed to interact specifically with the binding subsites of the active centre of the enzyme, emulating the natural polypeptide substrate residues (Figure 3).

Table 3. Basic classification of peptidases by the 'place of action' [108]

Endopeptidases	Cleave internal peptide alpha-bonds of polypeptide chain away from N-terminus or C-terminus.	
	Oligo-peptidases	Cleave shorter peptides and no proteins.[a]
Exopeptidases	Cleave peptide α-bonds adjacent to N-terminus or C-terminus of polypeptide chain.	
	Aminopeptidases	Cleave a single amino acid residue from N-terminus.
	Carboxypeptidases	Cleave a single amino acid residue from C-terminus.
	Dipeptidylpeptidases	Cleave a dipeptide from N-terminus.
	Tripeptidylpeptidases	Cleave a tripeptide from N-terminus.
	Peptidyldipeptidases	Cleave a dipeptide from C-terminus.
	Dipeptidases	Cleave dipeptides. Typically require both termini to be free.
Omegapeptidases	Cleave peptide α-bonds with no preference for N-terminus or C-terminus. They can also cleave isopeptide bonds.[b]	

Notes: Reprinted with permission from Elsevier, owner of the copyright [108]. [a] Peptide α-bonds are bonds where NH_2- or -COOH are directly attached to the α-carbon of the amino acid. [b] Isopeptide bonds are bonds where one or both of the NH_2- and -COOH groups are not directly attached to the α-carbon of the amino acid.

The problem with the EC classification system is that it groups together many enzymes which are very different from each other in terms of where and when they act, and, especially, in their evolutionary relationships. For this reason, Rawlings and Barrett [111], developed another classification system, MEROPS, in which taxonomic classification is performed (clan, family and unique peptidase) and the peptidases are grouped according to molecular structure, homology and function. It is based mainly on primary and tertiary structures of peptidases, and, according to some authors [108], is currently considered the most relevant approach to distinguish and group peptidases. "Merops" is the scientific name of a bird, the bee-eater (*Merops apiaster*), whose social behaviour inspired the authors to adopt the term "clan", as bee-eaters group their nests into families and clans. The inhabitants of each nest occupy a different part of the colony and have their own discrete area where the members hunt flying insects [112]. It has become clear that some proteolytic enzymes that depend upon the same residue for their catalytic activity belong to different families of proteins, sharing only a few general characteristics. Consequently, these families have been grouped into several "clans". A clan is defined as a group of families whose members have a common ancestor and that commonly have marked similarities, despite a lack of statistically significant correlation between their sequences (it mainly considers the catalytic centre and tertiary structure). Families of the same clan can be recognised because their peptidases have a similar protein folding pattern; a family is a group of enzymes with significantly high homology in the sequence, more than half of its residues, and which has evolutionary relationships. A unique peptidase is generally a single peptidase or a set of proteins, all of which display a particular type of peptidase activity, and are closely related by sequence. Clans normally include a single catalytic type although some include two or more [113-116] (Table 5).

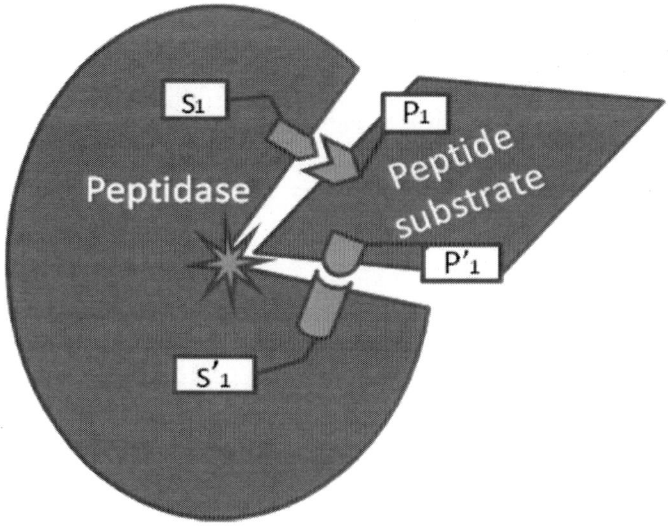

Figure 3. Diagram showing interaction between peptide substrate and the active site pocket of a peptidase.

1. Aspartic Peptidases

The aspartic peptidases are thought to mediate the peptide hydrolysis via general acid-base catalysis (without covalent complex formation). This often takes place through two aspartyl residues present in its active centre (Table 4). They are considered a highly conserved group compared to other peptidase classes, especially the cysteine and serine peptidases [54].

In *D. immitis* an aspartic peptidase is found mainly in the intestine, reproductive tissue and the developing eggs. It may thus be involved not only in nutrition but in gamete production and the development of the microfilarian [119].

In *A. cantonensis* a single aspartic peptidase is expressed after the stimuli of the digestion from the tissues of the intermediate host. It has been suggested that it may have a role in nutrition. This aspartic peptidase shows changes in the expression level of a putative aspartic peptidase in its development. The level increases from the infective larvae to the young adult, significantly decreases in male adults but increases in female adults. These findings could indicate differences in nutritional requirements at different developmental stages [75,120].

In extracts of hookworm *N. americanus* the digestion of haemoglobin and fibrinogen is inhibited by the aspartic peptidase inhibitor pepstatin A [83]. Two types of intestinal aspartic peptidases have been found in haematophage nematodes: those of cathepsin D type, which are lysosomal [83,121-123], and others more similar to mammalian pepsins [123]. This cathepsin

D has since been studied by Yang et al. [124] in *A. caninum* during embryonic development in the egg and the first stages of larval development, being found well distributed throughout all these stages, even in the infective larva and the E/S products. It has been suggested that it is involved in embryonic development, the same as other homologues in *C. elegans* [125], and that it may also play a part in host invasion and nutrient digestion, since, unlike the homologues of *C. elegans,* it is also found in L4 and the adult [124]. In *C. elegans* a cathepsin D (cathepsins are generally peptidases found in lysosomes and synthesized as inactive proenzymes) has been described which is thought to be involved in nutrient digestion and whose levels are modulated by the nutritional state of the nematode, decreasing during fasting [63].

Table 4. Basic classification of peptidases according to catalytic type[a] [108]

Serine Cysteine Threonine	S C T	The nucleophile attack during catalysis is facilitated by reactive group at amino acid side chain, a hydroxyl group (OH^-) of serine and TPs or a sulphydryl group (SH^-) of CPs.
		Catalytic triad of serine peptidases: Ser, Asp, His.
		Catalytic triad of CPs: Cys, His, Asn.
		Catalytic triad of TPs: Thr is conserved in active sites of all proteasomes [218]. All known threonine-type peptidases are N-terminal nucleophile peptidases [112].
Aspartic Metallo	A M	The nucleophile attack during catalysis is usually facilitated by an activated water molecule and followed by formation of tetrahedral intermediate.[b] The water molecule is bound by the side chains of aspartic residues of APs or by metal ions (e.g. one or two zinc ions, Zn^{2+}) of MPs.
		Catalytic dyad of APs: Asp and Asp.
		Catalytic zinc site of MPs is usually formed by His, Glu, Asp or Cys, which supply ligands for zinc.
Glutamic		The mechanisms of catalysis are similar to Aps, including an activated water molecule and tetrahedral intermediate. The water molecule is bound by the side chains of glutamic acid and glutamine residues.
		Catalytic dyad of APs: Gln and Glu.
Unknown		The peptidases of unknown catalytic type, that is, proteins where the sequence is known to belong to peptidases, but the mechanisms of catalysis are not determined [112].

Notes: Reprinted with permission from Elsevier, owner of the copyright [108]. [a] Catalytic type of peptidase is determined according to chemical mechanisms of catalysis related to the reactive group in the active site of peptidase. [b] Tetrahedral intermediate of AP catalysis is a formation that provides proton transfer from a water molecule to an aspartic acid dyad and another proton transfer from a dyad to a carbonyl oxygen of cleaved peptide bond [129]. A tetrahedral intermediate of MP is formed after attack of a zinc-bound water molecule on the carbonyl group of the cleaved peptide bond. This intermediate is further decomposed by transfer of glutamic acid proton to leaving the group [111]. AP, aspartic peptidase; CP, cysteine peptidase; MP, metallopeptidase; TP, threonine peptidase.

In *O. volvulus, A. caninum* and *C. elegans* a new family of aspartic peptidases was identified. This is closely related to the cathepsin D and E family, being more similar to cathepsins E. The latter are generally secreted and the adult female of *O. volvulus* is known to secrete specific aspartic peptidase inhibitors. It has thus been suggested that these peptidases are secreted by forming a complex with the inhibitor [126]. In *C. elegans* homogenate most activity is at an acidic pH and 90% is inhibited by the aspartic peptidase inhibitor pepstatin A, with evidence of up to 12 different aspartic peptidases being found [127].

In L4 and adults of *O. ostertagi* the presence of two aspartic peptidases which degrade mucins, haemoglobin and albumin has been reported, with one at pH 5.0 and the other at pH 3.0. The latter also degrades immunoglobulin G and would consequently be very useful for protecting the digestive tract of this nematode from attack by the immune system while it feeds on blood from the blood vessels irrigating the intestinal mucosa of the host [128].

In the insect parasitic nematode *S. carpocapsae* a secreted aspartic peptidase has been found which is a time dependent inducible gene, up-regulated in the gut of L3 and also expressed in the oesophagus. The expression of this peptidase can be induced in infective larvae with insect homogenate. Despite the high number of nematodes that are known to secrete aspartic peptidase, no clear function has been attributed to all of them, and the function of aspartic peptidase during the development stages remains unclear [98].

Table 5. Basic classification of peptidases according to molecular structure, homology and functions [108]

Clan[a]	Clan comprises a group of families probably sharing evolutionary ancestry, despite lack of statistically significant similarities in amino acid sequence. Distant relationship comes primarily from linear order of catalytic-site residues in polypeptide chains, and tertiary structure.[b] The name of each clan is formed from the letter for the catalytic type of peptidases (S, C, T, A, G, M or U, as for families) followed by second capital letter (e.g. CA).
Family	The family comprises peptidase members with evolutionary relationships based on primary structure similar to at least one another member of the family. Each family is named by a letter denoting the catalytic type (S, C, T, A, G, M or U) followed by number (e.g. C1).[c]
Unique peptidase	Unique peptidase is, in general, a single peptidase or a set of proteins all of which display a particular type of peptidase activity, and are closely related by sequence (e.g. C01.062).[d]

Notes: Reprinted with permission from Elsevier, owner of the copyright [108]. [a] The idea of using the terms 'family' and 'clan' for the groups of peptidases came from the ecological strategy of bee-eaters (*Merops apiaster*), which group their nests into families and clans. The inhabitants of each nest occupy a different part of the colony and have their own discrete area where the members hunt flying insects [112]. [b] Tertiary structure recognized by modelling is crucial for activity of many peptidases. For example, the polypeptide chain of papain forms two domains with a large cleft of the active site that blocks the pro-region part [117-118]. [c] Primary structure of peptidase determines statistically significant relationships in amino acid sequence with respect to a representative member, especially to its unit (peptidase unit is a part of the enzyme responsible for peptidase activity, [112]). [d] For example, cathepsin B-like peptidase, clan CA, family C1A, peptidase C01.062 (according to MEROPS classification).

2. Cysteine Peptidases

The cysteine peptidases usually contain a cysteine residue in their active centre, this being the principal residue of the typical catalytic triad (Cys/His/Asn) [114], which participates in covalent type catalysis. The sulfhydryl performs the nucleophilic attack, forming an intermediate covalent enzyme-substrate complex (Table 4).

Much is known about the cysteine peptidases in parasitic nematodes, since, due to their great diversity [16,130-132], they cover virtually all the functions in which peptides are involved in these animals. These enzymes are found in both larvae and adults and are thought to be involved in tissue penetration, nutrition and defense from the immune system of the host [77] as well as in moulting [66].

In *H. contortus* cysteine peptidases have been observed which produce extracorporeal degradation in vitro of tissues of an artificial extracellular matrix, presumably due to the E/S products, and which may help the parasite to reach and rupture the capillaries in the digestive tract mucosa in order to feed on the blood in vivo [3,133]. *Ostertagia ostertagi,* which feeds in the same way, exhibits cystein-like activity against mucins, suggesting that it may have a similar function [128].

In *A. suum* the cysteine peptidases have been related to the moulting process, as the presence of Z-Phe-Ala-FMK (cysteine peptidase inhibitor) prevents 77% of larvae from shedding the L3 cuticle, although their development continues [66]. In *B. pahangi* [4], *D. immitis* [134-135] and *O. volvulus* [136] they have also been related to moulting processes. In *Toxocara canis* a cathepsin Z has been related to moulting, although it expresses constitutively, suggesting that it must have another essential function [137]. It also expresses in *O. volvulus* and is again related to moulting [138]. In *C. elegans* it expresses at the same sites as in *O. volvulus* (in the epidermis of all the larval stages and in the oesophagus and gonads of the adult) and participates in moulting and other functions related to development such as spermatogenesis and embryonic development, suggesting it has a conserved function in nematodes [139]. In *N. americanus* a cysteine peptidase has been detected which is the most abundant in the fluids trapped between the L2 cuticle and that of the infective L3 before shedding of the old cuticle [140], suggesting it may also be involved in moulting.

In extracts of *H. contortus* [141] and in E/S products of *N. americanus* [83,221] cystein peptidases able to degrade haemoglobin, fibrinogen, collagen (only in *H. contortus*) and antibodies have been observed. In *H. contortus*, without considering the E/S products, the intestine has been shown to be the site where the highest concentration of cysteine peptidases is found [3].

Within the cysteine peptidase families there is one group which is widely distributed among the best studied parasites and one of the most important. This is the papain family, this also being the name of the clan to which they belong. The CA (papain type) clan is characterized by having the typical catalytic triad of cysteine peptidases in the active centre, consisting of three residues: cysteine, histidine and asparagine [15]. Analysis of the evolution, location and biological function of the cysteine peptidases belonging to this clan, in parasitic metazoa and protozoa, has increased understanding of the biochemistry and cellular function of this highly diverse family of enzymes, which take part in both intra- and extracellular processes [142]. For this reason, and given their diverse nature, they are regarded as attractive pharmacological targets, together with other cysteine peptidases, for the development of chemical agents to interfere with them. There are several reasons for this [15,143-145]:

- A sufficient number have different structural and biochemical properties from those of the host, including optimum pH and stability, alteration in peptide loops or domain extensions;
- The relative absence of redundancy with regard to the mammalian host. This means that the affinity spectrum for different substrates of the different cysteine peptidases in the hosts overlaps more than in the parasites;
- A higher concentration of cysteine peptidases in the intracellular compartments of the host than in the parasites. Cysteine peptidase activity of the host is generally more robust;
- Differentiated assimilation of inhibitors by parasites with regard to host, due in part to the great diversity in their cell location;
- The studies carried out to elucidate the mechanisms of resistance to cysteine peptidase inhibitors in parasite cultures show that they are independent of those of traditional antiparasitic agents.

Within the papain family there are two cathepsins which have been widely studied due to their presence in almost all organisms, cathepsins B and L. The term "cathepsin" was proposed by Willstätter and Bamann [146] to describe the peptidases which acted preferentially in an acidic medium and were later found in all cells and tissues and subsequently discovered to be of lysosomal origin. Thus, the term has traditionally referred to a lysosomal peptidase which can sometimes be secreted. They are widespread amongst the invertebrates and phyla of primitive protostome metazoa are known to express a wide range of cathepsins [77]. The papain family contains many cathepsins which are obviously cysteinic [147] and some of them are studied in nematodes [15]. They are organized in several multigenic families, some of which have been well characterized, such as those expressing in the intestinal tissues of the adults of *H. contortus* and *C. elegans.* This diversity reflects great functional specialization, the need for rapid protein digestion, or both, since all have a potential role in nutrient digestion [16,131-132,148].

In the cathepsins B of nematodes there is great variability in optimum pH and temperature, substrate affinity, etc. It is thought that their major role is nutrient digestion and that the high interspecific variability observed is due to the adaptation to its ecological niche undergone by each nematode [16]. Comparative phylogenetic analysis of some cathepsins B of both haematophagous and no-haematophagous helminths has been carried out. It has been shown that haematophagous worms' cathepsins B share some common motifs, known as hemoglobinase motifs, despite the fact that, phylogenetically, some of them may be more distant from each other, than from other cathepsins B belonging to the same group. For example cathepsins B of hookworms share this motif with those of schistosomes, whereas in nematodes such as *C. elegans* this does not occur. Interestingly, it appears in the intestinal nematode *A. suum*, but is not clearly understood since it is not known whether the cathepsin in question is expressed in all larval stages or during their migration in host tissues [149]. A cathepsin B with haemoglobinolytic activity from the intestine of *A. caninum* has been expressed in vitro, using a heterologous system. Optimal activity is achieved when it acts after a cathepsin D (aspartic peptidase) from the same parasite [56], and it is for this reason that both are placed in the enzymatic cascade that digests haemoglobin in this parasite. In *C. elegans* cathepsin B genes are expressed in the larval and adult digestive tube and also in the intestine of *H. contortus* [148,150-151], being related to nutrient digestion [150]. Furthermore, the gene promoter of this cathepsin B of *H. contortus* has been inserted into *C. elegans* and it has been shown that the mechanisms controlling spatial distribution of this peptidase are conserved in parasitic and free-living nematodes [152]. It had previously been suggested that, in addition to the structural similarity, there was a functional homology between the cathepsins B of the nematodes, which could play an important part in their nutrition [148]. They are also thought to have a role in embryonic and larval development as it has been shown that a cathepsin B is expressed in the egg and the infective larva of *A. caninum* [153]. In *A. cantonensis* a cathepsin B has been characterized which shows high specific expression in the worm stages which cause dysfunction of the blood–brain barrier of hosts. This cathepsin *Ac*CB displayed different expression profiles in non-permissive host (mice) and permissive host (rats) derived larval stages and was involved in the maturation of dendritic cells, suggesting a potential role in the central nervous system invasion and immunoregulation during parasite–host interactions [154].

Cathepsins L are less abundant in terms of their level of expression [77,155]. They also seem to be mainly involved in nutrient digestion, embryogenesis and moulting [2,7,156], although they have also been implicated in tissue invasion and immune response evasion in parasitic nematodes [77]. Homologues of many of these cathepsins L have been found in *C. elegans*, suggesting that they may be involved in conserved functions in different species of nematodes, although little is known of their precise functions [7]. In *B. pahangi* and *B. malayi* [45] a series of cathepsins L were studied. When compared with their corresponding cathepsins in *O. volvulus* and *C. elegans* [7,136,157] it was concluded that they are also involved in remodelling of the cuticle and the egg capsule in filarians. In *O. volvulus* cathepsins L and Z (closely related) have been described. These participate directly in moulting of L3 [136], later confirmed by RNAi experiments [138]. In *C. elegans* a cathepsin L (*Ce*CL_1), essential for development and embryogenesis, has been studied in depth. On inhibiting its expression, development of up to 98% of the embryos was detained [2]. It has also been found in the egg capsule and the cuticle of *O. volvulus* (*Ov*CL_1), suggesting that it is conserved in nematodes. From this it can be deduced that this peptidase is of great importance and probably has an identical function in many nematodes. It is expressed during embryonic and postembryonic development in the intestine, epidermis, oesophagus, gonads and the outer part of the egg, near the capsule. It's produced during the intermoult period, approximately 4 hours before the moult, and then, after moulting, production decreases to basal levels. Consequently, it is thought to be involved in processing products used in the synthesis of the egg capsule or in its degradation, the same in the cuticle, and also for other functions. It has been suggested that *Ce*CL_1 degrades the deutoplasmic proteins to leave the necessary amino acids available for embryonic development and for the synthesis or maturation of the egg capsule, and, later, for remodelling of the larval cuticle during moulting. It may also act in the adult, during growth after the last moult, by degrading the old cuticle and processing the new or digesting the proteins of the materials used to anchor the cuticle. Finally, it may also act indirectly, processing and activating other enzymes and hormones involved in moulting [7]. In *H. aduncum* the evolution of activity of both cathepsin L and B has been measured during its life cycle in vitro. The former showed higher activity and an acidic optimum pH and the latter, lower activity and a basic-neutral optimum pH. Having considered these activity variations and the optimum pH values, it has been suggested that cathepsin L has a role in digestive processes while cathepsin B could be involved in cuticle renewal, among other possible functions [158]. However, these data have been called into question since Rehman and Jasmer [16] reported that in *T. canis* and other nematodes such as hookworms there seemed to be modifications of the cathepsins L which caused them to show cathepsin B behaviour. This would suggest that both conventional substrates and inhibitors for this sub-class of peptidases would offer only provisional information, at least in nematodes.

3. Metallopeptidases

Within the metallopeptidases the endopeptidases and exopeptidases are equally usual. The name is due to the presence of a coordinated metal ion in their active centre. This is frequently a divalent cation of zinc but can also be cobalt, manganese, nickel or copper, and is able to polarize the peptide bond before it breaks, through the nucleophilic attack of hydroxyl from the water molecule activated by the ion. A common feature is the presence of catalytic residues of glutamate, histidine, aspartate and lysine, which maintain the metal cations in their positions and are thus essential for catalysis (Table 4). They act as exopeptidases when they possess only one metal cation and can act as exo- or endopeptidases when two of them are involved. Metallopeptidases exhibit a relatively broad range of specificity to peptide substrates, usually defined by the substrate residues of positions P1 and P'1 [159-160].

The metallopeptidases intervene in the invasion of host tissues by the parasite as they are able to degrade the extracellular matrix [46]. In *S. stercoralis* it has been shown that a zinc-endometallopeptidase secreted by the infective larva is involved in penetrating the skin of the mammalian host. This metallopeptidase is extremely potent and attacks the principal components of the dermal extracellular matrix (elastine, collagens and glycoproteins). Penetration of the skin is inhibited by specific inhibitors for metallopeptidases [9]. Furthermore, this peptidase conserves its proteolytic activity after the action of the antibodies, indicating that immunogenicity is absent from its active centre. This would represent a great evolutionary advantage for infection of the host [99]. In *Strongyloides venezuelensis*, the infective larvae have a zinc-metallopeptidase activity, which has been assumed to play a major role in skin penetration [161]. This metallopeptidase activity seems to be a homologue of Ss40 of the previous activity of *S. stercoralis*. More recently, an astacin-like metallopeptidase transcript in infective larvae of *S. stercoralis*, referred to as "strongylastacin" [162], has been identified and which shows high homology with another transcript only expressed in the infective larva of *S. venezuelensis* [163]. For this reason it has been suggested that if the expression of this transcript is specific for the infective larva stage it perfectly matches the metalloprotease activity previously reported [161]. In *H. aduncum* the presence of a metallopeptidase has been reported, with a similar optimum pH to those found in the host stomach where the worm presumably initiates invasion of the stomach tissues. It is inhibited by 1,10-phenanthroline and dithiothreitol, and, during its development in vitro it only appears in the infective L3 stage, suggesting that it may be employed by the parasite to undertake penetration of the host tissues [52]. In *A. cantonensis* some findings suggest a crucial role of metallopeptidases in the penetration of the intestinal wall of the host. Several metallopeptidase transcripts have been identified, induced after a pepsin-HCl treatment, among the most abundantly expressed transcripts in the infective larvae only. It has been related with the previously reported secreted peptidases with collagenolytic and elastinolytic activities [75], that are inhibited by serine peptidase or metallopeptidase inhibitors [164]. In the insect parasite *S. carpocapsae* an up-regulated astacin metallopeptidase has been reported during the parasitic stage, and strongly induced in vitro by insect tissues. This suggests that it plays a role in the parasitic process [165].

In *Trichuris suis* the presence of a secretory organ of a zinc-metallopeptidase has been reported and which has also been implicated in the degradation of host tissues during invasion [166]. In *H. contortus* a complex of several metallopeptidases associated with its intestinal microvilli, which, through immunization, offers moderate protection to sheep, has been described [227]. In addition to showing haemoglobinolytic activity, this complex may also be involved in the evasion of the immune system by degrading eotaxins, thus preventing the eosinophils from reaching the anchorage site of the parasite [167].

The first metalloendopeptidase activity described in nematodes was in the E/S products of *A. caninum* in 1983 [168]. Since it was able to degrade fibrinogen and plasminogen it was thought to prevent the host blood from coagulating. More recently, the cDNA of a metallopeptidase expressed in the intestinal lumen, *Ac*MEP$_1$, has been described and this is probably responsable for the activity mentioned previously [169]. In the same parasite there is another metallopeptidase, present both in the E/S products and extracts of larvae and adults, which also prevents coagulation by inhibiting fibrin precipitation and is also thought to be involved in tissue penetration due to its elastinolytic activity [168,170-171].

In *A. suum* an aminometallopeptidase released into the culture medium has been related to moulting as it reaches maximum concentration at the moment of passing from L3 to L4 [220]. The presence of amastatin (aminopeptidase inhibitor) prevents 34% of individuals from shedding the L3 cuticle. However, this does not halt development; the same aminopeptidase has later been found in the adult intestine, where it exhibits activity related with digestion [44]. In *B. pahangi* [4], *D. immitis* [134-135] and *O. volvulus* [136] aminopeptidases have also been detected in the moulting process. In the infective L3 of *H. contortus* there is a metalloendopeptidase which degrades an annular part of the L2 cuticle, so that it becomes detached, making way for the L3 cuticle [6].

In *H. aduncum* the expression of collagenase activity was studied during in vitro development and was observed to be of metallopeptidase type, principally inhibited by 1,10-phenanthroline. As L3 showed a different optimum pH to that of L4 and adult it was suggested that it has different functions in each developmental stage, with that of L3 possibly involved in migration of the parasite through the host tissues. Most activity was observed in the immature adult, after the final moult, suggesting that the collagenolytic activity may be involved in remodelling of the cuticle and in sexual maturity and digestion, as collagenolytic activity was shown to be significantly greater when no pepsin was added to the culture medium [172].

4. Serine Peptidases

This important group accounts for almost one third of all peptidases [108]. They are synthesized as inactive precursors (zymogens) and always consist of three domains: catalytic, substrate binding and zymogen activation. In order to be totally active they must be processed through the elimination of the zymogen domain (N-terminal extension) [173-174]. In their active centre the serine peptidases have an active serine residue, which, together with others of aspartate and histidine, form the catalytic triad His/Asp/Ser, which participates in covalent type catalysis. The hydroxyl group of the serine carries out the nucleophilic attack, giving rise to the intermediate covalent enzyme-substrate complex (Table 4).

The serine peptidases are also present in nematodes, and, together with other enzymes such as metallopeptidases, have an important role in invasion of host tissues by the parasite [9-10,175-176]. They have been identified in different parasitic nematodes such as *N. americanus* [140], filarians *D. immitis* [80] and *O. volvulus* [84], ascarids *A. suum* [177] and *T. canis* [178], anisakids *A. simplex* and *P. decipiens* [10,34], and *T. spiralis* [20]. An example of how serine peptidases are likely candidates for the degradation of host tissues was reported by Morris and Sakanari [17], who related an extracellular serine peptidase, present in the pathogenic bacteria *Dichelobacter nodosus*, which destroys host tissues (by degrading elastin, keratin and collagen), with a serine peptidase from the E/S products of the *A. simplex* larva in in vitro culture with similar activity. In the E/S products of the infective L3 of *N. americanus* serine peptidase activity able to degrade immunoglobulins G, A and M has been detected [140] which may help the parasite during entry by offering protection from the antibodies generated by the immune response of the host. When the hosts are insects, they can defend themselves from pathogens by melanotic encapsulation, a humoral immune response related to the production of reactive oxygen species, which can kill microorganisms or parasites. It has been shown that melanization in insects can be inhibited by a secreted serine peptidase of the insect parasitic nematode *S. carpocapsae,* presumably by disrupting the enzyme cascade of melanisation, inhibiting the phenoloxidase activity (a key enzyme for this process) [179]. In addition, there is experimental evidence that they can alter host haemocytes and their actin filaments [180]. Finally, a very clear and interesting example of how the serine peptidases participate in the invasion of host tissues by a parasite has been shown in this insect parasitic nematode. A secreted chymotrypsin-like serine peptidase is expressed only during the parasitic stages in the insect haemocoel, especially in the infective larvae (pre-invasive nematodes inside the mid-gut). This peptidase caused histolysis in the insect mid-gut, and in vitro assays demonstrated that it can digest extracellular proteins and induce apoptosis in cultured insect cells, thus suggesting that this peptidase is involved in the pathogenesis [181]. Another secreted serine peptidase of this nematode is involved in the parasitic process since is not expressed in the arrested stage. However, this gene was overexpressed in recovered nematodes, particularly those in the gut lumen preparing to invade the insect hemocoelium (infective larvae). In vitro assays showed that the expression started after stimulation with insect tissues, especially peritrophic membrane and hemolymph (peritrophic membrane lines the host gut lumen and the hemolymph represents the first lines of defense) [182].

In *H. aduncum* the expression of proteolytic activity during in vitro development has been studied. The serine peptidases did not appear until after moulting to the adult stage. This may well be related to the attainment of sexual maturity, appearing during the sexual maturation of the parasite [52,172]. Serine peptidases in the filarian *O. volvulus* have been related to moulting, embryogenesis and spermatogenesis [183]. *D. immitis* has an alternative splicing of furin (a subtilisin-like peptidase, which are serine peptidases), generating sex-specific isoforms. These sex-specific isoforms suggest that this peptidase processes targets involved in sex-determination and/or reproduction. The patterns of alternative splicing found in *D. immitis* are not conserved within the furin genes of *O. volvulus* and *B. malayi*. The alternative splicing observed within nematode furin genes probably represents a method for increasing diversity without increasing gene number. The subtilisin-like peptidase (subtilases) family is a group of serine peptidases involved in multiple processes such as the construction and maintenance of the cuticle, neural signalling and nematode development. At least two of these subfamilies of subtilases, kexin and pyrolysin, are present in nematodes. By cleaving or activating other proteins they play an essential part in the nematode biology. This makes these peptidases an attractive target for drug development. For example, cuticular collagen genes from *H. cortortus, A. suum* and *B. pahangi* have the conserved domain structure and kexin cleavage motifs found in *C. elegans* cuticular collagen genes, suggesting that this mechanism of collagen maturation is conserved among nematode species [184].

5. Peptidase Inhibitors in Parasitic Nematodes

Peptidase inhibitors have been studied in many groups of organisms including parasitic nematodes. For example, in *S. carpocapsae* it has been shown that the gene of a secreted cystatin is upregulated after induction by insect hemolymph and is thought to have a role among the parasitism genes expressed in *S. carpocapsae* throughout the parasitic cycle ([185]. In nematodes they would be expected to have a vital role in modulating proteolytic enzyme activity in, amongst others, development and parasite-host relationships. However, although not as widely studied as peptidases, their role is only clear in two cases:

- Protection of the digestive tract: in hookworms one of their functions seems to be to act as an anticoagulant, inhibiting several serine peptidases involved in host blood coagulation [186-188]. Consequently, some are promising pharmacological anticoagulants [189]. Aspartic peptidase inhibitors have been found in other groups of gastrointestinal parasites such as ascarids [190-191], and, more recently, trypsin-, chymotrypsin- and elastase-type serine peptidases have been found in *Ascaris* spp. [192-196] and *A. simplex* [17,225], where they are believed to be involved in protecting the parasite against host digestive enzymes and any other enzymes encountered while migrating through host tissues. In *Ancylostoma duodenale* an inhibitor acting on these enzymes has been characterized and shown to express itself in the oesophagus, intestine, and cuticle surface of the adult worms, suggesting that its role is to protect the exposed parts of the parasite from the host digestive environment [197]. Carboxypeptidase inhibitors which may have the same role have also been purified [198-199]. Some of these inhibitors are important allergens. Specifically, the aspartyl peptidase inhibitors from some parasitic nematodes (*A. caninum, Trichostrongylus colubriformis* and *O. volvulus*) are highly immunogenic, and have been proposed as potential vaccine antigens [200-202]. It has been reported that a single dose of the crude latex of *Carica papaya* provided comparable anthelmintic efficacy to the currently available synthetic anthelmintics, with respect to reductions in the egg output and worm burden, in pigs infected with the roundworm *A. suum* [203] and mice infected with *Heligmosomoides polygyrus*, one of the most common laboratory models for routine screening of potential drug candidates [204]. The mechanism of action of this phenomenon is related to the action of cysteine peptidases (such as papain) found in papaya latex [205], as shown by inhibiting the effect through addition of the cysteine peptidase inhibitor l-*trans*-epoxysuccinyl-leucylamido-4-guanidino-butane (E-64) [206]. In another ascaridoid,

A. simplex, as above, porcine pepsin has been shown to favor its development in in vitro culture [49] and resists it well, like *H. aduncum* [51], while papain and bovine trypsin had a negative effect on the former worm [67].

- Protection against host immune system: in the intestinal parasite *T. suis*, the inhibitors may be involved in a defense mechanism to escape the peptidase-mediated immune response of the mastocytes associated with the intestinal mucosa of the host [207]. In the filarian *B. malayi* it has been shown that certain inhibitors it produces are able to inhibit T-cell proliferation by preventing the processing of the antigen for its presentation by the antigen-presenting cells of the major class II histo-compatibility complex (MHC II) [208]. *N. brasiliensis* also modulates the immune response at the antigen processing level in the antigen-presenting cell [209]. In *A. cantonensis* certain cystatins are thought to be involved in host immunoregulation. It is surprising that in a non-permissive host, like the mouse, antibodies are generated against the cystatin of this nematode during an infection whereas in a permissive host, like the rat, no antibodies are generated [210]. For a non-permissive accidental host, human, a cystatin described in L3 of *A. simplex* is an important allergen which is expressed in the secretory gland and the basal layer of the cuticle [211]. In general, the cystatins in immune cells interfere with antigen processing and presentation, phagocytosis, expression of cytokines and nitric oxide, thus modifying the immune response. Consequently, it has been suggested that the cystatin-type molecules secreted from parasites down-modulate the host immune response [212]. In fact, it has been noted that in human and animal models helminth infections are concomitant with a reduction in the prevalence of inflammatory diseases and others related to hypersensitivity reactions. As a result, there is currently much interest in finding molecules from these parasites with antiallergic and/or anti-inflammatory properties [213-214].

Of course, peptidase inhibitors also take part in other regulatory processes, such as in the filarian *O. volvulus*, where they may have a regulatory role during moulting, embryogenesis and spermatogenesis [183].

Hosts have also developed peptidase inhibitors as a way of controlling nematode infections. For example, some plants such as the potato react to an infestation of the nematode *Globodera rostochiensis* by producing peptidase inhibitors [215]. Different plants have been created experimentally from transgenic crops (tomato and banana) in which there is heterologous expression of cystatin genes. These plants show, to a greater or lesser extent, resistance phenotypes against infestation by plant-parasitic nematodes (*M. incognita, Radopholus similis*) classified as serious pests [216-217]. The importance of the cysteine peptidases, the target of these cystatins, in nematodal development, is clear.

Conclusion

The nematodes are, after insects, the group of organisms with the largest number of species identified, indicating their great potential for adaptation and diversification. It includes members of medical, veterinary and agronomic significance, making this group one of the most important among animal and plant parasites. Peptidases, as in others organisms, play a major role in the biology of nematodes, but a great amount of the processes in which they are involved are not well known. Most effort is invested in investigation of tissue and cuticle renewal during development, host-tissue invasion, protein digestion, and, interestingly, in the role of parasitic peptidase inhibitors. Fortunately, unlike other groups of parasitic helminths, the nematodes include a species used as a model organism of great importance, *C. elegans,* enabling us to extract much relevant information by means of comparative studies, especially those related to development, processing of peptides and protein renewal by the proteasome, and, to a lesser extent, digestion. Although invertebrates appear to share several general features, digestion obviously depends largely on the ecological niche occupied by the nematode as well as the food source. However, such important processes and characteristics of parasitism as invasion and migration through host tissues or the host immune system's interaction with peptidases and inhibitors from both plants and animal parasites need the development of research adapted to each group of parasites or at least to groups that share a niche. This would serve as a reference for study in other closely related species.

ACKNOWLEDGMENTS

This work was partially funded by the Spanish Grant ACU01-027 from INIA and Research Group BIO-243 from Junta de Andalucía, Spain. David Malagón is a recipient of a postdoctoral fellowship of Ayudas del subprograma Estancias de movilidad posdoctoral en centros extranjeros from Ministerio de Educación of Spain Government. The English translation was by Mr. Robert Abrahams, BSc. Publishing of this work was supported by The Ministry of Education, Youth and Sports of the Czech Republic (KONTAKT II, Grant No. LH12096).

REFERENCES

[1] Munn E.A. & Munn P.D. (2002) Feeding and digestion. In: *The biology of nematodes,* (Lee D.L., ed.), pp. 211-232, Taylor & Francis, Florence, KY, USA.

[2] Britton C. & Murray L. (2002) A cathepsin L protease essential for *Caenorhabditis elegans* embryogenesis is functionally conserved in parasitic nematodes. *Molecular and Biochemical Parasitology,* 122: 21-33.

[3] Rhoads M.L. & Fetterer R.H. (1997) Extracellular matrix: A tool for defining the extracorporeal function of parasite proteases. *Parasitology Today,* 13: 119-122.

[4] Hong X., Bouvier J., Wong M.M., Yamagata G.Y. & McKerrow J.H. (1993) *Brugia pahangi*: identification and characterization of an aminopeptidase associated with larval molting. *Experimental Parasitology,* 76: 127-133.

[5] McKerrow J.H. (1994) Cysteine proteases of parasites: a remarkable diversity of function. *Perspectives in Drug Discovery and Design,* 2: 437-444.

[6] Gamble H.R., Purcell J.P. & Fetterer R.H. (1989) Purification of a 44 kilodalton protease which mediates the ecdysis of infective *Haemonchus contortus* larvae. *Molecular and Biochemical Parasitology,* 33: 49-58.

[7] Hashmi S., Britton C., Liu J., Guiliano D.B., Oksov Y. & Lustigman S. (2002) Cathepsin L is essential for embryogenesis and development of *Caenorhabditis elegans*. *The Journal of Biological Chemistry,* 277: 3477-3486.

[8] Cazzulo J.J. (2003) Amino acid and protein metabolism. In: *Molecular Medical Parasitology*, (Marr J.J., Nilsen T.W. & Komuniecki R.W., eds.), pp. 171-197, Academic Press, London.

[9] McKerrow J.H., Brindley P., Brown M., Gam A.A., Staunton C. & Neva F.A. (1990) *Strongyloides stercoralis*: identification of a protease that facilitates penetration of skin by the infective larvae. *Experimental Parasitology,* 70: 134-143.

[10] Sakanari J.A. & McKerrow J.H. (1990) Identification of the secreted neutral proteases from *Anisakis simplex*. *The Journal of Parasitology,* 76: 625-630.

[11] Lustigman S. (1993) Molting, enzymes and new targets for chemotherapy of *Onchocerca* volvulus. *Parasitology Today,* 9: 294-297.

[12] Criado-Fornelio A., de Armas-Serra C., Giménez-Pardo C., Casado-Escribano N., Jiménez-Gónzalez A. & Rodríguez-Caabeiro F. (1992) Proteolytic enzymes from *Trichinella spiralis* larvae. *Veterinary Parasitology,* 45: 133-140.

[13] Hawdon J.M., Jones B.F., Perregaux M.A. & Hotez P.J. (1995) *Ancylostoma caninum*: metalloprotease release coincides with activation of infective larvae in vitro. *Experimental Parasitology,* 80: 205-211.

[14] Delcroix M., Sajid M., Caffrey C.R., Lim K.C., Dvorak J., Hsieh I., Bahgat M., Dissous C. & McKerrow J.H. (2006) A multienzyme network functions in intestinal protein digestion by a platyhelminth parasite. *The Journal of Biological Chemistry,* 281: 39316-39329.

[15] Sajid M. & McKerrow J.H. (2002) Cysteine proteases of parasitic organism. *Molecular and Biochemical Parasitology,* 120: 1-21.

[16] Rehman A. & Jasmer D.P. (1999) Defined characteristics of cathepsin B-like proteins from nematodes: inferred functional diversity and phylogenetic relationships. *Molecular and Biochemical Parasitology,* 102: 297-310.

[17] Morris S.R. & Sakanari J.A. (1994) Characterization of the serine protease and serine protease inhibitor from the tissue-penetrating nematode *Anisakis simplex*. *The Journal of Biological Chemistry,* 269: 27650-27656.

[18] Lewert R.M. & Lee C.L. (1956) Quantitative studies of the collagenase-like enzyme of cercariae of *Schistosoma mansoni* and the larvae of *Strongyloides ratti*. *Journal of Infectious Diseases,* 99: 1-14.

[19] Hotez P.J. & Pritchard D.I. (1995) Hookworm infection. *Scientific American,* 272: 68-74.

[20] Todorova V.K. & Stoyanov D.I. (2000) Partial characterization of serine proteinases secreted by adult *Trichinella spiralis*. *Parasitology Research,* 86: 684-687.

[21] Page A.P. (2001). The nematode cuticle: synthesis, modification and mutants. In: *Parasitic nematodes: molecular biology, biochemistry and immunology,* (Kennedy M.W. & Harnett W., eds.), pp. 167-194, CABI Publishing, Wallingford-Oxon.

[22] Chvapil M., Boucek M. & Ehrlich E. (1970) Differences in the protocollagen hydroxylase activities from *Ascaris* muscle and hypodermis. *Archives of Biochemistry and Biophysics,* 140: 11-18.

[23] Page-McCaw A. (2008) Remodeling the model organism: matrix metalloproteinase functions in invertebrates. *Seminars in Cell and Developmental Biology,* 19: 14-23.

[24] Fraire-Zamora J.J. & Cardullo R.A. (2010). The physiological acquisition of amoeboid motility in nematode sperm: is the tail the only thing the sperm lost? *Molecular Reproduction and Development,* 77: 739-750.

[25] Bowerman B. & Kurz T. (2006) Degrade to create: developmental requirements for ubiquitin-mediated proteolysis during early *C. elegans* embryogenesis. *Development,* 133: 773-784.

[26] Colin J., Gaumer S., Guenal I. & Mignotte B. (2009) Mitochondria, Bcl-2 family proteins and apoptosomes: of worms, flies and men. *Frontiers in Bioscience,* 14: 4127-4137

[27] Platzer E.G. (1980) Biological control and genetics: Nematodes as biological control agents. *California Agriculture,* 34: pp 27

[28] Lee H.F., Chen I.L. & Lin R.P. (1973) Ultraestructure of the excretory system of *Anisakis* larva (Nematoda: Anisakidae). *The Journal of Parasitology,* 59: 289-298.

[29] Davey K.G. (1968) Molting in a parasitic nematode, *Phocanema decipiens.* IV. Ecdysis and its control. *Canadian Journal of Zoology-Revue Canadienne de Zoologie,* 46: 893-898.

[30] Sommerville R.I. (1982). The mechanics of moulting. In: *Aspects of parasitology,* (Meerovitch E., ed.), pp. 407-433, Montreal: the Institute of Parasitology, McGill University, Montreal.

[31] Morassutti A.L., Levert K., Pinto P.M., da Silva A.J., Wilkins P., Graeff-Teixeira C. (2012) Characterization of *Angiostrongylus cantonensis* excretory-secretory proteins as potential diagnostic targets. *Experimental Parasitology,* 130: 26-31.

[32] Hsu H.F. (1933) Study on the oesophageal glands of parasitic nematoda superfamily Ascaroidea. *Chinese Medical Journal,* 47: 1247-1288.

[33] Hoeppli R. (1933). On histolytic changes and extra-intestinal digestion in parasitic infection. *Lingnam Science Journal,* 12, (Supl.): 1-11.

[34] Matthews B.E. (1984). The source, release and specificity of proteolytic enzyme activity produced by *Anisakis simplex* larvae (Nematoda: Ascaridida) in vitro. *Journal of Helminthology,* 58: 175-185.

[35] Munn E.A. & Greenwood C.A. (1984). The occurrence of submicrovillar endotube (modified terminal web) and associated cytoskeletal structures in the intestinal epithelia of nematodes. *Philosophical Transactions of the Royal Society of London. Series B-Biological Sciences,* 306: 1-18.

[36] Barrett J. (1981) Nutrition and biosynthesis. In: *Biochemistry of parasitic helminths,* (Barrett J., ed.), pp. 149-244, Macmillan Publishers Ltd., London.

[37] van den Bossche H. & Borgers M. (1973) Subcellular distribution of digestive enzyme in *Ascaris suum* intestine. *International Journal for Parasitology,* 3: 59-65.

[38] Bush A.O., Fernández J.C., Esch G.W. & Seed R. (2001) Nematoda: the roundworms. In: *Parasitism. The diversity and ecology of animal parasites,* pp. 160-196, Cambridge University Press, Cambridge.

[39] Bird A.F. & Bird J. (1991) *The structure of nematodes,* 2nd edn., Academic Press, San Diego CA.

[40] Don T.A., Oksov Y., Lustigman S. & Loukas A. (2007) Saposin-like proteins from the intestine of the blood-feeding hookworm, *Ancylostoma caninum. Parasitology,* 134: 427-436.

[41] Schanbacher L.M. & Beames C.G., Jr. (1978) Factors in the movement of hexose across the intestine of *Ascaris suum*. *The Journal of Parasitology,* 64: 89-92.

[42] White R.R., Miyata S., Papa E., Spooner E., Gounaris K., Selkirk M.E. & Artavanis-Tsakonas K. (2011) Characterisation of the *Trichinella spiralis* deubiquitinating enzyme, TsUCH37, an evolutionarily conserved proteasome interaction partner. *PloS Neglected Tropical Diseases,* 5: e1340

[43] Rutherford T.A. & Webster J.M. (1974) Transcuticular uptake of glucose by the entomophilic nematode, *Mermis nigrescens*. *The Journal of Parasitology,* 60: 804-808.

[44] Rhoads M.L. & Fetterer R.H. (1998) Purification and characterisation of a secreted aminopeptidase from adult *Ascaris suum*. *International Journal for Parasitology,* 28: 1681-1690.

[45] Guiliano D.B., Hong X., McKerrow J.H., Blaxter M.L., Oksov Y., Liu J., Ghedin E. & Lustigman S. (2004) A gene family of cathepsin L-like proteases of filarial nematodes are associated with larval molting and cuticle and eggshell remodeling. *Molecular and Biochemical Parasitology,* 136: 227-242.

[46] Haffner A., Guilavogui A.Z., Tischendorf F.W. & Brattig N.W. (1998) *Onchocerca volvulus:* microfilariae secrete elastinolytic and males nonelastinolytic matrix-degrading serine and metalloproteases. *Experimental Parasitology,* 90: 26-33.

[47] Rutherford T.A., Webster J.M. & Barlow J.S. (1977) Physiology of nutrient uptake by the entomophilic nematode *Mermis nigrescens* (Mermithidae). *Canadian Journal of Zoology-Revue Canadienne de Zoologie,* 55: 1773-1781.

[48] Yasuraoka, K., Koyama, T. & Kato, K., (1967) Studies on the in vitro axenic development of *Anisakis* larvae (I). *Japanese Journal of Parasitology,* 16: 290–291 (in Japanese, cited by Sommerville & Davey [50]).

[49] Iglesias L., Valero A., Benítez R. & Adroher F.J. (2001) In vitro cultivation of *Anisakis simplex*: pepsin increases survival and moulting from fourth larval to adult stage. *Parasitology,* 123: 285-291.

[50] Sommerville R.I. & Davey K.G. (1976) Stimuli for cuticle formation and ecdysis in vitro of infective larva of *Anisakis* sp. (Nematoda: Ascaridoidea). *International Journal for Parasitology,* 6: 433-439.

[51] Iglesias L., Valero A., Gálvez L., Benítez R. & Adroher F.J. (2002) In vitro cultivation of *Hysterothylacium aduncum* (Nematoda: *Anisakidae*) from 3rd-stage larvae to egg-laying adults. *Parasitology,* 125: 467-475.

[52] Malagón D., Benítez R., Adroher F.J. & Díaz-López M. (2011) Proteolytic activity in *Hysterothylacium aduncum* (Nematoda: Anisakidae), a fish gastrointestinal parasite of worldwide distribution. *Veterinary Parasitology*, 183: 95-102.

[53] Rehman A. & Jasmer D.P. (1998) A tissue specific approach for analysis of membrane and secreted protein antigens from *Haemonchus contortus* gut and its application to diverse nematode species. *Molecular and Biochemical Parasitology,* 97: 55-68.

[54] Williamson A.L., Brindley P.J., Knox D.P., Hotez P.J. & Loukas A. (2003) Digestive proteases of blood-feeding nematodes. *Trends in Parasitology,* 19: 417-423.

[55] Nawaratna S.S., McManus D.P., Moertel L., Gobert G.N. & Jones M.K. (2011) Gene Atlasing of digestive and reproductive tissues in *Schistosoma mansoni*. *PLoS Neglected Tropical Diseases,* 5: e1043.

[56] Williamson A.L., Lecchi P., Turk B.E., Choe Y., Hotez P.J., McKerrow J.H., Cantley L.C., Sajid M., Craik C.S. & Loukas A. (2004) A multi-enzyme cascade of hemoglobin

proteolysis in the intestine of blood-feeding hookworms. *The Journal of Biological Chemistry,* 279: 35950-35957.

[57] Don T.A., Jones M.K., Smyth D., O'Donoghue P., Hotez P. & Loukas A. (2004) A pore-forming haemolysin from the hookworm, *Ancylostoma caninum. International Journal for Parasitology,* 34: 1029-1035.

[58] Knox D.P. (2011) Proteases in blood-feeding nematodes and their potential as vaccine candidates. *Advances in Experimental Medicine and Biology,* 712: 155-176.

[59] Bethony J.M., Loukas A., Hotez P.J. & Knox D.P. (2006) Vaccines against blood-feeding nematodes of humans and livestock. *Parasitology,* 133: 63-79

[60] Ekoja S.E. & Smith W.D. (2010) Antibodies from sheep immunized against *H. contortus* with H-gal-GP inhibit the haemoglobinase activity of this protease complex. *Parasite Immunology,* 32: 731–738.

[61] Sarkis G.J., Ashcom J.D., Hawdon J.M. & Jacobson L.A. (1988) Decline in protease activities with age in the nematode *Caenorhabditis elegans. Mechanisms of Ageing and Development,* 45: 191-201.

[62] Sarkis G.J., Kurpiewski M.R., Ashcom J.D., Jen-Jacobson L. & Jacobson L.A. (1988) Proteases of the nematode *Caenorhabditis elegans. Archives of Biochemistry and Biophysics,* 261: 80-90.

[63] Hawdon J.M., Emmons S.W. & Jacobson L.A. (1989) Regulation of proteinase levels in the nematode *Caenorhabditis elegans*. Preferential depression by acute or chronic starvation. *The Biochemical Journal,* 264: 161-165.

[64] Brindley P.J., Kalinna B.H., Wong J.Y., Bogitsh B.J., King L.T., Smyth D.J., Verity C.K., Abbenante G., Brinkworth R.I., Fairlie D.P., Smythe M.L., Milburn P.J., Bielefeldt-Ohmann H., Zheng Y. & McManus D.P. (2001) Proteolysis of human hemoglobin by schistosome cathepsin D. *Molecular and Biochemical Parasitology,* 112: 103–112.

[65] Sajid M., McKerrow J.H., Hansell E., Mathieu M.A., Lucas K.D., Hsieh I., Greenbaum D., Bogyo M., Salter J.P., Lim K.C., Franklin C., Kim J.H. & Caffrey C.R. (2003) Functional expression and characterization of *Schistosoma mansoni* cathepsin B and its trans-activation by an endogenous asparaginyl endopeptidase. *Molecular and Biochemical Parasitology,* 131: 65–75.

[66] Rhoads M.L., Fetterer R.H. & Urban J.F., Jr. (1998) Effect of protease class-specific inhibitors on in vitro development of the third- to fourth-stage larvae of *Ascaris suum. The Journal of Parasitology,* 84: 686-690.

[67] Dziekońska-Rynko J., Rokicki J. & Jabłonowski Z. (1997) The impact of animal and plant proteases on nematoda 3rd stage *Anisakis simplex* larvae. *Oceanological Studies,* 1: 103-107.

[68] Iglesias L., Valero A. & Adroher F.J. (1997) Some factors which influence the in vitro maintenance of *Anisakis simplex* (Nematoda). *Folia Parasitologica,* 44: 297-301.

[69] Rogers W.P. (1961) The physiology of hatching of eggs of *Ascaridia galli. Journal of Helminthology,* R.T. Leiper Supplement: 151-156.

[70] Rogers W.P. (1962) *The nature of parasitism: The relationship of some metazoan parasites to their hosts,* Academic Press, New York.

[71] Rogers W.P. & Sommerville R.I. (1968) The infectious process, and its relation to the development of early parasitic stages of nematodes. *Advances in Parasitology,* 6: 327-348.

[72] Jabłonowski Z., Romaniuk K., Piechocki D., Żołtowska K., Łukaszewicz-Babečka J. & Dziekońska-Rynko I.J. (1995) [Effect of hydrochloric acid on invasion of *Ascaris suum;* enzyme activity in the digestive system and serum of newborn piglets] (original in Polish, abstract in English). *Wiadomosci Parazytologiczne,* 41: 413-419.

[73] Wang L.C., Chao D. & Chen E.R. (1989) Acquired immunity in rats against *Angiostrongylus cantonensis* infection. *International Journal for Parasitology,* 19: 617-620.

[74] Wang L.C., Chao D. & Chen E.R. (1991) Experimental infection routes of *Angiostrongylus cantonensis* in mice. *Journal of Helminthology,* 65: 296-300.

[75] Chang S.H, Tang P. & Wang L.C. (2011) A transcriptomic study on the pepsin-activated infective larvae of *Angiostrongylus cantonensis. Molecular and Biochemical Parasitology,* 179: 47- 50.

[76] http://www.faculty.pnc.edu/jcamp/parasit/nematode.html (accessed 1st of December of 2011). Dr. Joe Camp (Purdue University North Central).

[77] Tort J., Brindley P.J., Knox D.P., Wolfe K.H. & Dalton J.P. (1999) Proteinases and associated genes of parasitic helminths. *Advances in Parasitology,* 43: 161-266.

[78] Knox D.P., Skuce P.J., Newlands G.F. & Redmond D.L. (2001) Nematode gut peptidases, proteins and vaccination. En: *Parasitic nematodes: molecular biology, biochemistry and immunology* (Kennedy M.W. & Harnett W., eds.) pp. 247-268, CABI Publishing, Wallingford-Oxon.

[79] Dalton J.P., Brindley P.J., Knox D.P., Brady C.P., Hotez P.J., Donnelly S., O'Neill S.M., Mulcahy G. & Loukas A. (2003) Helminth vaccines: from mining genomic information for vaccine targets to systems used for protein expression. *International Journal for Parasitology,* 33: 621-640.

[80] Tamashiro W., Rao M. & Scott A. (1987) Proteolytic cleavage of IgG and other protein substrates by *Dirofilaria immitis* microfilarial enzymes. *The Journal of Parasitology,* 73: 149-154.

[81] Monroy F.G., Cayzer C.J., Adams J.H. & Dobson C. (1989) Proteolytic enzymes in excretory-secretory products from adult *Nematospiroides dubius. International Journal for Parasitology,* 19: 129-131.

[82] Kumar S. & Pritchard D.I. (1992) Secretion of metalloproteases by living infective larvae of *Necator americanus. The Journal of Parasitology,* 78: 917-919.

[83] Brown A., Burleigh J.M., Billett E.E. & Pritchard D.I. (1995) An initial characterization of the proteolytic enzymes secreted by the adult stage of the human hookworm *Necator americanus. Parasitology,* 110: 555-563.

[84] Lackey A., James E.R., Sakanari J.A., Resnick S.D., Brown M., Bianco A.E. & McKerrow J.H. (1989) Extracellular proteases of *Onchocerca. Experimental Parasitology,* 68: 176-185.

[85] Drake L.J., Bianco A.E., Bundy D.A. & Ashall F. (1994) Characterization of peptidases of adult *Trichuris muris. Parasitology,* 109: 623-630.

[86] Young C.J., McKeand J.B. & Knox D.P. (1995) Proteinases released in vitro by the parasitic stages of *Teladorsagia circumcincta*, an ovine abomasal nematode. *Parasitology,* 110: 465-471.

[87] Read A.F. & Skorping A. (1995) The evolution of tissue migration by parasitic nematode larvae. *Parasitology,* 111: 359-371.

[88] Balbuena J.A., Karlsbakk E., Kvenseth A.M., Saksvik M. & Nylund A. (2000) Growth and emigration of third-stage larvae of *Hysterothylacium aduncum* (Nematoda: Anisakidae) in larval herring *Clupea harengus*. *The Journal of Parasitology,* 86: 1271-1275.

[89] Sukhdeo M.V.K. & Sukhdeo S.C. (1994) Optimal habitat selection by helminths within the host environment. *Parasitology,* 109: 41-55.

[90] Rhoads M.L., Fetterer R.H. & Urban J.F., Jr. (2001) Release of hyaluronidase during in vitro development of *Ascaris suum* from the third to fourth larval stage. *Parasitology Research,* 87: 693-697.

[91] Wilson R., Ainscough R., Anderson K., Baynes C., Berks M., Bonfield J., Burton J., Connell M., Copsey T., Cooper J., Coulson A., Craxton M., Dear S., Du Z., Durbin R., Favello A., Fraser A., Fulton L., Gardner A., Green P., Hawkins T., Hillier L., Jier M., Johnston L., Jones M., Kershaw J., Kirsten J., Laisster N., Latreille P., Lightning J., Lloyd C., Mortimore B., O'Callaghan M., Parsons J., Percy C., Rifken L., Roopra A., Saunders D., Shownkeen R., Sims M., Smaldon N., Smith A., Smith M., Sonnhammer E., Staden R., Sulston J., Thierry-Mieg J., Thomas K., Vaudin M., Vaughan K., Waterston R., Watson A., Weinstock L., Wilkinson-Sproat J. & Wohldman P. (1994) 2.2 Mb of contiguous nucleotide sequence from chromosome III of *C. elegans*. *Nature,* 368: 32-38.

[92] Hotez P.J., Capello M., Hawdon J., Beckers C. & Sakanari J.A. (1994) Hyaluronidases of the gastrointestinal invasive nematodes *Ancylostoma caninum* and *Anisakis simplex*: possible functions in the pathogenesis of human zoonoses. *The Journal of Infectious Diseases,* 170: 918-926.

[93] Murrell K.D., Eriksen L., Nansen P., Slotved H.C. & Rasmussen T. (1997) *Ascaris suum:* a revision of its early migratory path and implications for human ascariasis. *The Journal of Parasitology,* 83: 255-260.

[94] Anderson R.C. & Bartlett C.M. (1993) The significance of precocity in the transmission of the nematode parasites of vertebrates. *Canadian Journal of Zoology-Revue Canadienne de Zoologie,* 71: 1917-1922.

[95] Roberts L.S. & Janovy J., Jr. (2005) Nematodes: Ascaridida, intestinal large roundworms. In: Gerald D. Schmidt & Larry S. Roberts' *Foundations of Parasitology* 7ª edn. (Roberts L.S. & Janovy J., Jr., eds.) pp. 431-444, McGraw-Hill, New York.

[96] Balbuena J.A., Karlsbakk E., Saksvik M., Kvenseth A.M. & Nylund A. (1998) New data on the early development of *Hysterothylacium aduncum* (Nematoda, Anisakidae). *The Journal of Parasitology,* 84: 615-617.

[97] Klimpel S. & Rückert S. (2005) Life cycle strategy of *Hysterothylacium aduncum* to become the most abundant anisakid fish nematode in the North Sea. *Parasitology Research,* 97: 141-149.

[98] Balasubramanian N., Toubarro D., Nascimento G., Ferreira R. & Simões N. (2012) Purification, molecular characterization and gene expression analysis of an aspartic protease (Sc-ASP113) from the nematode *Steinernema carpocapsae* during the parasitic stage. *Molecular and Biochemical Parasitology,* 182: 37-44.

[99] Brindley P.J., Gam A.A., McKerrow J.H. & Neva F.A. (1995) Ss40: the zinc endopeptidase secreted by infective larvae of *Strongyloides stercoralis*. *Experimental Parasitology,* 80: 1-7.

[100] Castagnone-Sereno P., Deleury E., Danchin E.G.J., Perfus-Barbeoch L. & Abad P. (2011) Data-mining of the *Meloidogyne incognita* degradome and comparative analysis of proteases in nematodes. *Genomics*, 97: 29–36.

[101] Southan S. (2001) A genomic perspective on human proteases. *Federation of European Biochemical Societies (FEBS) Letters*, 498: 214–218.

[102] Wu Y., Wang X., Liu X., & Wang Y. (2003) Data-mining approaches reveal hidden families of proteases in the genome of malaria parasite. *Genome Research,* 13: 601–616

[103] Opperman C.H., Bird D.M., Williamson V.M., Rokhsar D.S., Burke M., Cohn J., Cromer J., Diener S., Gajan J., Graham S., Houfek T.D., Liu Q., Mitros T., Schaff J., Schaffer R., Scholl E., Sosinski B.R., Thomas V.P. & Windham E. (2008) Sequence and genetic map of *Meloidogyne hapla*: a compact nematode genome for plant parasitism. *Proceedings of the National Academy of Sciences of the United States of America* (PNAS), 105: 14802–14807.

[104] Ghedin E., Wang S., Spiro D., Caler E., Zhao Q., Crabtree J., Allen J.E., Delcher A.L., Guiliano D.B., Miranda-Saavedra D., Angiuoli S.V., Creasy T., Amedeo P., Haas B., El-Sayed N.M., Wortman J.R., Feldblyum T., Tallon L., Schatz M., Shumway M., Koo H., Salzberg S.L., Schobel S., Pertea M., Pop M., White O., Barton G.J.,. Carlow C.K, Crawford M.J., Daub J., Dimmic M.W., Estes C.F., Foster J.M., Ganatra M., Gregory W.F., Johnson N.M., Jin J., Komuniecki R., Korf I., Kumar S., Laney S., Li B.W., Li W., Lindblom T.H., Lustigman S., Ma D., Maina C.V., Martin D.M.A., McCarter J.P., McReynolds L., Mitreva M., Nutman T.B., Parkinson J., Peregrín-Alvarez J.M., Poole C., Ren Q., Saunders L., Sluder A.E., Smith K., Stanke M., Unnasch T.R., Ware J., Wei A.D., Weil G., Williams D.J., Zhang Y., Williams S.A., Fraser-Liggett C., Slatko B., Blaxter M.L. & Scott A.L. (2007) Draft genome of the filarial nematode parasite *Brugia malayi. Science,* 317: 1756–1760.

[105] Grassmann, W. & Dyckerhoff, H. (1928) Über die Proteinase und die Polypeptidase der Hefe. 13. Abhandlung über Pflanzenproteasen in der von R. Willstätter und Mitarbeitern begonnenen Untersuchungsreihe. *Hoppe-Seyler's Zeitschrift für physiologische Chemie,* 179: 41-78

[106] García-Carreño F.L. (1992) Protease inhibition in theory and practice. *Biotechnology Education,* 3: 145-150.

[107] http://www.chem.qmul.ac.uk/iubmb/enzyme/EC3/4/ (accessed 1st of December of 2011). *Committee of the International Union of Biochemistry and Molecular Biology* (NC-IUBMB).

[108] Kašný M., Mikeš L., Hampl V., Dvořák J., Caffrey C.R., Dalton J.P., & Horák P. (2009) Peptidases of Trematodes. *Advances in Parasitology,* 69: 205-297.

[109] Núñez de Castro I. (2001) Catálisis enzimática. In: *Enzimología,* pp. 69-99, Pirámide, Madrid.

[110] Schechter I. & Berger A. (1967) On the size of the active site in proteases. I. Papain. *Biochemical and Biophysical Research Communications,* 27: 157-162.

[111] Rawlings N.D. & Barrett A.J. (1993) Evolutionary families of peptidases. *The Biochemical Journal,* 290: 205–218.

[112] Rawlings N.D., Morton F.R., Kok C.Y., Kong J. & Barrett A.J. (2008) MEROPS: The peptidase database. *Nucleic Acids Research,* 36: 320–325.

[113] Rawlings N.D. & Barrett A.J. (1994) Families of serine peptidases. *Methods in Enzymology,* 244: 19-61.

[114] Rawlings N.D. & Barrett A.J. (1994) Families of cysteine peptidases. *Methods in Enzymology,* 244: 461-486.

[115] Rawlings N.D. & Barrett A.J. (1995) Families of aspartic peptidases, and those of unknown catalytic mechanism. *Methods in Enzymology,* 248: 105-120.

[116] Rawlings N.D. & Barrett A.J. (1995) Evolutionary families of metallopeptidases. *Methods in Enzymology,* 248: 183-228.

[117] Illy C., Quraishi O., Wang J., Purisima E., Vernet T. & Mort J.S, (1997) Role of the occluding loop in cathepsin B activity. *The Journal of Biological Chemistry,* 272: 1197-1202.

[118] Musil D., Zucic D., Turk D., Engh R.A., Mayr I., Huber R., Popovic T., Turk V., Towatari T. & Katunuma N. (1991) The refined 2.15 A X-ray crystal structure of human liver cathepsin B: The structural basis for its specificity. *EMBO Journal* 10: 2321-2330.

[119] Sato K., Yamaguchi H., Waki S., Suzuki M. & Nagai Y. (1995) *Dirofilaria immitis*: immunohistochemical localization of acid proteinase in the adult worm. *Experimental Parasitology,* 81: 63-71.

[120] Hwang K.P., Chang S.H. & Wang L.C. (2010) Alterations in the expression level of a putative aspartic protease in the development of *Angiostrongylus cantonensis. Acta Tropica,* 113: 289-294.

[121] Harrop S.A., Prociv P. & Brindley P.J. (1996) Acasp, a gene encoding a cathepsin D-like aspartic protease from the hookworm *Ancylostoma caninum. Biochemical and Biophysical Research Communications,* 227: 294-302.

[122] Williamson A.L., Brindley P.J., Abbenante G., Prociv P., Berry C., Girdwood K., Pritchard D.I., Fairlie D.P., Hotez P.J., Dalton J.P. & Loukas A. (2002) Cleavage of hemoglobin by hookworm cathepsin D aspartic proteases and its potential contribution to host specificity. *FASEB Journal,* 16: 1458-1460.

[123] Williamson A.L., Brindley P.J., Abbenante G., Datu B.J., Prociv P., Berry C., Girdwood K., Pritchard D.I., Fairlie D.P., Hotez P.J., Zhan B. & Loukas A. (2003) Hookworm aspartic protease, Na-APR-2, cleaves human hemoglobin and serum proteins in a host-specific fashion. *Journal of Infectious Diseases,* 187: 484-494.

[124] Yang Y., Wei H., Qin W. & Zheng J. (2009) Expression and characterization of aspartic protease gene in eggs and larvae stage of *Ancylostoma caninum. Parasitology Research,* 104: 1327–1333.

[125] Tcherepanova I, Bhattacharyya L, Rubin CS & Freedman JH. (2000) Aspartic proteases from the nematode *Caenorhabditis elegans*. Structural organization and developmental and cell-specific expression of asp-1. *The Journal of Biological Chemistry,* 275: 26359-26369.

[126] Jolodar A. & Miller D.J. (1998) Identification of a novel family of non-lysosomal aspartic proteases in nematodes. *Biochimica et Biophysica Acta,* 1382: 13-16.

[127] Geier G., Banaj H.J., Heid H., Bini L., Pallini V. & Zwilling R. (1999) Aspartyl proteases in *Caenorhabditis elegans*. Isolation, identification and characterization by a combined use of affinity chromatography, two-dimensional gel electrophoresis, microsequencing and databank analysis. *European Journal of Biochemistry*, 264: 872-879.

[128] Geldhof P., Claerebout E., Knox D.P., Agneessens J. & Vercruysse J. (2000) Proteinases released in vitro by the parasitic stages of the bovine abomasal nematode *Ostertagia ostertagi*. *Parasitology,* 121: 639-647.

[129] Polgár L. (1987). The mechanismof action of aspartic proteases involves 'push–pull' catalysis. *FEBS Letters,* 219: 1-4.

[130] Pratt D., Cox G.N., Milhausen M.J. & Boisvenue R.J. (1990) A developmentally regulated cysteine protease gene family in *Haemonchus contortus*. *Molecular and Biochemical Parasitology,* 43: 181-192.

[131] Jasmer D.P., Roth J. & Myler P.J. (2001) Cathepsin B-like cysteine proteases and *Caenorhabditis elegans* homologues dominate gene products expressed in adult *Haemonchus contortus* intestine. *Molecular and Biochemical Parasitology,* 116: 159-169.

[132] Shompole S. & Jasmer D.P. (2001) Cathepsin B-like cysteine proteases confer intestinal cysteine protease activity in *Haemonchus contortus*. *The Journal of Biological Chemistry,* 276: 2928-2934.

[133] Karanu F.N., Rurangirwa F.R., McGuire T.C. & Jasmer D.P. (1993) *Haemonchus contortus:* identification of proteases with diverse characteristics in adult worm excretory-secretory products. *Experimental Parasitology,* 77: 362-371.

[134] Richer J.K., Sakanari J.A., Frank G.R. & Grieve R.B. (1992) *Dirofilaria immitis*: proteases produced by third- and fourth-stage larvae. *Experimental Parasitology,* 75: 213-222.

[135] Richer J.K., Hunt W.G., Sakanari J.A. & Grieve R.B. (1993) *Dirofilaria immitis*: effect of fluoromethyl ketone cysteine protease inhibitors on the third- to fourth-stage molt. *Experimental Parasitology, 76*: 221-231.

[136] Lustigman S., McKerrow J.H., Shah K., Lui J., Huima T., Hough M. & Brotman B. (1996) Cloning of a cysteine protease required for the molting of *Onchocerca volvulus* third stage larvae. *The Journal of Biological Chemistry,* 271: 30181-30189 (Erratum in: *The Journal of Biological Chemistry* (1997) 272: 4645).

[137] Falcone F.H., Tetteh K.K., Hunt P., Blaxter M.L., Loukas A. & Maizels R.M. (2000) The new subfamily of cathepsin-Z-like protease genes includes Tc-cpz-1, a cysteine protease gene expressed in *Toxocara canis* adults and infective stage larvae. *Experimental Parasitology*, 94: 201-207.

[138] Lustigman S., Zhang J., Liu J., Oksov Y. & Hashmi S. (2004) RNA interference targeting cathepsin L and Z-like cysteine proteases of *Onchocerca volvulus* confirmed their essential function during L3 molting. *Molecular and Biochemical Parasitology,* 138: 165-170.

[139] Hashmi S., Zhang J., Oksov Y. & Lustigman S. (2004) The *Caenorhabditis elegans* cathepsin Z like cysteine protease, Ce-CPZ-1, has a multifunctional role during the worms' development. *The Journal of Biological Chemistry,* 279: 6035-6045.

[140] Kumar S. & Pritchard D.I. (1992) The partial characterization of proteases present in the excretory/secretory products and exsheathing fluid of the infective (L3) larva of *Necator americanus*. *International Journal for Parasitology,* 22: 563-572.

[141] Knox D.P., Redmond D.L. & Jones D.G. (1993) Characterization of proteinases in extracts of adult *Haemonchus contortus*, the ovine abomasal nematode. *Parasitology,* 106: 395-404.

[142] Bond J.S. & Butler P.E. (1987) Intracellular proteases. *Annual Review of Biochemistry,* 56: 333-364.

[143] McKerrow J.H., Engel J.C. & Caffrey C.R. (1999) Cysteine protease inhibitors as chemotherapy for parasitic infections. *Bioorganic and Medicinal Chemistry,* 7: 639-644.

[144] McKerrow J.H. (1999) Development of cysteine protease inhibitors as chemotherapy for parasitic diseases: insights on safety, target validation, and mechanism of action. *International Journal for Parasitology,* 29: 833-837.

[145] Selzer P.M., Pingel S., Hsieh I., Ugele B., Chan V.J., Engel J.C., Bogyo M., Russell D.G., Sakanari J.A. & McKerrow J.H. (1999) Cysteine protease inhibitors as chemotherapy: Lessons from a parasite target. *Proceedings of the National Academy of Sciences of the United States of America (PNAS),* 96: 11015-11022.

[146] Willstätter R. & Bamann E. (1929) Über die proteasen der magenschleimhaut. Erste abhandlung über die enzyme der leukozyten. *Hoppe-Seyler's Zeitschrift für Physiologische Chemie,* 180: 127-143.

[147] Berti P.J. & Storer A.C. (1995) Alignment/phylogeny of the papain superfamily of cysteine proteases. *Journal of Molecular Biology,* 246: 273-283.

[148] Pratt D., Armes L.G., Hageman R., Reynolds V., Boisvenue R.J. & Cox G.N. (1992) Cloning and sequence comparisons of four distinct cysteine proteases expressed by *Haemonchus contortus* adult worms. *Molecular and Biochemical Parasitology, 51*: 209-218.

[149] Baig S., Damian R.T. & Peterson D.S. (2002) A novel cathepsin B active site motif is shared by helminth bloodfeeders. *Experimental Parasitology,* 101: 83-89.

[150] Ray C. & McKerrow J.H. (1992) Gut-specific and developmental expression of a *Caenorhabditis elegans* cysteine protease gene. *Molecular and Biochemical Parasitology,* 51: 239-249.

[151] Britton C., McKerrow J.H. & Johnstone I.L. (1998) Regulation of the *Caenorhabditis elegans* gut cysteine protease gene cpr-1: requirement for GATA motifs. *Journal of Molecular Biology,* 283: 15-27.

[152] Britton C., Redmond D.L., Knox D.P., McKerrow J.H. & Barry J.D. (1999) Identification of promoter elements of parasite nematode genes in transgenic *Caenorhabditis elegans. Molecular and Biochemical Parasitology,* 103: 171-181

[153] Yang Y., Qin W., Wei H., Ying J. & Zhen J. (2011) Characterization of cathepsin B proteinase (AcCP-2) in eggs and larvae stages of hookworm *Ancylostoma caninum. Experimental Parasitology,* 129: 215–220.

[154] Han Y.P., Li Z.Y., Li B.C., Sun X., Zhu C.C., Ling X.T., Zheng H.Q., Wu Z.D. & Lv Z.Y. (2011) Molecular cloning and characterization of a cathepsin B from *Angiostrongylus cantonensis*. *Parasitology Research*, 109: 369–378.

[155] Loukas A., Selzer P.M. & Maizels R.M. (1998) Characterisation of Tc-clp-1, a cathepsin L-like cysteine protease from *Toxocara canis* infective larvae. *Molecular and Biochemical Parasitology,* 92: 275-289.

[156] Koiwa H., Shade R.E., Zhu-Salzman K., D'Urzo M.P., Murdock L.L., Bressan R.A. & Hasegawa P.M. (2000) A plant defensive cystatin (soyacystatin) targets cathepsin L-like digestive cysteine proteinases (DvCALs) in the larval midgut of western corn rootworm (*Diabrotica virgifera virgifera*). *FEBS Letters,* 471: 67-70.

[157] Lizotte-Waniewski M., Tawe W., Guiliano D.B., Lu W., Liu J., Williams S.A. & Lustigman S. (2000) Identification of potential vaccine and drug target candidates by expressed sequence tag analysis and immunoscreening of *Onchocerca volvulus* larval cDNA libraries. *Infection and Immunity,* 68: 3491-3501.

[158] Malagón D., Díaz-López M., Benítez R. & Adroher F.J. (2010) Cathepsin B- and L-like cysteine protease activities during the in vitro development of *Hysterothylacium aduncum* (Nematoda: Anisakidae), a worldwide fish parasite. *Parasitology International,* 59: 89-92.

[159] Auld D.S. (2004) Catalytic mechanisms of metallopeptidases. In: Barrett, Rawlings & Woessner (Eds.), *Handbook of Proteolytic Enzymes*, 2nd edn. Elsevier. Amsterdam, The Netherlands pp. 268-289.

[160] Lowther W.T. & Matthews B.W. (2002) Metalloaminopeptidases: Common functional themes in disparate structural surroundings. *Chemical Reviews,* 102: 4581–4608

[161] Maruyama H., Nishimaki A., Takuma Y., Kurimoto M., Suzuki T., Sakatoku Y., Ishikawa M. & Ohta N. (2006). Successive changes in tissue migration capacity of developing larvae of an intestinal nematode, *Strongyloides venezuelensis. Parasitology,* 132: 411-418.

[162] Gómez Gallego S., Loukas A., Slade R.W., Neva F.A., Varatharajalu R., Nutman T.B. & Brindley P.J. (2005) Identification of an astacin-like metallo-proteinase transcript from the infective larvae of *Strongyloides stercoralis. Parasitology International,* 54: 123–133.

[163] Yoshida A., Nagayasu E., Nishimaki A., Sawaguchi A., Yanagawa S. & Maruyama H. (2011) Transcripts analysis of infective larvae of an intestinal nematode, *Strongyloides venezuelensis. Parasitology International,* 60: 75–83.

[164] Lee J.D. & Yen C.M. (2005) Protease secreted by the infective larvae of *Angiostrongylus cantonensis* and its role in the penetration of mouse intestine. *American Journal of Tropical Medicine and Hygiene*, 72: 831–836.

[165] Jing Y., Toubarro D., Hao Y.J. & Simões N. (2010) Cloning, characterisation and heterologous expression of an astacin metalloprotease, Sc-AST, from the entomoparasitic nematode *Steinernema carpocapsae. Molecular and Biochemical Parasitology*, 174: 101-108.

[166] Hill D.E., Gamble H.R., Rhoads M.L., Fetterer R.H. & Urban J.F., Jr. (1993) *Trichuris suis:* a zinc metalloprotease from culture fluids of adult parasites. *Experimental Parasitology,* 77: 170-178.

[167] Culley F.J., Brown A., Conroy D.M., Sabroe I., Pritchard D.I. & Williams T.J. (2000) Eotaxin is specifically cleaved by hookworm metalloproteases preventing its action in vitro and in vivo. *Journal of Immunology,* 165: 6447-6453.

[168] Hotez P.J. & Cerami A. (1983) Secretion of a proteolytic anticoagulant by *Ancylostoma* hookworms. *The Journal of Experimental Medicine,* 157: 1594-1603.

[169] Jones B.F. & Hotez P.J. (2002) Molecular cloning and characterization of Ac-mep-1, a developmentally regulated gut luminal metalloendopeptidase from adult *Ancylostoma caninum* hookworms. *Molecular and Biochemical Parasitology,* 119: 107-106.

[170] Hotez P.J., Trang N.L., McKerrow J.H. & Cerami A. (1985) Isolation and characterization of a proteolytic enzyme from the adult hookworm *Ancylostoma caninum. The Journal of Biological Chemistry,* 260: 7343-7348.

[171] Hotez P.J., Haggerty J., Hawdon J., Milstone L., Gamble H.R., Schad G. & Richards F. (1990) Metalloproteases of infective *Ancylostoma* hookworm larvae and their possible functions in tissue invasion and ecdysis. *Infection and Immunity,* 58: 3883-3892.

[172] Malagón D., Adroher F.J., Díaz-López M. & Benítez R. (2010) Collagenolytic activity related to metalloproteases (and serine proteases) in the fish parasite *Hysterothylacium aduncum* (Nematoda: Anisakidae). *Diseases of Aquatic Organisms,* 90: 129-34.

[173] Hedstrom L. (2002) Serine protease mechanism and specificity. *Chemical Reviews,* 102: 4501–4523.

[174] Rawlings N.D. & Barrett A.J. (2004) Introduction: Serine peptidases and their clans. In: Barrett, Rawlings & Woessner (Eds.) *Handbook of Proteolytic Enzymes,* 2nd edn. Elsevier, Amsterdam, The Netherlands. pp. 1417–1439.

[175] McKerrow J.H. (1989) Parasite proteases. *Experimental Parasitology,* 68: 111-115.

[176] Cohen F.E., Gregoret L.M., Amiri P., Aldape K., Railey J. & McKerrow J.H. (1991) Arresting tissue invasion of a parasite by protease inhibitors chosen with the aid of computer modeling. *Biochemistry,* 30: 11221-11229.

[177] Knox D.P. & Kennedy M.W. (1988) Proteinases released by the parasitic larval stages of *Ascaris suum,* and their inhibition by antibody. *Molecular and Biochemical Parasitology,* 28: 207-216.

[178] Robertson B.D., Bianco A.T., McKerrow J.H. & Maizels R.M. (1989) *Toxocara canis:* proteolytic enzymes secreted by the infective larvae in vitro. *Experimental Parasitology,* 69: 30-36.

[179] Balasubramanian N., Hao Y.J., Toubarro D., Nascimento G. & Simões N. (2009) Purification, biochemical and molecular analysis of a chymotrypsin protease with prophenoloxidase suppression activity from the entomopathogenic nematode *Steinernema carpocapsae. International Journal for Parasitology,* 39: 975-984.

[180] Balasubramanian N., Toubarro D. & Simões N. (2010) Biochemical study and in vitro insect immune suppression by a trypsin-like secreted protease from the nematode *Steinernema carpocapsae. Parasite Immunology,* 32: 165-175.

[181] Toubarro D., Lucena-Robles M., Nascimento G., Costa G, Montiel R., Coelho A.V. & Simões N. (2009) An apoptosis-inducing serine protease secreted by the entomopathogenic nematode *Steinernema carpocapsae. International Journal for Parasitology,* 39: 1319-1330.

[182] Toubarro D., Lucena-Robles M., Nascimento G., Santos R., Montiel R., Veríssimo P., Pires E., Faro C., Coelho A.V. & Simões N. (2010) Serine protease-mediated host invasion by the parasitic nematode *Steinernema carpocapsae. The Journal of Biological Chemistry*, 285: 30666-30675.

[183] Ford L., Guiliano D.B., Oksov Y., Debnath A.K., Liu J., Williams S.A., Blaxter M.L. & Lustigman S. (2005) Characterization of a novel filarial serine protease inhibitor, Ov-SPI-1, from *Onchocerca volvulus*, with potential multifunctional roles during development of the parasite. *The Journal of Biological Chemistry,* 280: 40845-40856.

[184] Poole C.B., Jin J. & McReynolds L.A. (2007) Subtilisin-like proteases in nematodes. *Molecular and Biochemical Parasitology,* 155: 1-8.

[185] Hao Y.J., Montiel R., Nascimento G., Toubarro D. & Simões N. (2008) Identification, characterization of functional candidate genes for host–parasite interactions in entomopathogenetic nematode *Steinernema carpocapsae* by suppressive subtractive hybridization. *Parasitology Research,* 103: 671-683.

[186] Capello M., Clyne L.P., McPhedran P. & Hotez P.J. (1993) *Ancylostoma* factor Xa inhibitor: Partial purification and its identification as a major hookworm-derived anticoagulant in vitro. *The Journal of Infectious Diseases,* 167: 1474-1477.

[187] Capello M., Vlasuk G.P., Bergum P.W., Huang S. & Hotez P.J. (1995) *Ancylostoma caninum* anticoagulant peptide: A hookworm-derived inhibitor of human coagulation factor Xa. *Proceedings of the National Academy of Sciences of the United States of America (PNAS),* 92: 6152-6156.

[188] Stassens P., Bergum P.W., Gansemans Y., Jespers L., Laroche Y., Huang S., Maki S., Messens J., Lauwereys M., Cappello M., Hotez P.J., Lasters I. & Vlasuk G.P. (1996) Anticoagulant repertoire of hookworm *Ancylostoma caninum. Proceedings of the National Academy of Sciences of the United States of America (PNAS),* 93: 2149-2154.

[189] Lee A.Y. & Vlasuk G.P. (2003) Recombinant nematode anticoagulant protein c2 and other inhibitors targeting blood coagulation factor VIIa/tissue factor. *Journal of Internal Medicine,* 254: 313-321.

[190] Keilová H. & Tomásek V. (1972) Effect of pepsin inhibitor from *Ascaris lumbricoides* on cathepsin D and E. *Biochimica et Biophysica Acta,* 284: 461-464.

[191] Abu-Erreish G.M. & Peanasky R.J. (1974) Pepsin inhibitors from *Ascaris lumbricoides.* Isolation, purification, and some properties. *The Journal of Biological Chemistry,* 249: 1558-1565.

[192] Babin D.R., Peanasky R.J. & Goos S.M. (1984) The isoinhibitors of chymotrypsin/elastase from *Ascaris lumbricoides*: the primary structure. *Archives of Biochemistry and Biophysics,* 232: 143-161.

[193] Peanasky R.J., Bentz Y., Homandberg G.A., Minor S.T. & Babin D.R. (1984) The isoinhibitors of chymotrypsin/elastase from *Ascaris lumbricoides*: the reactive site. *Archives of Biochemistry and Biophysics,* 232: 135-142.

[194] Gronenborn A.M., Nilges M., Peanasky R.J. & Clore G.M. (1990) Sequential resonance assignment and secondary structure determination of the *Ascaris* trypsin inhibitor, a member of a novel class of proteinase inhibitors. *Biochemistry,* 29: 183-189.

[195] Martzen M.R., McMullen B.A., Smith N.E., Fujikawa K. &y Peanasky R.J. (1990) Primary structure of the major pepsin inhibitor from the intestinal parasitic nematode *Ascaris suum. Biochemistry,* 29: 7366-7372.

[196] Hawley J. & Peanasky R.J. (1992) *Ascaris suum*: are trypsin inhibitors involved in species specificity of ascarid nematodes? *Experimental Parasitology,* 75: 112-118.

[197] Jin X., Deng L., Li H., Zhang Z., He Q., Yang C., Jiang H., Zhu X.Q. & Peng L. (2011) Identification and characterization of a serine protease inhibitor with two trypsin inhibitor-like domains from the human hookworm *Ancylostoma duodenale. Parasitology Research,* 108: 287–295.

[198] Homandberg G.A. & Peanasky R.J. (1976) Characterization of proteins from *Ascaris lumbricoides* which bind specifically to carboxypeptidase. *The Journal of Biological Chemistry,* 251: 2226-2233.

[199] Homandberg G.A., Litwiller R.D. & Peanasky R.J. (1989) Carboxypeptidase inhibitors from *Ascaris suum*: the primary structure. *Archives of Biochemistry and Biophysics,* 270: 153-161.

[200] Delaney A., Williamson A., Brand A., Ashcom J., Varghese G., Goud G.N. & Hawdon J.M. (2005) Cloning and characterisation of an aspartyl protease inhibitor (API-1) from *Ancylostoma* hookworms. *International Journal for Parasitology,* 35: 303-313.

[201] Burbelo P.D., Leahy H.P., Iadarola M.J. & Nutman T.B. (2009) A four-antigen mixture for rapid assessment of *Onchocerca volvulus* infection. *PLoS Neglected Tropical Diseases,* 3: e438.

[202] Shaw R.J., McNeill M.M., Maass D.R., Hein W.R., Barber T.K., Wheeler M., Morris C.A. & Shoemaker C.B. (2003) Identification and characterisation of an aspartyl protease inhibitor homologue as a major allergen of *Trichostrongylus colubriformis. International Journal for Parasitology,* 33: 1233-1243.

[203] Satrija F, Nansen P, Bjørn H, Murtini S & He S. (1994) Effect of papaya latex against *Ascaris suum* in naturally infected pigs. *Journal of Helminthology,* 68: 343-346.

[204] Satrija F., Nansen P., Murtini S. & He S. (1995) Anthelmintic activity of papaya latex against patent *Heligmosomoides polygyrus* infections in mice. *Journal of Ethnopharmacology*, 48: 161-164.

[205] Stepek G., Buttle D.J., Duce I.R. & Behnke J.M. (2006) Human gastrointestinal nematode infections: are new control methods required? *International Journal of Experimental Pathology,* 87: 325-341.

[206] Stepek G., Buttle D.J., Duce I.R., Lowe A., & Behnke J.M. (2005) Assessment of the anthelmintic effect of natural plant cysteine proteinases against the gastrointestinal nematode, *Heligmosomoides polygyrus*, in vitro. *Parasitology,* 130: 203-211.

[207] Rhoads M.L., Fetterer R.H., Hill D.E. & Urban J.F.Jr. (2000) *Trichuris suis:* a secretory chymotrypsin/elastase inhibitor with potential as an immunomodulator. *Experimental Parasitology,* 95: 36-44.

[208] Manoury B., Gregory W.F., Maizels R.M. & Watts C. (2001) Bm-CPI-2, a cystatin homolog secreted by the filarial parasite *Brugia malayi*, inhibits class II MHC-restricted antigen processing. *Current Biology,* 11: 447-451.

[209] Dainichi T., Maekawa Y., Ishii K., Zhang T., Nashed B.F., Sakai T., Takashima M. & Himeno K. (2001) Nippocystatin, a cysteine protease inhibitor from *Nippostrongylus brasiliensis*, inhibits antigen processing and modulates antigen-specific immune response. *Infection and Immunity,* 69: 7380-7386.

[210] Liu Y.H., Han Y.P., Li ZY., Wei J., He H.J., Xu C.Z., Zheng H.Q., Zhan X.M., Wu Z.D. & Lv Z.Y. (2010) Molecular cloning and characterization of cystatin, a cysteine protease inhibitor, from *Angiostrongylus cantonensis*. *Parasitology Research,* 107: 915-922.

[211] Rodríguez-Mahillo A.I., González-Muñoz M., Gómez-Aguado F., Rodríguez-Pérez R., Corcuera M.T., Caballero M.L. & Moneo I. (2007) Cloning and characterisation of the *Anisakis simplex* allergen Ani s 4 as a cysteine-protease inhibitor. *International Journal for Parasitology,* 37: 907-917.

[212] Zavasnik-Bergant T. (2008) Cystatin protease inhibitors and immune functions. *Frontiers in Bioscience,* 13: 4625-4637.

[213] Schnoeller C., Rausch S., Pillai S., Avagyan A., Wittig B.M., Loddenkemper C., Hamann A., Hamelmann E., Lucius R. & Hartmann S. (2008) A helminth immunomodulator reduces allergic and inflammatory responses by induction of IL-10-producing macrophages. *Journal of Immunology,* 180: 4265-4272.

[214] Klotz C., Ziegler T., Daniłowicz-Luebert E. & Hartmann S. (2011) Cystatins of parasitic organisms. *Advances in Experimental Medicine and Biology,* 712: 208-221.

[215] Turrà D., Bellin D., Lorito M. & Gebhardt C. (2009) Genotype-dependent expression of specific members of potato protease inhibitor gene families in different tissues and in response to wounding and nematode infection. *Journal of Plant Physiology,* 166: 762-774.

[216] Atkinson H.J., Grimwood S., Johnston K. & Green J. (2004) Prototype demonstration of transgenic resistance to the nematode *Radopholus similis* conferred on banana by a cystatin. *Transgenic Research,* 13: 135-142.

[217] Chan Y.L, Yang A.H., Chen J.T., Yeh K.W. & Chan M.T. (2010) Heterologous expression of taro cystatin protects transgenic tomato against *Meloidogyne incognita* infection by means of interfering sex determination and suppressing gall formation. *Plant Cell Reports,* 29: 231–238.

[218] Guerra-Sá R., Castro B.W., Evangelista E.A., Kettelhut I.C. & Rodrigues V. (2005) *Schistosoma mansoni*: Functional proteasomes are required for development in the vertebrate host. *Experimental Parasitology,* 109: 228-236.

[219] Caffrey C.R. & Ryan M.F. (1994) Characterisation of proteolytic activity of excretory-secretory products from adult *Strongylus vulgaris. Veterinary Parasitology,* 52: 285-296.

[220] Rhoads M.L., Fetterer R.H. & Urban J.F., Jr. (1997) Secretion of an aminopeptidase during transition of third- to fourth-stage larvae of *Ascaris suum. The Journal of Parasitology,* 83: 780-784.

[221] Kumar S. & Pritchard D.I. (1993) New observations on proteases of the human hookworm *Necator americanus. Biochemistry and Molecular Biology International,* 30: 13-19.

[222] Køie M. & Fagerholm H.P. (1995) The life cycle of *Contracaecum osculatum* (Rudolphi, 1802) sensu stricto (Nematoda, Ascaridoidea, Anisakidae) in view of experimental infections. *Parasitology Research,* 81: 481-489.

[223] Dziekońska-Rynko J., Rokicki J. & Jabłonowski Z. (1997) Wpływ zarażenia larwami III stadium *Anisakis simplex* na aktywnośćproteaz w przewodzie pokarmowym świnek morskich [in Polish; abstract in English]. *Wiadomosci Parazytologiczne,* 43: 425-430.

[224] Dziekońska-Rynko J., Rokicki J. & Jabłonowski Z. (2003) Activity of selected hydrolases in excretion-seccretion products and homogenates from L3 and L4 larvae of *Anisakis simplex* (Nematoda: Anisakidae) parasiting herring. *Acta icthyologica et piscatoria*, 33: 125-135.

[225] Nguyen T.T., Qasim M.A., Morris S.R., Lu C.C., Hill D., Laskowski M., Jr. & Sakanari J.A. (1999) Expression and characterization of elastase inhibitors from the ascarid nematodes *Anisakis simplex* and *Ascaris suum. Molecular and Biochemical Parasitology*, 102: 79-89.

[226] Ruitenberg E.J. & Loendersloot H.J. (1971) Histochemical properties of excretory organ of *Anisakis* sp. larva. *The Journal of Parasitology*, 57: 1149-1150.

[227] Smith S.K., Pettit D., Newlands G.F., Redmond D.L., Skuce P.J., Knox D.P. & Smith W.D. (1999) Further immunization and biochemical studies with a protective antigen complex from the microvillar membrane of the intestine of *Haemonchus contortus. Parasite Immunology*, 21: 187-199.

In: Parasites
Editor: Gilmar S. Erzinger

ISBN: 978-1-62257-692-0

Chapter 5

FASCIOLA: PARASITE BIOLOGY, DISEASE AND CONTROL

Luke Norbury[1], Kamil Januszkiewicz[1,2] and Peter M. Smooker[3,4]
[1]Witold Stefański Institute of Parasitology PAS, Warsaw, Poland
[2]Central Forensic Laboratory of the Police, Warsaw, Poland
[3]School of Applied Sciences, RMIT University, Bundoora, Victoria, Australia
[4]Health Innovations Research Institute, RMIT University, Bundoora, Victoria, Australia

ABSTRACT

The trematodes *Fasciola hepatica* and *F.gigantica* can infect a broad range of hosts, and cause the disease Fasciolosis. Fasciolosis is a disease of economic importance, with livestock infection having a considerable impact on the agriculture industry. Human infection is now considered of public health importance and is hyperendemic in some regions. The incidence of animal and human infections is increasing in parts of the world.

This review will provide an overview of our current understanding of *Fasciola*, from distribution, diversity and disease development, to current control measures and progress towards an effective vaccine.

Fasciola excrete and secrete a broad range of molecules while residing in the host. The repertoire of molecules changes as the parasite matures, reflecting the different needs of the fluke during parasite development. These molecules, which are involved in important parasite functions, are often found in the parasites excretory-secretory material, and function at the host-parasite interface. Common among these molecules are antioxidents as well as a variety of proteases, including cathepsin proteases. The roles of these key molecules during *Fasciola* infection, including their ability to modulate host immune responses will be a focus of the review. The findings from recent transcriptome and proteome analysis will also be a discussed, as will the potential for RNA interference in characterisation of this parasite.

Triclabendazole is the drug of choice for control of *Fasciola*, however resistance has been reported in a number of countries. Recent studies of population genetics to determine genetic lineages and diversity will be discussed, as will future directions for control of this important parasite, particularly the successes and failures in the development of an effective *Fasciola* vaccine.

.

INTRODUCTION

The flatworms *Fasciola hepatica* and *F. gigantica* are commonly known as liver flukes and are the two main species of the Fasciolidae family, which belong to the trematode subclass known as digenea. *Fasciola* parasitize sheep and other ruminants, as well as a wider range of mammalian hosts, including humans, causing the disease known as Fasciolosis, an economically and medically important food- and water-borne parasitic zoonosis. Recent years has seen an increased interest in these parasites due to increases in domestic animal infections in Europe, and revelations of the extent of human infection in parts of the world. The development and application of molecular and proteomic techniques means that the last few years has seen major improvements of our understanding of fluke biology.

1. ECOLOGY

Life Cycle and Development

The life cycle of *Fasciola* (Figure 1) is complex, but is typical of a Digenean lifecycle involving several phases and two hosts; a mammalian primary host, with certain *genera* of amphibious snails acting as the intermediate host, reviewed in [1-2].

Thin-shelled, unembryonated ovoid eggs laid by the adult parasite in the bile duct system of the primary host are subsequently passed in the host's faeces. The eggs must be separated from the faeces and encounter fresh water for the fluke life cycle to continue. Egg viability is influenced by temperature and moisture, with temperatures above 30ºC and dryness detrimental [3-4], however, eggs have been shown to remain viable for years if kept refrigerated [5]. Embryonic development occurs in water, and is regulated by temperature, oxygen tension and moisture. In laboratory conditions, the optimum temperature for egg development is 30ºC, and it generally takes one to two weeks before the operculum opens and the fully developed miracidium escapes, although development may take 6 months under non-optimal conditions [1, 6]. Light also stimulates the hatching process [7-8].

The motile ciliated miracidium must locate and actively penetrate a suitable snail host within 24 hours. Chemotactic structures and phototropic tendencies are believed to aid the miracidium in locating the snail host [9]. The miracidium penetrates the snail using a retractile papilliform process, possibly facilitated by the secretion of proteolytic enzymes [10], and a series of asexual developmental changes occur within the snail intermediate host.

Once in the snail, the miracidium metamorphoses into a rounded sporocyst which migrates to the snail's digestive gland. The sporocyst, upon rupturing, gives rise to several motile rediae. These rediae can cause serious damage to the snail host, even causing death. Germ balls in the brood chamber of the rediae develop into cercariae, or under certain conditions give rise to subsequent generations of rediae [1, 11]. The cercariae actively emerge from the snail and become free swimming. The whole process, from snail infection to cercaria emergence usually lasts four to seven weeks, and a single miracidium is able to produce over a thousand cercariae [12-13].

Within two hours of leaving the snail, the cercariae attach to various objects such as submerged aquatic vegetation, and encyst as metacercariae. The cyst, almost immediately infective, darkens and hardens over a few days. Metacercariae are able to survive for up to a year in appropriate conditions, but show reduced survivability at increasing temperatures and in dry conditions [14-15].

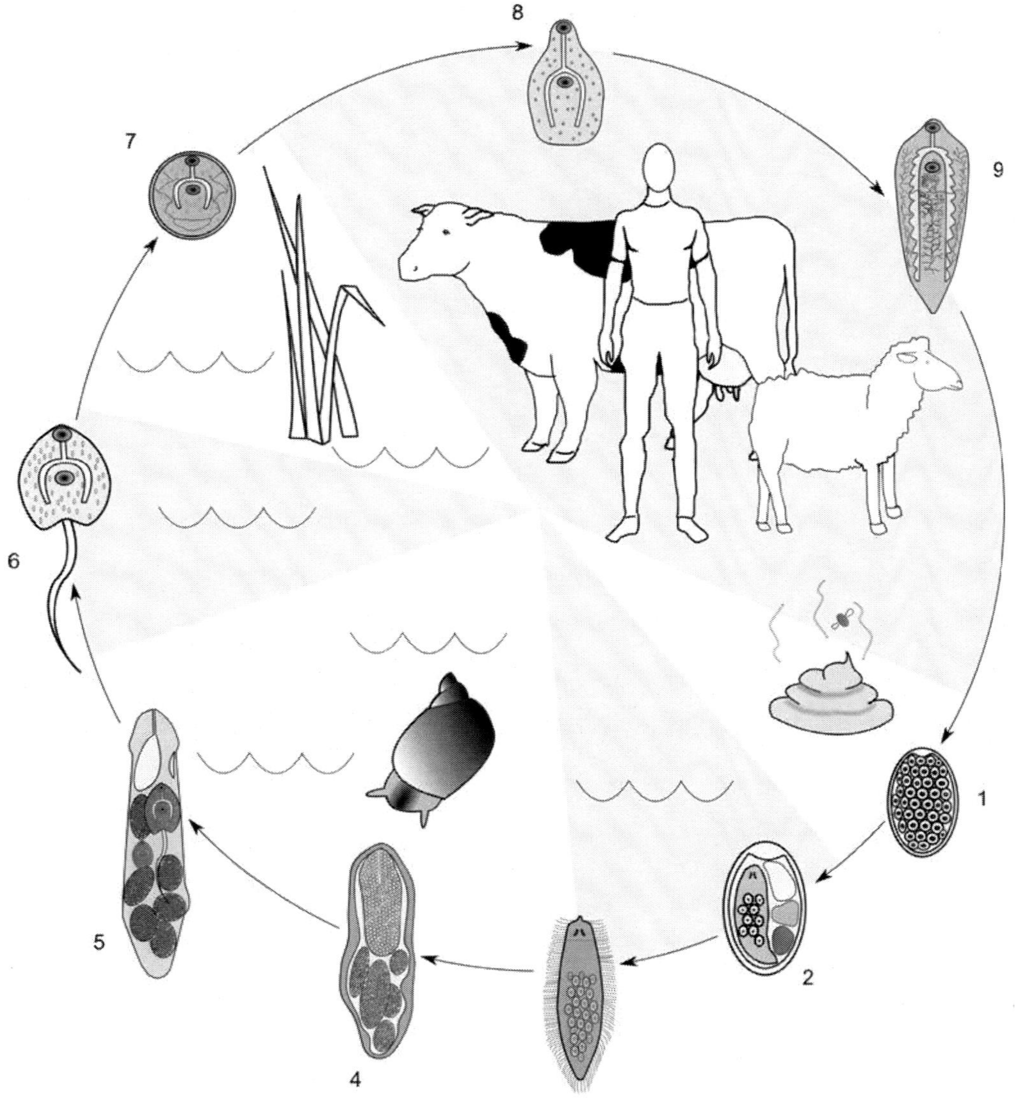

Figure 1. Stages in the *Fasciola* life cycle include 1) passage of the unembryonated egg in faeces, 2) egg embryonation in water, 3) miracidia, and snail asexual phases; 4) sporocyst, 5) redia, 6) cercaria and emergence from the snail host, 7) metaceraria excystment and definitive host infection, 8) newly excysted juvenile and 9) adult fluke within the definitive host. Figure courtesy of Dr Sapna Thoduka.

Upon being swallowed by the definitive host, metacercariae are stimulated to excyst in the duodenum. Excystment *in vitro* is stimulated by high carbon dioxide concentrations, reducing conditions and temperature at 39°C, the normal ruminant body temperature (reviewed in [1]). The emergence of the parasite is also stimulated by the presence of bile, bile salts and gastric juices [16-17]. Newly excysted juveniles (NEJ) penetrate the intestinal wall, and enter the peritoneal cavity within 24 hours. In sheep, the liver is penetrated four to six days post infection. The juvenile flukes then migrate through the liver parenchyma, consuming liver cells and blood for several weeks.

Seven to eight weeks post infection they reach the bile ducts where they attain sexual maturity, generally within 12 weeks of the initial infection [5]. Flukes attach to the bile duct wall using their suckers, with their spines securing them in place, and abrading the host epithelia and rupturing blood vessels, providing the parasite with additional nutrition [18]. Adult *F. hepatica* can live for up to 11 and 13 years in sheep and humans respectively [19-21], and produce up to 25,000-50,000 eggs per day in sheep [22].

Fasciola are hermaphrodites. Sexual reproduction is the most common form of procreation, however self-fertilisation can occur, and facultative gynogenesis may be widespread in this parasite [23] with asexually reproducing diploid and triploid *Fasciola* present, particularly in Asia [24-26].

There may be some variation with the *Fasciola* lifecycle, particularly within different primary hosts, which influences adult fluke size, egg size and infectivity and time course of infection, and fluke lifespan [27-30]. Genetic differences among fluke strains also result in phenotypic fluke variations [31]. Occasionally ectopic infections can occur, with immature flukes observed migrating to various organs [32-45] while prenatal infections have also been observed [46-47].

The main factors affecting the life cycle however tend to be the requirement for suitable temperature and sufficient moisture. *F. gigantica* are present in tropical areas where temperature and moisture are not generally limiting factors, and their life cycle is essentially identical to that of *F. hepatica*. However variations do exist, which can dramatically influence the epidemiology and pathology of these two parasites. The eggs of *F. gigantica* are generally larger in size than those of *F. hepatica*, though this is not always the case [48]. The periods of *F. gigantica* egg embryonation and intramolluscan larval development are also slightly longer [2]. In some cases the miracidium of *F. gigantica*, unlike that of *F. hepatica*, are actively repelled by light [49] which has implications for the types of intermediate hosts they are likely to infect, and the fluke species have adapted to different intermediate hosts. Additionally, immature *F. gigantica* take longer to migrate through the host liver and are slower to develop to sexual maturity compared to *F. hepatica*. As their name suggests, *F. gigantica* flukes are considerably larger than *F. hepatica*, and the adult stages of the two species are morphological distinct, though some overlap is evident and intermediate forms exist. Differences in the biochemistry between the two species means disease development and pathology differ in definitive hosts, with certain hosts more susceptible or resistant to each species.

Fasciola Intermediate Hosts

The intermediate hosts of *Fasciola* are the amphibious freshwater snails of the family *Lymnaeidae*. Different affinities are observed by *Fasciola* species for different snail species, though some snail species, such as *Pseudosuccinea columella* are susceptible to both *Fasciola* species [50], while resistant populations of this species have also been identified [51-52].

Taxonomic confusion among the Lymnaeids complicates the host-parasite interactions, however, studies suggest an old parasite-host relationship; *F. hepatica* with *Galba* (*Fossaria*) and *F. gigantica* with *Radix* [53]. This suggests a parallel evolution of each liver fluke species with different lymnaeid branches towards an increasing specificity [50].

Fasciola, particularly *F. hepatica* have shown to be able to adapt to various localised lymnaeid species, with a recent study indicating that *F. hepatica* is able to infect widely across the phylogeny of the Lymnaeidae family, while species involved in *F. gigantica* transmission are more phylogenetically clustered [54]. This has implications for the spread of Fasciolosis, as the intermediate snail hosts of *F. hepatica* and *F. gigantica* differ in their environmental preferences, thus affecting their distribution and ultimately the distribution of the *Fasciola* parasites.

Along with the susceptibility of a range of local lymnaeids to *F. hepatica*, the capacity for ecological niche widening by *Galba Truncatula*, the typical intermediate host for *F. hepatica* in Europe and parts of Asia has facilitated the worldwide distribution of *F. hepatica* [55]. Conversely, the more restricted distribution of *F. gigantica* is in part linked to the more limited capacity of its intermediate snail hosts to exploit wider ecological niches, with the absence of *F. gigantica* in the USA linked to the scarcity of *Radix* lymnaeids in that part of the world [56].

Fasciola Primary Hosts

Important primary hosts of *F. hepatica* are generally ruminants such as cattle, sheep and goats. Alternatively cattle and buffalo are the important domesticated hosts for *F. gigantica*, as well as goats, sheep and donkeys. However, *Fasciola* species have a wide range of primary hosts, and have shown an ability to expand from infecting important European livestock to parasitising many animals indigenous to other continents. As well as humans and domesticated animals, primary hosts can also include various wild animals such as rodents, otters, deer, bison, and can include species that have only relatively recently been encountered such as marsupials, llamas, alpacas, capybara and camels [57-62]. While higher avian body temperatures (42°C) are thought to confer resistance to *Fasciola* infection [63], *F. hepatica* has also been discovered in emus and rheas [64-65].

Fasciola species are better adapted to certain hosts, and infestation levels of primary hosts vary among sites and among host species, as does the range and level of infection of host species in different regions. The ability to infect such a wide range of hosts has undoubtedly facilitated the worldwide spread of these parasites, and influences transmission patterns and the dissemination of disease.

2. Epidemiology

2.1. Distribution and Abundance

Fasciola are dependent on a consistent set of suitable environmental conditions to survive, requiring acceptable moisture and temperature conditions both for the intermediate host and for their own growth and development. *F. hepatica* infection occurs in temperate regions, and though the parasite originated in Europe, it now has a cosmopolitan distribution, having successfully spread to all continents with the exception of Antarctica. *F. gigantica* has a more restricted distribution and is largely confined to the tropics, and appears only in Africa, Eastern Europe, the Middle East and Asia. There is overlap among their ranges, and while both species are endemic to many countries, *F. hepatica* often predominates at higher altitudes, due to its lower temperature requirements. Over 500 million animals are proposed to be infected by *Fasciola*, and levels of infection among domestic animals can reach over 90%, with highest prevalences generally observed in tropical regions with *F. gigantica* [66].

As mentioned earlier, fasciolosis has a great capacity to spread to new areas, which is facilitated by the export of infected livestock and the ability of *F. hepatica* to infect a large range of primary hosts and adapt to new hosts which then contribute to further transmission. Additionally, the colonizing ability and introduction of intermediate-host snail species such as *Galba truncatula* and *P. columella* and the parasites' ability to adapt to numerous local intermediate hosts is vital to the parasites spread [55].

Europe has seen increases in infections of domestic livestock over the past decades, with some regions experiencing significant increases in recent years. Changes in climate can influence *Fasciola* development and transmission rates, and can introduce *Fasciola* to new areas. Several recent reports indicate climate change (in combination with drug resistance) will facilitate increases in infection rates in some regions. This will be due to warmer temperature and increased rainfall creating friendly climactic conditions for the spread and survival of intermediate host species and free-living *Fasciola* stages [55, 67-68].

While the veterinary importance of *Fasciola* has long been recognised, the emergence or re-emergence of human fasciolosis has only been acknowledged in the last couple of decades. The number of detected human cases has increased significantly, with increasing reports of human fasciolosis in 51 countries [69]. Estimates of human infections range from 2.4 [70] to 17 million [71], with a further 91 [72] to 180 million at risk of infection [56].

Seven distinct human infection classifications have been proposed, including; autochthonous isolated, nonconstant cases; imported cases; hypoendemic; mesoendemic; hyperendemic; epidemics in nonhuman endemic areas but in animal endemic areas; and epidemics in human endemic areas [73].

Ecologically, human infection occurs most frequently in sheep and cattle raising regions [74]. However, there is only a very basic correlation between human and livestock infection, and high or low prevalence do not seem to be related between humans and animals [69, 73, 75].

Fasciola infection is regarded as a serious health problem in several regions and several human endemic areas exist, most notably in the Andean countries of South America, including Bolivia, Peru, Chile and Ecuador [76]. A striking description of a human hyperendemic area is the Bolivian altiplano where over 360,000 humans are estimated to be infected, and in which high intensity infections, and high infection prevalences have been recorded [77-83]. Human fasciolosis is relatively rare in the U.S.A and Australia, with only sporadic cases reported, while no cases have been reported in New Zealand [84-85]. It is also generally not prevalent in Europe, although France, Portugal and Spain have recurrent outbreaks and possess endemic areas [86-87] and see [56, 73, 75-76]. Both *F. hepatica* and *F. gigantica* infections are prevalent in livestock in the region around the Caspian Sea, which has also been suggested to contain widespread human endemic areas [56], with endemic regions and large disease outbreaks occurring in the north of Iran [88-89], and increasing detection of human fasciolosis in Turkey [90-93]. Several islands in the Caribbean also report sporadic cases of human fasciolosis, with several outbreaks in Cuba having occurred in the last century [56, 94-96]. Reporting of human cases is limited in Asia, however, the number of cases is considered to be increasing, particularly in Vietnam which has seen a recent rise in infection levels [97-98]. An estimated 830,000 people are infected in Egypt, and hyperendemic regions are present in the Nile delta region, with prevalences up to 19% reported for some villages [85, 99].

2.3. Transmission and Infection

Temperature and climatic factors, such as rainfall, are crucial to *Fasciola* transmission. In many areas of the world there are yearly cycles of livestock infection, with disease outbreaks usually relating to peak pasture contamination by metacercariae. In some warmer areas, where winter temperatures do not halt the fluke life cycle, continuous contamination of pastures results in a year-round possibility of infection. Both domestic and wild animals act as reservoir hosts, with the range of hosts infected and their influence on the spread of disease differing strongly among regions. Sheep, cattle and goats are generally the predominant animal reservoirs. Humans can also play an important role in transmission of the parasite, at least in hyperendemic areas [73, 76].

Mas-Coma and colleagues have described a number of epidemiological patterns for human fasciolosis transmission in hypo- and hyperendemic areas that relate to different environments on different continents. These include Andean high altitude, Caribbean insular, Afro-Mediterranean lowland, and Caspian area patterns [56, 76]. A transmission pattern specific for Vietnam and Southeast Asia has also been proposed [100].

Human infections occur through ingestion of contaminated watercress and other freshwater associated plants that are eaten raw in salads. These differ in geographic regions, for example the 'water morning glory' in Southeast Asia, and spearmint in Europe and the Middle East (reviewed in [100-102]). Water has also been cited as a source of human infection through direct drinking or contamination of food or utensils [75, 84, 101-102]. Irrigation developments have been linked to increased *Fasciola* infections among animals and humans [103], while appropriate water filtration can reduce risk of infection [101]. Experimental evidence in mice and pigs also suggests that people who consume raw dishes prepared from fresh livers infected with immature flukes can also become infected [104].

2.4. Fluke Diversity and Population Genetics

Some controversy exists in regards to the Fasciolidea family. Despite morphological differences, *Fasciola* species show considerable overlap. In regions where both species coexist, inconsistencies in morphological features mean that differentiation of *Fasciola* species cannot always be accurately undertaken, while the existence of 'intermediate' *Fasciola*, and the presence of aspermic diploid and triploid specimens, particularly in Asia, further complicates the issue [23-26, 105-109].

Recent years have seen the use of modern phenotypic techniques [108, 110-112] and various molecular approaches developed and applied to the identification and differentiation of *Fasciola*, and a number of studies have identified novel 'intermediate' *Fasciola* and indicate hybridization between the two species [26, 106-107, 113-119].

Fasciola molecular epidemiology has developed in recent years, with a number of different DNA regions and alternative molecular assays utilised for analysis and characterisation of *Fasciola*. These studies not only identify and differentiate between *F. hepatica*, *F. gigantica* and in some cases 'intermediate' *Fasciola*, but also enable comparisons to determine genetic diversity of *Fasciola* species infecting individual animals or herds at single or multiple locations [120-135], and reviewed in [136-137].

This has revealed multiple genetic *Fasciola* lineages and has enabled the spread of *Fasciola* to be traced [121-122, 133, 138-141], reviewed in [100].

The presence of diversity both within and between infrapopulations has been observed, with the presence of diverse fluke populations within a single definitive host [121-123, 138, 142] and also between herds even at relatively close geographical proximity (~100km) reported [122]. These fluke populations have also been shown to change relatively rapidly in a given area [122]. To date, a clear lack of genetic differentiation in fluke from different hosts has been observed, even between domestic and wild animals [120-121, 123-124, 131, 133, 142] however different patterns of fluke genetic structure dependant on the host has recently been reported, with greater gene flow and greater genetic variation in cattle and sheep respectively [125].

The results from these studies lead to a greater understanding of *Fasciola* epidemiology and population biology, which has implications for understanding and predicting the development and spread of drug resistant strains and developing effective control programs. The diversity seen within *Fasciola* populations and the high level of gene flow described in some of the aforementioned studies suggests relatively rapid response or adaptation to environmental changes, which indicates among other things that development of drug resistance is likely to occur relatively frequently. Additionally, the presence of the same fluke populations in hares and cattle in Holland suggests some wild animals may be able to facilitate the spread of triclabendazole resistance [123].

Additionally, the application of molecular epidemiology in relation to *Fasciola* infection of intermediate hosts has also contributed to our understanding of fluke epidemiology, helping to determine transmission dynamics and facilitate risk assessment [143-146].

A future application for *Fasciola* molecular epidemiology will be the linkage of genetic diversity to practical phenotypic differences and identification of important traits such as infectivity or drug resistance. A number of studies have begun to characterise such phenotypic differences between genetically distinct strains [31, 147-151].

3. Disease

3.1. Disease Development and Pathology

Animal Fasciolosis

Development of *Fasciola* infection within the definitive host can be divided into two phases, known as parenchymal and biliary, reviewed in [152].

The parenchymal phase coincides with the migratory stages of fluke development, where fluke penetrate the intestinal wall, traverse the abdominal cavity and migrate through liver tissue, to which they have a strong predilection. During this phase, pathology results from mechanical damage to tissue caused by migrating flukes. Clinical signs are generally not associated with prehepatic stages of penetration, though eosinophilia and some fibrosis may be evident. However, as the fluke grow, they cause increasing damage and elicit a greater inflammatory response. The migrating fluke can cause severe liver damage which can even result in death. Fluke excretory/secretory material and the host inflammatory response also contribute to disease pathology.

The biliary stage coincides with fluke residence in the bile ducts, with symptoms resulting from the feeding action of the parasites. Symptoms can also occur as a result of biliary obstruction and inflammation, though biliary obstruction is generally rare in sheep [152]. Symptoms include severe anaemia, and generalized oedema.

There is a large variation in susceptibility of definitive hosts to *Fasciola* infection, while host susceptibility to the different *Fasciola* species can also vary. In regard to infection with *F. hepatica*, relatively few fluke survive to maturity in cattle and humans, while sheep, rabbits, and mice appear more permissive, with the biliary stage of the disease more common. In general, the disease progresses faster in experimental animals than in sheep, with pathology exacerbated in smaller animals due to their smaller livers.

Disease severity can be classified as acute or chronic, and among definitive hosts this has been most thoroughly described in sheep, with four clinical categories detailed, relating to the infective dose of metacercariae [152]. The pathology of acute fasciolosis occurs when large numbers of metacercariae are ingested in a short amount of time. Acute fasciolosis (type II) in sheep occurs when greater than 1000 infective metacercariae are ingested, with sheep showing pallor, loss of condition and ascites before death, while an infective dose of greater than 5000 (type I) results in sudden death generally without previous clinical signs. Weight loss is a dominant feature of subacute fasciolosis, which generally results from trickle infections resulting in an accumulated infective dose of greater than 800 metacercaria, Sheep may also die, while other symptoms include lethargy and anaemia. Chronic fasciolosis occurs with infected doses of greater than 200 and is identified with emaciation and gradual development of oedema or swelling under the jaw (bottle jaw) and ascites.

Acute Fasciolosis does not usually occur, except in sheep and goats. The chronic form of the disease is generally more common among hosts, though considerable variation exists between animals. Calves for instance require infection with greater than 10,000 metacercaria for death to occur, while over 1000 are required to cause clinical Fasciolosis. Chronic Fasciolosis, though rarely fatal, is of considerable economic importance. In general the level of infection affects pathology, however, disease severity also depends on the nutritional status of the animals. There is also considerable variation in susceptibility to infection and in immune responses against liver fluke [153] with certain hosts possessing innate resistance against one or both *Fasciola* species. The prime example of this is Indonesian thin tailed (ITT) sheep, which are susceptible to *F. hepatica* infection (which they do not normally encounter), but have innate and acquired resistance to *F. gigantica* infection, and hence represent a valuable tool to examine potential mechanisms involved in parasite resistance [154-157].

What degree of the variability seen in fluke development is due to genetic differences among the liver fluke infecting different hosts, or results from phenotypic plasticity is unknown. Fluke growth, fecundity, metacercariae infectivity, egg numbers and developmental mortality of flukes, and biochemical secretions have all been shown to be affected by the fluke's choice of definitive host [27-30, 158-160].

Immunomodulation of the hosts immune response during *Fasciola* infection can leave the host susceptible to co-infection with other pathogens and aggravation of disease development due to repression of the required protective response [161-162]. This is discussed further in following sections. An often fatal complication of fasciolosis, known as black disease, is caused by secondary infection with *Clostridium oedematiens* or *C. novyi.* The bacteria proliferate in necrotic lesions, and cause hepatitis.

Human Fasciolosis

Human fasciolosis follows a similar course to that in animals as described above, and is reviewed in [75, 102, 163]. Infection does not generally cause mortality in humans, who are not considered highly susceptible with most fluke trapped in the liver parenchyma. The invasive or acute phase lasts 2-4 months, corresponding with fluke migratory life stages and the mechanical damage they inflict, particularly to the liver, and the induction of immune responses. The time of onset of this stage from ingestion of metacercaria can vary considerably, depending on the individual and infection dose of metacercariae, but can be anywhere from a few days to over 3 months. Major symptoms of this phase include fever, abdominal pain, hepatomegaly, anaemia, urticaria, eosinophilia and a range of gastrointestinal disturbances, while respiratory symptoms, splenomegaly, ascites, and jaundice have also been reported [75].

The chronic phase can develop months or years after the initial infection, and may be preceded by a latent phase. Adult fluke residing in the bile duct cause inflammation and hyperplasia of the epithelium, resulting in dilation and thickening of the duct wall, which combined with the fluke bodies can result in obstruction. The chronic phase is generally less severe than the acute phase, and symptoms are often sub-clinical or indistinguishable from other hepatic conditions, such as cholangitis, cholecystitis and cholelithiasis. The long life span of *Fasciola* in humans, between 9-13.5 years [19-20], indicates that in endemic areas, repeat infections likely result in acute phases overlapping with the chronic phase of disease.

Ectopic infections can also occur, with fluke isolated to various tissues other than the liver, though most frequently to the gastrointestinal tract. Resulting associated pathology is due to the migratory tracks, and can involve inflammation and fibrosis [75].

Additionally, in some regions, particularly the Middle East, *Fasciola* has been reported to cause a condition known as halzoun, with young worms, ingested with raw liver, proposed to attach to buccal or pharyngeal membranes, causing pain, irritation, hoarseness, and coughing [74].

Economic Impact

As liver flukes infect a wide range of domestic animals worldwide, the disease causes considerable impact on the economy of the livestock industry. Clinical fasciolosis effects can range from sudden death to chronic underperformance, while subclinical infections often go unnoticed, yet also produce marked economic effects (reviewed in [6, 66] and [164]).

Fasciola infection of sheep results in impaired live weight gain and wool production and infection of buffalo and cattle leads to weight gain reductions with higher burdens leading to increased impairment [165-169, 170 , 171]. Milk quality and yields are also affected by fluke infection and fertility rates have also been reported to be lower [164, 172-175]. In addition, reduced draught capacity of water buffalo resulting from *F. gigantica* infection is a significant adverse effect, as these animals are the main source of draught power in parts of Asia [176]. Significantly, infections with *Fasciola* can also lead to increased susceptibility to other pathogens which can further debilitate animals [161, 177-179].

Economic losses include costs of treatment, labour, liver condemnation; and losses in production. This, along with the public health significance of human fasciolosis means that even if a majority of animals infected have modest burdens, worldwide economic losses are still in the billions of dollars (reviewed in [6, 66]), with annual losses to the worldwide agriculture industry estimated at over US$3 billion [66]. Additionally, human fasciolosis leads to significant economic costs that can be estimated in terms of absenteeism, hospitalisation, treatments and disability [70].

Immunology of Infection

Fasciola infection is accompanied with down-regulation of Th1 and inflammatory immune responses, and induction of a Th2 and T-regulatory type immune response [153, 161, 180-182], characterised in sheep by production of IL-4, IL-5, IL-10, elevated IgE and IgG1 levels, eosinophilia and alternative activation of macrophages [157, 171, 183-188]. The development of the Th2 response is associated with chronic disease, while alternatively a Th1 response has been associated with protection [181, 184, 188-190].

During *Fasciola* infection there is a complex and dynamic contest between host immune responses and parasite defence and immunomodulatory systems. The immune mechanisms employed by various hosts differ, as does fluke susceptibility to the different killing mechanisms and as a result there is a wide variation in susceptibility to *Fasciola* infection and ability to acquire resistance amongst definitive hosts [153, 155, 184, 191-193]. For example, mice have poor resistance to *Fasciola*, and sheep and goats neither possess innate resistance nor are able to develop acquired immunity to *F. hepatica*; alternatively rats and cattle have been demonstrated to acquire resistance to *F. hepatica* [153, 155, 194-195]

While *Fasciola* species possess a large range of immune evasion and modulation techniques through ES secretions and tegumental shedding [155, 182, 196-197], there are differences between the two species, including their ability to modulate and evade host immune responses, which is reflected by the different susceptibility and disease progression observed in certain hosts [155, 171, 185-186, 193, 198]; the relative resistance of Indonesian thin tailed sheep against *F. gigantica* compared to *F. hepatica* is a prime example [184, 199].

Protective Immune Responses

Recent years have seen increases in the understanding of the protective immune responses that can be mounted by some host species against *Fasciola*, although there is still some way to go before a full understanding of host-parasite interactions and the immunology of *Fasciola* infection is understood. Protective immune responses have been most thoroughly elucidated in rats and sheep, reviewed in [155].

Rats

A pro-inflammatory response during early stages of fluke infection is important to the generation of a protective response in rats [200-201]. The protective immune responses in immune rats are believed to target juvenile fluke with the gut wall and peritoneal cavity the sites of parasite killing [155, 202-203]. The gut wall is the initial site of resistance, and a much smaller percentage of fluke are found to survive gut wall penetration and migration through the peritoneal cavities of immune rats when compared to naïve rats [202, 204].

During early stages of *F. hepatica* infection the peritoneal areas of rats are infiltrated by neutrophils, eosinophils and CD4+ lymphocytes, and elevated nitric oxide (NO) from eosinophils and superoxide expression in macrophages and neutrophils have been observed [205]. In resistant rats, NEJ entering the peritoneal cavity after penetrating the gut wall are coated with antibodies and eosinophils, neutrophils, macrophages and mast cells [206-207]. Thus resistance to fluke in rats appears to involve both an effective antibody response and lymphoid cells that target juvenile fluke, with eosinophils and IgG1 shown to correlate with protection at the gut wall [207-208].

Eosinophil killing of NEJ has not been demonstrated *in vitro* however, and antibody-dependent cell-mediated cytotoxicity (ADCC) dependent on macrophages, rather than eosinophils, and reactive nitrogen species, as opposed to reactive oxygen species, have been identified as an effector mechanism for killing juvenile *F. hepatica* in naïve rats [155, 209-210].

Juvenile *F. hepatica*, are relatively resistant to cell mediated killing by reactive oxygen species [211], however juvenile *F. gigantica* are more susceptible to killing by reactive oxygen and nitrogen species, and are able to be killed even in the absence of specific antibody by reactive nitrogen species [155]. This difference in susceptibility of *F. gigantica* to rat ADCC correlates with low infectivity observed by *F. gigantica* in rats [155].

Sheep

Sheep have generally been considered to possess no innate immunity and do not naturally develop acquired immunity [153]. However, while ITT sheep are susceptible to *F. hepatica*, they have an immunologically based resistance to infection with *F. gigantica*, thus represent an ideal model to investigate mechanisms of resistance [157, 184, 199, 212]. Induction of an early local Th1 immune response appears to be associated with ITT sheep's ability to resist infection to *F. gigantica* [184]. Similar to rats, targeting of NEJ in the peritoneal cavity appears important in resistance.

Effective ADCC in ITT sheep differs from that observed in rats, with both macrophages and eosinophils able to mediate killing with superoxide anion radicals, as opposed to nitric oxide [193]. This effector mechanism in ITT sheep is able to effectively kill juvenile *F. gigantica*, but is ineffective against juvenile *F. hepatica* [193]. The relative resistance of *F. hepatica* to superoxide-mediated killing may be due to increased levels of superoxide radical defence enzyme, superoxide dismutase, being expressed by *F. hepatica* [193, 211]. This in part explains the different susceptibilities to the two fluke species observed for this sheep breed [193].

Cattle

Cattle are relatively resistant to primary infections and can develop an acquired immunological resistance to *Fasciola*. Cattle acquire resistance much later than is observed in rats, targeting fluke after they have reach the liver, and fibrosis of the liver also plays a role in development of a non-immune resistance to *Fasciola* reinfection among cattle [153]. However, vaccine studies show that cattle can mount an alternate protective response that target fluke prior to liver invasion [190, 213]

The protective immune responses of cattle against fasciolosis are not as well elucidated as that of rats or ITT sheep, and the ADCC effector molecules involved in killing juvenile *Fasciola* in cattle is unknown. However, through vaccine studies, protection has been linked to a Th1 or mixed Th1/2 response [190].

Fasciola Immune Evasion & Modulation

Fasciola are able to survive for long periods inside hosts, and are relatively safe from the host immune system while residing in bile ducts. However, during migration juvenile and immature fluke encounter various host immune responses and utilise a variety of molecules to help penetrate defensive barriers and avoid and modulate the host immune system.

F. hepatica modulates host immunity towards a Th2 type response and down-regulates the Th1 response [161, 181, 214-216], with infection producing alternatively activated macrophages [183, 215, 217]. Immune modulation occurs in both the early stages of infection [161, 181, 214] as well as during chronic fasciolosis [218], with both migrating fluke, and adult flukes residing in the bile ducts, secreting molecules that help elicit and maintain the Th2-polarised host immune response. It appears that several immunomodulatory molecules are secreted that are effective at both actively driving Th2 responses as well as down-regulating both the induction and effector stages of the Th1 response [161, 216]. Further, *Fasciola* immunomodulation has been shown to influence host susceptibility to other pathogens [161, 181, 219-220].

Fasciola ES material and tegumental proteins function at the host-parasite interface and are exposed to the host immune response where they play vital roles in host-parasite interaction. In recent years, several of these molecules have been identified that contribute directly and indirectly to effective immune modulation and evasion, and the mechanisms through which they act is increasingly becoming elucidated.

Tegument

The fluke's tegument, the outer component of the parasite, is in constant contact with host matter, and as such is believed to play an important role in evasion of the host immune response. Migrating flukes make alterations to the outer tegument, secrete granules and are constantly shedding the tegument, only slowing down these processes once in the relative safety of the bile duct, reviewed in [221-222].

As the fluke matures, different granules are secreted which changes the composition of the tegument, which is proposed to promote production of irrelevant antibody. Further, the constant sloughing of the tegument, with turnover every 2-3 hours during migration of juveniles, may help protect against specific antibody and eosinophil attachment, while the shed antigen acts as a decoy for the host immune response [155, 223-226].

The tegument of *F. hepatica* also possesses potent anti-inflammatory properties, being capable of suppressing murine dendritic cell maturation and function, resulting in suppression of a protective Th1 proinflammatory response [227]. An initial study has also implicated it in alternate activation of ovine macrophages [228], which are involved in promoting Th2 responses and suppressing Th1-driven inflammation. Thus the tegument of *F. hepatica* is not only involved in evading host immune responses but also modulating them.

Excretory/Secretory Material

As *Fasciola* matures and migrates, encountering distinct environments, the composition of the ES changes as different functions are required (expanded upon in the last section). While the molecules in *Fasciola* ES facilitate many functions, ES has been implicated in immune invasion and modulation through several mechanisms with several different molecules contributing directly and indirectly.

Early studies indicated defence roles for *Fasciola* ES through digestion of host immunoglobulins [229] which would aid evasion of antibody-mediated effector functions. *Fasciola* ES can prevent peritoneal inflammatory cells killing NEJ *in vitro*, and has been shown to also kill lymphocytes [230], and induce early apoptosis in eosinophils [231]. Additionally, *Fasciola* ES is also able to modulate cell proliferation [198, 232-235] and the functions of a range of immune cell types in various hosts, reducing phagocytic abilities, and/or suppressing free radical production [180, 236-240]. *Fasciola* ES also induces suppression of delayed-type hypersensitivity [180]. Further, recent studies have linked *Fasciola* ES with alternative activation of macrophages [183, 241] and modulation of dendritic cell maturation and function [197]. These findings clearly demonstrate the ability of *Fasciola* ES to suppress Th1 responses and drive Th2 immune responses during *Fasciola* infections.

Cathepsin Proteases

The predominant proteins present in *Fasciola* ES are the cathepsin proteases, while an array of antioxidant enzymes are also secreted. *Fasciola* temporally express a variety of cathepsin L proteases that have overlapping but different substrate specificities, and are proposed to have roles in several important parasite processes (expanded upon in fluke biology).

Fasciola cathepsin L proteases have subsequently been shown to aid immune evasion, with juvenile and adult-specific cathepsins able to cleave immunoglobulins [242-244], facilitating prevention of *in vitro* antibody-mediated eosinophil attachment to NEJ [245-246]. Additionally, *F. hepatica* cathepsin L2 (FhCatL2) is able to cleave fibrinogen to produce a fibrin clot which may prevent excessive bleeding at feeding points in the bile ducts, alternatively the clot may prevent access of immune effector cells to the parasite's surface [247-248].

Cathepsin L also has an immunomodulatory function, with *F. hepatica* cathepin L shown to suppress a protective Th1 immune response in mice [162]. *F. hepatica* cathepsin L (FhCatL5) can suppress sheep T lymphocyte proliferation, and is able to reduce CD4 expression on human and sheep T cells [235]. *F. hepatica* cathepsin L (FhCatL1) has been shown to suppress macrophage-driven Th1 inflammation by inhibiting macrophage activation by LPS through degradation of toll-like receptor 3 [249], and can also activate dendritic cell phenotypes that suppress development of Th17 cells and resultant inflammation in mice [182]. Thus *F. hepatica* cathepsin L proteases are involved in host immune modulation through suppression of the protective Th1 proinflammatory response and induction of a Th2 response through a number of mechanisms (reviewed in [250]).

The cathepsin B proteases are not as abundant as the cathepsin L's in *Fasciola* ES, and are predominantly expressed by juvenile fluke. In *F. gigantica*, cathepsin B has been localised to the tegument, where it has been proposed to assist in preventing immune cell attachment, and aid tegumental sloughing [251].

Antioxidant Enzymes

Fasciola ES contains a vast repertoire of antioxidant defence enzymes, including several fatty acid-binding proteins (FABPs), several glutathione-S-transferases (GSTs), peroxiredoxin, thioredoxin, superoxide dismutase and a protein-disulfide isomerase [252-253]. These antioxidant defence enzymes, which provide protection against oxidative stress, have been shown to protect *Fasciola* from both reactive nitrogen and oxygen species released by host immune cells (reviewed in [155]). Higher expression levels of specific antioxidant defence enzymes (SOD and/or GST) observed in adult and juvenile *F. hepatica* compared to NEJ *F. hepatica*, and in juvenile *F. hepatica* compared to juvenile *F. gigantica* or *Schistosoma mansoni* have been associated with increased resistance to killing by free radicals [193, 211]. Interestingly, *Fasciola* GST levels are restricted in resistant hosts (cattle and rats), which may be linked with host immune responses [159].

Fasciola antioxidants have also been shown to be capable of affecting the functionality of immune cells, modulating the host immune response. *F. hepatica* GST has been shown to suppress the ability of rat macrophages to produce nitric oxide, and inhibits the proliferative response of rat spleen cell [254]. Further, like cathepsin L but through a different mechanism, GST can activate dendritic cell phenotypes that suppress development of Th17 cells and resultant inflammation in mice [182].

Peroxiredoxins (also known as thioredoxin peroxidase or thiol-specific antioxidant protein) have also been shown to have immunomodulatory capabilities, playing a key role driving host Th2 immune responses via the recruitment and alternative activation of peritoneal macrophages [183, 215, 217, 255].

Recently, another group of molecules, helminth defence molecules (HDM), have also been identified as possessing immunomodulatory capabilities. *F. hepatica* HDM appears to modulate the host immune response away from an excessive inflammatory response via molecular mimicry of host defence peptides, preventing activation of macrophages [256].

4. MANAGEMENT

Diagnosis

Detection and diagnosis of *Fasciola* infection is vital not just for determining the presence of fluke to ensure timely treatment, but is required to determine treatment efficacy as well as identify drug resistance. As such development of simple, sensitive and specific cost-effective diagnostics is essential.

Diagnosis of fasciolosis in both humans and animals has traditionally been determined through coproparasitological examination [22] and detection of eggs is probably still the most widely employed method though other tests, including indirect immunological tests and PCR have become available.

Several methods can be used to detect eggs in faecal matter, including Kato-Katz smear [257-258], sedimentation or FLOTAC techniques [259-260]. However, identification requires the fluke to reach sexual maturity and release eggs. Thus positive identification is only possible after several months of infection and by this stage, disease pathology due to fluke migration has occurred. Additionally, weak infections can pass undetected, while a lack of egg release or sporadic egg release during chronic infections can also hinder detection. Ectopic infections also cannot be detected, and quantitative determination of infecting fluke is not possible.

Early detection of *Fasciola* infection is required for timely chemotherapeutic intervention and greatly facilitates epidemiological studies. An array of serological and immunological detection methods have been developed and reported in numerous studies over the last 2 decades utilising immunogenic *Fasciola* antigens. Most of these assays focus on the detection of circulating antibodies, however, the presence of *Fasciola* antigens in the blood and faeces of infected hosts has resulted in several antigen detection assays also being developed. Antibody and antigen detection assays have been shown to detect infection far earlier than traditional methods and can be more sensitive, but specificity is sometimes a problem.

Many of the initial assays to detect circulating antibodies utilised crude homogenate or ES extracts from either *F. hepatica* or *F. gigantica*, though specificity was often found to be poor due to cross-reactivity and subsequently a range of purified native antigens or in some cases recombinant antigens have been used [261-268]. However, the cathepsin L proteases, due to both their abundance [253, 269] and immunodominance during infections of both animals and humans [270-272], arguably appear the most promising antigens and have been extensively applied to *Fasciola* detection in humans and a variety of definitive hosts. Assays utilising purified native cathepsins have been show to possess generally high sensitivity, and specificity (approaching 100% in both cases) when screening human populations for fasciolosis [82, 88, 273-278], while assays screening a number of ruminant species, including sheep, cattle, buffalo, goat and alpacas, have also given promising results, with antibody generally detected as early as 2-4 weeks post infection, highlighting the ability of these assays to identify pre-patent infection [271, 279-288]. The use of recombinant cathepsin L proteins and peptides can provide comparable sensitivity and specificity to purified natural antigen, while improving reproducibility and enabling large scale production of antigen [270, 281, 287, 289-296].

However, the high sensitivity and specificity reported in experimental settings often drops under the field situation of natural infections, and false positives caused by cross-reactivity with other parasite antigens is an issue with some assays [270, 297]. Restricting detection to IgG4 has been shown to enhance specificity and accuracy in diagnosis of human *Fasciola* infection when using a recombinant cathepsin L antigen [291, 293, 295].

Assays based on the use of a monoclonal antibody (MM3) that binds to *Fasciola* cathepsin L epitopes [298] have also been developed that detect serum (or milk) antibodies [299-302] within 3-4 weeks of infection with specificities and sensitivities nearing 100%. These are available commercially.

Despite issues with specificity with some assays, detection of antibodies in sera for diagnosis of fasciolosis in both humans and animals looks promising. However, due to the persistence of circulating antibodies post infection [303], these assays have limited application in determining active infections and are unreliable when inferring drug efficacy.

Detection of *Fasciola* infection though the presence of antibodies in milk has also been applied in recent years, through analysis of individual and bulk milk samples [172, 301-302, 304-309]. Although it is not always as sensitive as detection using sera, a high correlation between sera and milk detection has been found [301-302, 304, 306, 310] and detection of milk antibodies is possible in cattle harbouring less than 10 fluke [301-302]. Bulk milk tank testing has been used for epidemiology studies and in herd prevalence estimation [307-308, 311-313], and has been effective when herd infection prevalences are above 12.5% [302, 304-305].

Several assays utilising purified polyclonal or monoclonal antibodies that bind to *Fasciola* antigens have been reported and also show high sensitivity and specificity [314-320]. Assays have detected circulating *Fasciola* antigens from human [320-322], sheep [314, 323] and cattle sera [316-317, 324-325], as well as coproantigens [314, 316-318, 320-323, 326-331].

Detection of infection via circulating antigens is possible earlier than through detection of antibodies, and is typically possible from 1 week post infection [323, 332], but has been reported as early as one day post infection when detecting tegumental antigen in infected mice [319]. However, antigen detection in sera is often only practical during the acute phase of infection, with antigen levels decreasing as infection progresses [314, 319, 321, 332].

Coproantigen assays cannot detect infection as early, but can still detect prepatent infections several weeks prior to the fecal egg counts and also detect antigens during patent infection [299-300, 316, 321, 323, 330, 332]. One coproantigen assay was sensitive enough to detect infections of 1 or 2 fluke in sheep and cattle respectively [330] and out-performed sera antibody or antigen detecting ELISAs in regard to sensitivity and specificity has been reported [317, 320, 322]. Additionally, coproantigen levels can also correlate to worm burden and may lead to the development of semi-quantitative assays [314, 317, 323, 326-327, 329-330, 333]. Further, coproantigen assays indicate current infections, and have been utilised to confirm effective flukicide treatment [299, 316, 326, 328, 331, 334].

Differentiation between *F. hepatica* and *F. gigantica* cannot generally be undertaken with the aforementioned diagnostic assays, which has implications in areas where the species overlap. However, as outlined in the section on genetic diversity, numerous PCR-based assays have been described to detect and differentiate between the two species, and in some cases intermediate forms [114, 129, 134-135, 335-337]. These assays can be applied to diagnosis and epidemiological surveys of definitive hosts and humans, or for epidemiological studies of *Fasciola* infection in snails to determine transmission dynamics and facilitate risk assessment.

Control Strategies and Treatment Options

There are a range of methods available for the control and prevention of fasciolosis in animals, though the type of control program utilized and its effectiveness depends on the species of *Fasciola*, local epidemiological factors, animal husbandry and climatic conditions and socio-economic factors (reviewed in [6, 338-339]). Complete eradication of infection is generally not practical, and most strategies focus on minimising exposure to the parasite and its increasing prevalence in the environment. Anthelminthics, pasture management regimes, molluscicides and biological competitors have all been shown to be effective in certain situations, while the use of resistant animals has also been proposed. However, the availability and low cost of anthelmintics means that *Fasciola* infections in developed countries (particularly those caused by *F. hepatica*) are nearly always controlled by a combination of anthelmintic drugs and pasture management.

Effective human control programmes rely on reducing infection in animals and preventing contamination of food and water with metacercarie, with widespread or focused anthelmintic treatments and education programs forming part of combined efforts.

Modelling Systems

The strong link between climate and *Fasciola* infection levels means several forecasting systems to predict high risk periods and infection outbreaks have been developed. These models allow prediction of Fasciolosis incidence at local and regional scales, facilitating development and implementation of suitable effective control strategies. Several systems have been developed and implemented in areas where Fasciolosis is a problem [145, 340-342]. A number of parameters and techniques can be used, and the current trend for these systems is towards Geographic Information Systems (GIS).

Anthelmintic Treatment

Administration of anthelmintics is the main method used to control *F. hepatica* in livestock. There are a range of different anthelmintic drugs available for the treatment of animal fasciolosis, including albendazole, closantal, clorsulon, rafoxanide, nitroxynil and triclabendazole; which differ in availability, price, efficacy and safety, however, most products are efficacious against both *F. hepatica* and *F. gigantica* (see [6, 343]). Despite their equivalent efficacy, the use of anthelmintics in the control of *F. gigantica* has not been as widely adopted as with *F. hepatica*, due mainly to economic differences in their zones of distribution.

Triclabendazole (TCBZ) is the only drug effective against both the adult flukes and early stage parasites, and as a result has become the drug of choice for treating fluke infections in livestock for over 25 years and has more recently been used successfully to treat human cases [344].

Treatment regimes range from the targeted treatment of infected animals showing clinical signs, to strategic drenching of herds for fluke control. However, knowledge of the local, specific epidemiological requirements is required for prophylaxis to be most effective, with the optimal intervals and timing of drenching dependent on numerous factors such as the species of animal to be treated and animal husbandry practices, to climatic conditions and timing and length of seasonal transmission, resulting in varying recommendations from 1 to 5 treatments annually [6]. Rapid recontamination of pastures can also occur through exposure to natural wildlife reservoirs and contaminated irrigation, limiting the effectiveness of drenching.

The recommended oral dosage of TCBZ is 10 mg/kg for sheep and goats and 12 mg/kg for cattle respectively, while higher oral doses (24 mg/kg), or intraruminal injections, are required for buffalo due to lower serum concentrations [6, 345-347].

TCBZ has been used in recent years to treat human fasciolosis in a number of countries where the disease is endemic including in Egypt, Iran, Bolivia, Peru, Vietnam, Georgia and Yemen. It has been shown to be highly effective and is generally well tolerated, with the most common side effects including nausea, vomiting and abdominal pain. Single oral doses of 10 mg/kg give cure rates of about 80%, while double dose cure rates range from 92% to 100% [94, 348-350] and reviewed in [56, 102, 351].

TCBZ action may be due to the combined and sequential effects of different metabolites, and it is believed to act against microtubules (as supported by morphological analysis of fluke tegument, testis and vitelline cells), disrupting microtubule-based processes, though the precise mode of action is still undetermined. These actions inhibit fluke mobility and secretory functions, which affects tegument synthesis and turnover, and leads to rapid cessation of egg production [344, 352-356]

The time scales of action *in vivo* on adult and juvenile flukes are comparable. In studies conducted in sheep, most major tissue changes occur from 48-72 hr post treatment (pt), with severe tegument necrosis at 72 hr pt and fluke death within 96 hr pt. Adult fluke shows rapid effects on reproductive organs, with egg production cessation by 24 hr pt [357-362] reviewed in [356].

Drug Resistance

Drug resistance to a number of fasciolicides has been identified and while it is not yet a major problem, it may pose a serious problem for fasciolosis control in the future [343, 363]. Resistance to TCBZ in *Fasciola* infected livestock was first reported in Australia [364], just 12 years after its commercial release in Ireland. While the incidence of resistance is low, it is now widespread within the veterinary sector, with subsequent resistance observed in several other countries, including Holland, Spain, Ireland, the UK, and recently Argentina [365-373]. To date, anthelmintic resistance has not been reported for *F.gigantica*, probably due to the lower levels of anthelmintic use against this parasite.

Regarding TCBZ resistance, despite microtubules being the most likely targets for drug action, the resistance mechanism does not appear to be linked to tubulin mutations. Instead, it appears that upregulation of drug efflux and drug metabolism processes are more important [374-376] [377-385] and reviewed in [355-356, 386].

Several treatment options, including the use of other flukicides and rotation of usage, or synergistic drug combinations are being used to treat resistant strains or to prevent development of resistance, and while several novel chemotherapeutic agents are also being trialled they require more research, with commercial production likely to be many years away [343, 356].

Alternative Anthelmintics

While some less effective flukicides, such as bithionol, are no longer commercially available [94] others are still produced, including clorsulon, closantel, nitroxynil, clorsulon, oxyclozanide, rafoxanide and albendazole. Although their activity is restricted to the adult stage, they are still important in fluke control, as some (closantel, clorsulon, albendazole, oxyclozanide and nitroxynil) have been shown to be active against adult TCBZ-resistant flukes *in vivo*, indicating the possibility of use in areas where TCBZ resistance is present [365, 387-389]

New Drugs

In humans, TCBZ resistance has not represented a problem to date (although there have been reports of patients failing to respond to triclabendazole [390-391]), though the extent of resistance observed in animal fasciolosis suggests this is not likely to continue. Since the emergence of resistance, a number of alternative drugs have been assessed for their effectiveness in combating fasciolosis both in animals and humans.

Compound Alpha, a TCBZ derivative has shown promise as a flukicide and is effective against all stages of fluke [151, 392-394], reviewed by [355-356], although it does not appear to be efficacious against TCBZ-resistant fluke *in vivo* [151], presumably due to its similar mode of action.

Artemisinin derivatives have given promising initial results *in vitro* and *in vivo*, in rats, sheep and humans, including promising results against TCBZ-resistant fluke [395-403]. However, artemether has been shown to not be effective for treatment of human fasciolosis in a recent phase II trial in Egypt [404]. As initial studies observed synergistic interactions between artemether and other artemisinin derivatives in combination with TCBZ [405], their role in combination chemotherapy warrants further investigation.

Nitazoxanide has been reported to be effective against human fasciolosis with cure rates of 82.4% and 97% observed treating human fasciolosis cases in Egypt [406-407]. However, it showed limited efficacy in a trial in Peru, with reported cure rates of 60% in adults and 40% in children [408].

Myrrh (Mirazid), has been registered in Egypt for the treatment of human fasciolosis, and while most studies report successful treatment of human fasciolosis [409-414], there are also some conflicting reports as to its effectiveness in both sheep and human fasciolosis treatment [415-416] which indicate more study of this compound is required.

Other drugs, and various natural products have also been trialled and are reviewed in [355].

Drug Combinations

Drug combinations are often used for treatment of parasite infections, and the use of synergistic drug combinations has been proposed at the veterinary level to slow development of drug resistance [343].

A number of combinations have been trialled against animal *Fasciola* infections including combinations involving nitroxynil, clorsulon and invermectin, TCBZ-ivermectin, nitroxynil-closantel, and combinations of TCBZ with clorsulon or luxabendazole either with or without closantel. All these combinations show synergy, and are seen as viable avenues to pursue even though some combinations have not been effective against TCBZ-resistant fluke *in vivo*, while others still require their efficacy against TCBZ resistant fluke populations to be investigated [343, 374, 387-388, 417-420]. However, the TCBZ-luxabendazole and TCBZ-clorsulon combinations have already been shown to have synergy against TCBZ-resistant fluke in sheep [343, 421].

Co-administration of TCBZ with metabolite and Pgp inhibitors has also been shown to enhance the drug's efficacy [422-423], and TCBZ in combination with ketoconazole, piperonyl butoxide or methimazole has been shown to be effective *in vitro*, and in rats against TCBZ resistant fluke [377-385], however effects in sheep still need to be elucidated.

Prevention and Control

A number of non-anthelmintic strategies are available to minimise infections of animals. Their use and effectiveness in combination with each other and/or with anthelmintic treatments depend on local situations. These approaches generally focus on the minimisation of exposure of the animals to infective metacercariae, either through avoidance of high risk areas through grazing management, or through snail control, either with the use of molluscicides, through biological means, or through drainage or other management practices. Additionally, quarantine strategies to avoid introduction of metacercariae or infected animals to *Fasciola*-free regions play an important role.

Management Strategies

Grazing management to deny access to metacercariae, or minimisation of snail habitats from developing are effective measures towards reducing fasciolosis. Strategies include drainage of fields or wetlands, improved drainage techniques for ditches and fields, rotation of pastures, and fencing off or denying access to wet areas at high risk occasions. However, while these practices can be effective, they may not always be feasible, for example drainage can be expensive, and may not be considered environmentally friendly.

Snail Control

The control of snails has been shown to be able to minimise *Fasciola* infections. Besides management regimes, a broad range of control methods focusing on the intermediate snail host including the use of organic, inorganic or plant-derived molluscicides have been suggested. However, the use of molluscides requires the right topographical habitat to be effective, and is thus most effective for treatment of dams or irrigated rice fields, and it is often considered environmentally unacceptable due to toxicity towards other aquatic fauna. A number of natural molluscicides have been reported [424-428], and treatment at levels that kill *Fasciola* redia and cercaria but are sublethal to snails have been proposed [429], but application to field situations is generally lacking and further study is required.

A number of biological control measures have also been suggested or trialled, including the use of competitor snails, predators, snail parasites and commensals (reviewed by [9, 66, 339, 430]).. A convenient method of sterilisation of *L. rubiginosa* through natural infection with *E. revolutum* has recently been shown to reduce the prevalence of *F. gigantica* in the field [431]. Additionally, the discovery of strains of *Pseudosuccinea columella* and *Lymnaea schirazensis* resistant to *Fasciola* may lead to future applications in control strategies [51-52, 432-433]. While some methods have been shown to be effective in certain situations, their widespread use is often not practical. Thus, despite the range of snail control strategies, they are often not practical or sufficiently effective to eliminate liver fluke disease.

Resistant Livestock

The use of resistant livestock has been reported as another option. While the large host range available to *Fasciola* limits this to some degree, a number of resistant animal breeds have been identified and breeding of relatively resistant livestock, especially in developing countries where *Fasciola* is endemic may be a viable means to reduce the impact of *Fasciola* infections, providing a long term solution [156, 339, 434].

Human Education

Since human infections occur in animal endemic areas, reducing infection of animals is a major component to an effective control regime, and the methods mentioned above can form part of an effective integrated control program. However, along with anthelmintic treatment, education programmes stressing the importance of separating grazing animals from food production areas, discouraging consumption of edible water plants that can be contaminated, and knowledge of disease transmission can be important aspects of control. Additionally, waterborne infections have been implicated in several human endemic areas, and water treatment, such as water filtering can also reduce disease transmission [101, 103, 435].

Summary

Implementation of integrated parasite control programmes, utilising several of the varied options outlined above, and generally including anthelmintic treatment, have been shown to reduce fluke infections of definitive hosts, and represent effective means to control *Fasciola*. However, for programmes to be effective requires consideration of local epidemiology factors and other conditions, meaning tailoring to individual farms of regions. This has been facilitated in recent years with increasingly sophisticated programs for prediction of outbreaks and times of high risk.

Further, successful treatment and control of human fasciolosis through well designed programmes is realistically achievable [436] and is facilitated by the fact that endemic regions tend to be relatively small and geographically isolated.

Despite the varied fluke control options and their reported successes, none have been widely adopted, and anthelmintic treatment, often with the aid of management strategies, continues to be the favoured means of control, and looks set to remain so, at least until a sufficiently safe and effective vaccine is produced.

Vaccine Development

The need for cost effective and alternate means of controlling *Fasciola* infections, especially since the emergence of drug resistance, means there is a significant focus towards developing an efficacious vaccine. Despite several decades of research, there is no commercially available vaccine, and while several hurdles still remain it appears an achievable goal with several promising vaccine candidates identified that provide protection through reducing worm burdens and/or egg production and viability.

The cathepsin L proteases are the most thoroughly studied *Fasciola* proteins. Their exposure to the immune system and their importance to parasite invasion and survival make them leading vaccine candidates. Native purified proteins [213, 437-440], and a range of recombinant cathepsin L isoforms (FhCatL1, -L2, -L3 and -L5)[441-444], derived from both adult and juvenile developmental stages and belonging to various phylogenetic clades, have been able to induce protection through reduced worm burdens in a range of experimental and ruminant hosts, while anti-fecundity and reduced egg counts have also been reported [445]. Choice of adjuvant is important with Freund's complete adjuvant generally eliciting strongest responses and highest levels of protection, however, this is not a commercially usable adjuvant. Nevertheless, the ability of these antigens to induce protection has been observed in several species, through various systems and modes of delivery, including as recombinant protein [441-444, 446-447], DNA vaccines [446, 448], peptides [289, 449], or mimotopes [450], and through injection, intranasal or oral administration. Recently, plasticity has been observed among the FhCatL1-like clades, which has implications for the use of these proteases as vaccine candidates [451]. Consequently, it has been suggested that other cathepsin L proteases, such as FhCatL2, -L3 and -L5, with their observed relative lack of variability and plasticity, may subsequently offer better options as vaccine candidates.

Another leading candidate are FABP's, components of the first defined purified antigen fraction ($FhSm_{III}$(M)) trialled as a *Fasciola* vaccine. Both native (nFh12), and recombinant (rFh15) FABP conferred variable, but generally moderate, protection through reducing fluke number and viability, egg production and/ or disease pathology in experimental and ruminant hosts [452-459]. They also provide cross protection against schistosomiasis [460-462]. The converse is also true: the recombinant schistosome protein (rSm14) also protects against *Fasciola* infection [463-466].

The highly heterologous cellular detoxification GST family of proteins can also confer protection in sheep and cattle, as well as reduce egg counts [467-468], however, the high variability between vaccine trials (reviewed in [469]) and poor protection elicited with the recombinant protein in cattle [470] means the focus has generally shifted away from this family of antigens, although protection in mice (up to 84%) with a recombinant *F. gigantica* GST was recently reported [471].

Numerous other candidate antigens have been trialled in several hosts, with varying degrees of success, including paramyosin [472], phosphoglycerate kinase [473], haemoglobin [190, 438, 474], saposin [475-478], peroxiredoxin [479-480], and thioredoxin-glutathione reductase [481]. Arguably the most promising of these candidates are the leucine aminopeptidases (LAPs), with the native antigen reported to confer almost 90% protection in sheep [439], even higher than when used in a combined vaccine with cathepsin L, while recombinant LAP induced 78% protection in rabbits [482] and 86.7 % in sheep [483].

Despite the promise seen with the mentioned antigens, further work still needs to be undertaken before a commercial vaccine is available. The need to ensure the correct choice of recombinant protein to provide protection comparable to native antigen needs to be considered, as does the choice of appropriate commercially acceptable adjuvant, with adjuvant choice vital for maximal protection. Another issue to overcome to ensure maximum efficacy is to determine the correct mode of delivery. The variability in protection levels induced in different hosts means it is possible that different formulations will be required for specific species.

The use of DNA vaccines may overcome the adjuvant issue, and has already given promising results in experimental animals [446, 448, 473, 484], however, the ability of this form of vaccine to produce a protective response in larger hosts is questionable [485]. The availability of a commercial DNA vaccine for horses (West Nile Innovator-DNA) indicates this may not be a limitation, although this is an antiviral vaccine, with the effector mechanisms considerably different to that required for protection against a large parasite.

Several other methods to induce protective immunity are also being assessed, such as an adjuvant adaptation system [457-459, 486-487], the use of inclusion bodies [444, 447] and targeting the mucosal immune response through oral vaccination with plant-expressed recombinant protein [488].

A majority of vaccine studies focus on *F. hepatica*, and it was initially assumed extrapolation of results toward *F. gigantica* would be possible. However, several studies have shown that antigens and formulations that work against *F. hepatica* are not necessarily effective against *F. gigantica* [472, 489-491], and most studies to date, using native or recombinant proteins or peptides, including cathepsin L, paramyosin, peroxiredoxin, FABP, GST and LAP have reported little or no protection [472, 489-493]. The best results have been observed with rGST in mice (up to 84% reduced worm burden) [471], a combination of rFABP/rGST in buffalo (35 % reduced worm burden)[491] and rFABP in buffalo (reduced worm burden (36%) and egg count (98.6%))[494].

Two current trends in the search for the development of an efficacious *Fasciola* vaccine are the use of antigens expressed during early stages of infection, and also the use of multivalent vaccines. While the adult parasite residing in the bile duct is relatively safe from the host immune response, the migratory stages, particularly prior to reaching the liver, are exposed to the host immune response and are considered most vulnerable. Natural protective immune responses also seem to target this stage of the disease process. With this in mind juvenile stage specific proteins are being looked at and trialled. Validation of this approach has been achieved with FhCatL3 and FhCatB2, with both antigens showing considerable promise [440, 442-443, 449].

A view going forward towards generating required protection is the use of multivalent or 'cocktail' vaccines. In studies undertaken to date, these have generally resulted in increased levels of protection when compared to single antigen vaccines. Dalton (1996) elicited >72% protection in cattle, when vaccinating with native FhCatL2 and haemoglobin [438], while FhCatL1 and FhCatL2 in combination, and FhCatL1 and haemoglobin have also been trialled [190, 438-439]. The combination of FhCatL5 and FhCatB2 conferred 83% protection in rats [442]. A combination of FABP/GST in buffalo elicited better protection against *F. gigantica* than either antigen alone [491]. Although hurdles still exist, the future for *Fasciola* vaccines looks promising, with several high efficacy candidates identified in trials with mice, rats, cattle and sheep. Additionally, a *Fasciola* vaccine need not be 100% protective to be commercially viable [495]. It has been suggested that even moderately effective vaccines, reducing fluke burdens by as little as 50%, would be enough to be cost-effective in areas with low infection levels, by reducing infection levels below the threshold required to have economic impacts [496]. Apart from reducing fluke burdens, the anti-fecundity aspect observed with some vaccines would also be expected to reduce transmission rates [495], while the reduced pathology resulting from smaller flukes (as a consequence of vaccine-induced retarded development) would lead to more productive hosts.

The identification of new target antigens is a major challenge for future research and will be discussed in the following section.

5. The Current State of *Fasciola* Biology

Recent years has seen significant developments in the understanding of both fundamental and functional *Fasciola* biology. Advances in a range of techniques and the availability of public *Fasciola* transcriptomes has facilitated many of these findings, including the identification and characterisation of the temporal and spatial expression profiles, and functionality of individual molecules or groups of molecules that function at the host-parasite interface.

The important roles these molecules play in facilitating parasitism and survival and their exposure to the host immune system means these ES and tegumental molecules are considered important diagnostic antigens, drug targets or vaccine candidates. For this reason much of the proteomic, transcriptomic, and functional genomic approaches that have recently been applied to *Fasciola* have focused on examining these important parasite molecules.

Jefferies and colleagues (2001) undertook the initial proteomic (2D electrophoresis combined with mass spectroscopy) approach towards investigating adult *F. hepatica* ES, and identified a relative abundance of protective proteins, specifically cathepsin L proteases and antioxidant enzymes [252]. However, their approach also highlighted the lack of genetic sequences available for *Fasciola*, with only 29 of 60 prominent adult ES proteins able to be identified. Subsequently, Morphew *et al.* (2007) compared the ES produced by *F. hepatica in vitro* and *in vivo* (within sheep bile). The *in vitro* ES composition was similar to that observed previously (ie. predominance of cathepsin L's and antioxidant enzymes), however, the only *Fasciola* proteins identified in the sheep bile, and recognized by immune sera, were cathepsin L's, further emphasising their predominance in adult ES and status as leading vaccine candidates [269]. These results highlighted concerns about extrapolating conclusions from *in vitro* analysis.

The last few years has seen the number of *Fasciola* transcripts rise significantly from just a few dozen to the current situation where a number of adult *Fasciola* next-generation EST databases are publicly available. An initial EST dataset was released by the Wellcome Trust Sanger Institute (ftp://ftp.sanger.ac.uk/pub4/pathogens/Fasciola/hepatica/ESTs/) and consisted of 6819 unannotated adult *F. hepatica* derived ESTs. This initial database spurred several integrated studies incorporating database sequences with phylogenetic and proteomic analysis and greatly facilitated the identification and characterisation of a number of *Fasciola* genes and their evolutionary relationships [451, 497-500].

Several studies have applied proteomics to *Fasciola* over the last few years, examining different aspects of the parasites biology, and allowing comparison between different life stages, parasite species, and between TCBZ-resistant and susceptible fluke strains. Comparison of adult and NEJ *F. hepatica* proteomes with adult *S. mansoni* and *S. bovis* schistosomula respectively, have been undertaken to identify common antigenic parasite proteins that are situated at the host parasite interface, in a search for novel drug and vaccine targets [501-502]. Proteomic comparison of the responses from TCBZ-resistant and susceptible fluke strains upon exposure to TCBZ-SO has revealed differential expression for GST and FABP, indicating study of these enzymes in regard to TCBZ-resistance in *Fasciola* is warranted [503]. Further, an initial study specifically focused towards elucidating the makeup of *Fasciola*'s tegument and its secretions also utilised proteomic techniques [504]. This proteomic study was the first to integrate data from the large next-generation *Fasciola* transcriptome databases that have recently become available.

While several studies have undertaken proteomic analysis of either somatic extracts or the secretomes from various *F. hepatica* life stages, including eggs, sporocysts, metacercaria, NEJ, 3-week old juveniles, and adults [253, 505-507], these studies relied on homology screening against the Sanger adult EST database, or sequences from other parasite species due to the limited sequence data available for these *Fasciola* life stages. Despite this, they were able to further the understanding of *Fasciola* biology throughout its life cycle, and Moxon *et al.* (2010) identified a promising diagnostic antigen [506]. Robinson *et al.* (2009) analysed the secretome of *F. hepatica* at four distinct time points during fluke development within the primary host [253]. This study, integrating transcriptomic and proteomic data, was able to lead to a deeper understanding of the sequential temporal expression patterns of the proteases and antioxidant enzymes secreted by the fluke as it matures and migrates through the host, encountering distinct host environments and requiring divergent functions.

A *F. hepatica* NEJ EST database has subsequently been constructed, and though consisting of only 1684 high quality sequences (resulting in approximately 500 contigs), it provided an initial glimpse of the NEJ transcriptome, and identified several novel genes [508]. When compared with the Sanger adult *F. hepatica* transcriptome, differential expression of several genes, including cathepsins was observed [508].

When discussing *Fasciola* biology, the focus generally turns towards the cathepsin proteases, and several reviews on *Fasciola* cathepsin L and cathepsin B are available [509-513]. *F. hepatica* cathepsin L's are probably the most thoroughly characterised group of trematode proteins. They comprise the major component of *Fasciola* ES in adult fluke, are the immunodominant antigens during infection, and undertake important roles for fluke survival at the host-parasite interface, making them ideal diagnostic antigens and vaccine candidates, as has already been discussed in this chapter.

Fasciola express a large number of cathepsin L and B proteases, with recent studies identifying further variants [253, 451, 514-515]. Transcript analysis and several proteomic studies have demonstrated that distinct protease repertoires are expressed at different life stages, with FhCatL3 and cathepsin B proteases abundant in metacercaria extracts and NEJ ES [253, 514]. As the fluke mature, expression and secretion of these proteases is down regulated, while expression of FhCatL1, -L2 and -L5 is upregulated. These cathepsin L's comprise over 80% of ES components in the adult fluke of which about 5-10% is FhCatL5, about 20-25% is FhCatL2, with the remainder FhCatL1-like proteases [253, 451, 497].

Phylogenetic analysis shows that the cathepsin L's are a monophyletic group, evolutionarily divided into clades by duplication events that coincided with landmarks in the evolution of the mammalian hosts of the parasite [516]. These clades reflect substrate specificity as well as temporal expression profiles [497, 516]. The substrate specificity of these proteases are largely dependent on the preference of the S2 subsite. This specificity is able to be altered by a single amino acid substitution at residues lining the subsite, including at amino acid 69 [244, 517-518].

Fasciola cathepsins have been shown to cleave a range of basement membrane and extracellular matrix proteins [244, 519]. The adult expressed cathepsins have overlapping but distinct substrate specificities, that combined with temporal expression patterns and biochemical analysis provide clues to the functions of the proteases at the host-parasite interface [244, 517-518, 520-522]. The adult specific cathepsin L's have a preference for substrates with aliphatic residues at P_2, and FhCatL1 is proposed to have evolved to digest haemoglobin, indicating a role in nutrient acquisition for the parasite [521]. FhCatL2 can additionally cleave substrates with a proline at P_2 (something FhCatL1 and –L5 cannot do), enabling it to cleave fibrinogen to produce clots, and also cleave collagen implicating it in migration through the liver [248, 518]. FhCatL2 has also been implicated in generating liver fibrosis through activation of hepatic stellate cells [523]. FhCatL5 has a unique preference for P_2 aspartate, though what functionality this may confer has not been elucidated (Norbury et al., in press). The juvenile specific FhCatL3 (and the *F. gigantica* homologue (FgCatL1G)) have a distinct preference at P_2 for glycine and proline enabling them to also cleave collagen [244, 520, 522]. The temporal expression profile of FhCatL3 implicates roles in excystment as well as migration through the gut wall, and excystment of metacercaria is significantly reduced in the presence of the cysteine protease inhibitor E64 [253]. Interestingly, while

FgCatL1G can cleave immunoglobulins, FhCatL3 cannot, indicating functional differences between these apparent orthologues [244, 520].

Further, the development and application of RNAi to *Fasciola* [524-525] offers the opportunity to characterise gene function, and has already been used to demonstrate that cathepsin L and cathepsin B are important in juvenile fluke migration and intestinal wall penetration [524]. Incubation of NEJ in the presence of a specific cathepsin B inhibitor also reduced fluke motility [526]. The application of RNAi to *Fasciola* offers a means to evaluate the function of genes, as well as assess their suitability as drug and vaccine targets, and will likely aid in selection of such molecules in the future.

As mentioned briefly earlier in this section, a large number of sequences are now available for *Fasciola*. Application of large-scale next-generation sequencing has been used to produce a normalised adult *F. hepatica* transcriptome, and a non-normalised adult *F. gigantica* transcriptome [515, 527]. These annotated, compiled databases have been made available to the public by the Gasser laboratory (http://bioinfosecond.vet.unimelb.edu.au /wblast2.html). Recently, a second normalised adult *F. hepatica* transcriptome database became available, hosted by Helminth Genomics at the University of Liverpool (http://worm1.liv.ac.uk/). Interestingly this site is also hosting a draft *F. hepatica* genome, the dataset composed of *F. hepatica* shotgun genome sequences. The transcriptomes provide an invaluable resource that are capable of giving insight into *Fasciola* biology and evolution and provide a starting point for future fundamental and applied studies towards understanding fluke processes, and identification of novel targets for parasite control. The paucity of transcript data from juvenile fluke means that transcriptomes from developing fluke stages would provide additional important insights into those stages, while the long awaited *Fasciola* genome also looks capable of significantly transforming our understanding of *Fasciola*.

The utilisation of modern high-throughput techniques, especially the use of a combined and integrated or 'systems biology' approach, has helped develop our understanding of the complex biology of *Fasciola*. This enables comparisons between developmental stages and among fluke strains that allows elucidation of molecular pathways, and has particularly improved our understanding of *Fasciola* virulence factors and in the next few years appears likely to play an increasingly powerful role in the identification of novel therapeutic targets.

The development of proteomic techniques, availability of transcriptome and genome datasets and the ability to undertake functional genomics means that it is currently a time of great opportunity to readily characterise and unravel vital features of *Fasciola* biology, and identify and implement novel means to control this parasite.

REFERENCES

[1] Andrews SJ. The life cycle of *Fasciola hepatica*. In: Dalton JP, editor. *Fasciolosis.* Wallingford: CABI Publishing; 1999. p. 1-30.

[2] Kumar V. *Trematode infections and diseases of man and animals.* Dordrecht: Kluwer academic publishers; 1999.

[3] Rowcliffe SA, Ollerenshaw CB. Observations on the bionomics of the egg of *Fasciola hepatica. Ann Trop Med Parasitol.* 1960 Jun;54:172-81.

[4] Suhardono, Roberts JA, Copeman DB. Variations in the survival of *Fasciola gigantica* eggs in bovine dung stored in the sun as opposed to the shade. *Trop Anim Health Prod.* 2006 Jul;38(5):379-82.

[5] Boray JC. *Experimental fascioliasis in Australia. Adv Parasitol. 1969*;7:95-210.

[6] Torgerson PR, Claxton J. *Epidemiology and control.* In: Dalton JP, editor. Fasciolosis. Wallingford: CABI Publishing; 1999. p. 113-50.

[7] Rowan WB. *The mode of hatching the egg of Fasciola hepatica. Exp Parasitol.* 1956 Mar;5(2):118-37.

[8] Erasmus DA. *The biology of Trematodes.* London: Edward Arnold; 1972.

[9] Graczyk TK, Fried B. Development of *Fasciola hepatica* in the intermediate host. In: Dalton JP, editor. *Fasciolosis.* Wallingford: CABI Publishing; 1999. p. 31-46.

[10] Smythe JD, Halton DW. *The Physiology of Trematodes.* 2nd ed. Cambridge ; New York: Cambridge University Press; 1983.

[11] Rondelaud D, Belfaiza M, Vignoles P, Moncef M, Dreyfuss G. Redial generations of *Fasciola hepatica*: a review. *J Helminthol.* 2009 Sep;83(3):245-54.

[12] Krull WH. The number of cercariae of *Fasciola hepatica* developing in snails infected with a single miracidium. In: Christie JR, editor. P Helm *Soc Wash: Proceedings of the Helminthological Society of Washington.;* 1941. p. 55-8.

[13] Lee CG, Cho SH, Lee CY. Metacercarial production of Lymnaea viridis experimentally infected with Fasciola hepatica. *Vet Parasitol.* 1995 Jul;58(4):313-8.

[14] Boray JC, Enigk K. Laboratory Studies on the Survival and Infectivity of Fasciola Hepatica- and F. Gigantica-Metacercariae. *Z Tropenmed Parasitol.* 1964 Oct;15:324-31.

[15] Suhardono, Roberts JA, Copeman DB. The effect of temperature and humidity on longevity of metacercariae of *Fasciola gigantica. Trop Anim Health Prod.* 2006 Jul;38(5):371-7.

[16] Dixon KE. The physiology of excystment of the metacercaria of *Fasciola hepatica* L. *Parasitology.* 1966 Aug;56(3):431-56.

[17] Sukhdeo MV, Mettrick DF. The behavior of juvenile *Fasciola hepatica. J Parasitol.* 1986 Aug;72(4):492-7.

[18] Dawes B. Some observations of *Fasciola hepatica* L. during feeding operations in the hepatic parenchyma of the mouse, with notes on the nature of liver damage in this host. *Parasitology.* 1963;53:135-43.

[19] Chatterjee KD. *Fasciola hepatica.* In: Chatterjee KD, editor. *Parasitology (protozoology and helminthology).* 10th ed ed. Calcutta: S. N. Guha Ray At Sree Saraswaty Press; 1975. p. 146-8.

[20] Dan M, Lichtenstein D, Lavochkin J, Stavorowsky M, Jedwab M, Shibolet S. Human fascioliasis in Israel. An imported case. *Isr J Med Sci.* 1981 Jun;17(6):430-2.

[21] Durbin CG. Longevity of the liver fluke, *Fasciola* sp. in sheep. *P Helm Soc Wash.* 1952;19:120.

[22] Happich FA, Boray JC. Quantitative diagnosis of chronic fasciolosis. 2. The estimation of daily total egg production of *Fasciola hepatica a*nd the number of adult flukes in sheep by faecal egg counts. *Aust Vet J.* 1969 Jul;45(7):329-31.

[23] Fletcher HL, Hoey EM, Orr N, Trudgett A, Fairweather I, Robinson MW. The occurrence and significance of triploidy in the liver fluke, *Fasciola hepatica. Parasitology.* 2004 Jan;128(Pt 1):69-72.

[24] Terasaki K, Moriyama-Gonda N, Noda Y. Abnormal spermatogenesis in the common liver fluke (*Fasciola* sp.) from Japan and Korea. *J Vet Med Sci.* 1998 Dec;60(12):1305-9.

[25] Terasaki K, Noda Y, Shibahara T, Itagaki T. Morphological comparisons and hypotheses on the origin of polyploids in parthenogenetic *Fasciola* sp. *J Parasitol.* 2000 Aug;86(4):724-9.

[26] Itagaki T, Sakaguchi K, Terasaki K, Sasaki O, Yoshihara S, Van Dung T. Occurrence of spermic diploid and aspermic triploid forms of *Fasciola* in Vietnam and their molecular characterization based on nuclear and mitochondrial DNA. *Parasitol Int.* 2009 Mar;58(1):81-5.

[27] Dixon KE. The Relative Suitability of Sheep and Cattle as Hosts for the Liver Fluke, *Fasciola Hepatica* L. *J Helminthol.* 1964;38:203-12.

[28] Valero MA, Darce NA, Panova M, Mas-Coma S. Relationships between host species and morphometric patterns in Fasciola hepatica adults and eggs from the northern Bolivian Altiplano hyperendemic region. *Vet Parasitol.* 2001 Dec 3;102(1-2):85-100.

[29] Rondelaud D, Dreyfuss G. *Fasciola hepatica* : the influence of the definitive host on the characteristics of infection in the snail Lymnaea truncatula. *Parasite.* 1995;2(3):275-80.

[30] Abrous M, Comes AM, Gasnier N, Rondelaud D, Dreyfuss G, Chauvin A, et al. Morphological variability in *Fasciola hepatica* eggs in ruminants, rodents and lagomorphs. *J Helminthol.* 1998 Dec;72(4):313-7.

[31] Hanna RE, Gordon AW, Moffett D, Edgar HW, Oliver LF, McConnell S, et al. Fasciola hepatica: comparative effects of host resistance and parasite intra-specific interactions on size and reproductive histology in flukes from rats infected with isolates differing in triclabendazole sensitivity. *Vet Parasitol.* 2011 Jun 10;178(3-4):251-63.

[32] Arjona R, Riancho JA, Aguado JM, Salesa R, Gonzalez-Macias J. Fascioliasis in developed countries: a review of classic and aberrant forms of the disease. *Medicine* (Baltimore). 1995 Jan;74(1):13-23.

[33] Le TH, De NV, Agatsuma T, Blair D, Vercruysse J, Dorny P, et al. Molecular confirmation that *Fasciola gigantica* can undertake aberrant migrations in human hosts. *J Clin Microbiol.* 2007 Feb;45(2):648-50.

[34] Dalimi A, Jabarvand M. *Fasciola hepatica* in the human eye. Trans R Soc Trop Med Hyg. 2005 Oct;99(10):798-800.

[35] Naresh G, Gomez PA, Salmah B, Suryati MY. Fasciolosis (liver fluke) of the breast in a male patient: a case report. *Breast.* 2006 Feb;15(1):103-5.

[36] Zali MR, Ghaziani T, Shahraz S, Hekmatdoost A, Radmehr A. Liver, spleen, pancreas and kidney involvement by human fascioliasis: imaging findings. *BMC Gastroenterol.* 2004 Aug 4;4:15.

[37] Prociv P, Walker JC, Whitby M. Human ectopic fascioliasis in Australia: first case reports. *Med J Aust.* 1992 Mar 2;156(5):349-51.

[38] Cho SY, Yang HN, Kong Y, Kim JC, Shin KW, Koo BS. Intraocular fascioliasis: a case report. *Am J Trop Med Hyg.* 1994 Mar;50(3):349-53.

[39] Park CI, Kim H, Ro JY, Gutierrez Y. Human ectopic fascioliasis in the cecum. *Am J Surg Pathol.* 1984 Jan;8(1):73-7.

[40] Vatsal DK, Kapoor S, Venkatesh V, Vatsal P, Husain N. Ectopic fascioliasis in the dorsal spine: case report. *Neurosurgery.* 2006 Sep;59(3):E706-7; discussion E-7.

[41] Ying M, Xiaosu H, Wang B. A case of ectopic parasitism: *Fasciola hepatica* larvae burrow through a human brain and mimic cerebral aneurysm. *Trans R Soc Trop Med Hyg.* 2007 Oct;101(10):1051-2.

[42] Yi-Zhu X, Zhi-Bang Y. A case of ectopic fascioliasis in the skin. *Trop Doct.* 2010 Oct;40(4):253-4.

[43] Rashed AA, Khalil HH, Morsy AT. Zoonotic ectopic fascioliasis: review and discussion. *J Egypt Soc Parasitol.* 2010 Dec;40(3):591-608.

[44] Yoshihara S, Hung NH, Hung NP, Goto N. An erratic parasitism found in the lungs of sheep during experimental infection with *Fasciola gigantica* (Japanese strain). *J Helminthol.* 1998 Sep;72(3):227-30.

[45] Tanir G, Karaman A, Tufekci SB, Erdogan D, Tuygun N, Ozkan AT. A case of ectopic intraabdominal fascioliasis presented with acute abdomen. *Turk J Gastroenterol.* 2011 Jun;22(3):347-50.

[46] Rees JB, Sykes WE, Rickard MD. Prenatal infection with *Fasciola hepatica* in calves. *Aust Vet J.* 1975 Nov;51(11):497-9.

[47] Vermeer J, Augustijn C, Berger J, Berndsen F, Charpentier G, Eshuis J, et al. [An intrauterine and prenatal *Fasciola hepatica* infection in a dairy herd in South Netherlands]. *Tijdschr Diergeneeskd.* 1993 Jan 15;118(2):41-2.

[48] Valero MA, Perez-Crespo I, Periago MV, Khoubbane M, Mas-Coma S. Fluke egg characteristics for the diagnosis of human and animal fascioliasis by *Fasciola hepatica* and *F. gigantica*. *Acta Trop.* 2009 Aug;111(2):150-9.

[49] Taylor EL. Fascioliasis and the Liver Fluke. *FAO Agricultural Studies.* Rome1964. p. 234.

[50] Bargues MD, Vigo M, Horak P, Dvorak J, Patzner RA, Pointier JP, et al. European Lymnaeidae (Mollusca: Gastropoda), intermediate hosts of trematodiases, based on nuclear ribosomal DNA ITS-2 sequences. *Infect Genet Evol.* 2001 Dec;1(2):85-107.

[51] Gutierrez A, Pointier JP, Yong M, Sanchez J, Theron A. Evidence of phenotypic differences between resistant and susceptible isolates of *Pseudosuccinea columella* (Gastropoda: Lymnaeidae) to *Fasciola hepatica* (Trematoda: Digenea) in Cuba. *Parasitol Res.* 2003 Jun;90(2):129-34.

[52] Gutierrez A, Pointier JP, Fraga J, Jobet E, Modat S, Perez RT, et al. Fasciola hepatica: identification of molecular markers for resistant and susceptible Pseudosuccinea columella snail hosts. *Exp Parasitol.* 2003 Nov-Dec;105(3-4):211-8.

[53] Bargues MD, Mas-Coma S. Phylogenetic analysis of Lymnaeid snails based on 18S rDNA sequences. *Mol Biol Evol.* 1997 May;14(5):569-77.

[54] Correa AC, Escobar JS, Durand P, Renaud F, David P, Jarne P, et al. Bridging gaps in the molecular phylogeny of the Lymnaeidae (Gastropoda: Pulmonata), vectors of Fascioliasis. *BMC Evol Biol.* 2010;10:381.

[55] Mas-Coma S, Valero MA, Bargues MD. Climate change effects on trematodiases, with emphasis on zoonotic fascioliasis and schistosomiasis. *Vet Parasitol.* 2009 Aug 26;163(4):264-80.

[56] Mas-Coma S, Bargues MD, Valero MA. Fascioliasis and other plant-borne trematode zoonoses. *Int J Parasitol.* 2005 Oct;35(11-12):1255-78.

[57] Shimalov VV, Shimalov VT. Findings of *Fasciola hepatica* Linnaeus, 1758 in wild animals in Belorussian Polesie. *Parasitol Res.* 2000 Jun;86(6):527.

[58] Haridy FM, Morsy TA. Camel: a new Egyptian host for *Fasciola gigantica. J Egypt Soc Parasitol.* 2000 Aug;30(2):451-4.

[59] Santarem VA, Tostes RA, Alberti H, Sanches OD. *Fasciola hepatica* in capybara. *Acta Trop.* 2006;98(3):311-3.

[60] Spratt DM, Presidente PJ. Prevalence of *Fasciola hepatica* infection in native mammals in southeastern Australia. *Aust J Exp Biol Med Sci.* 1981 Dec;59(Pt 6):713-21.

[61] Valero MA, Panova M, Comes AM, Fons R, Mas-Coma S. Patterns in size and shedding of *Fasciola hepatica* eggs by naturally and experimentally infected murid rodents. *J Parasitol.* 2002 Apr;88(2):308-13.

[62] Menard A, Agoulon A, L'Hostis ML, Rondelaud D, Collard S, Chauvin A. Myocastor coypus as a reservoir host of Fasciola hepatica in France. *Vet Res.* 2001 Sep-Oct;32(5):499-508.

[63] Kawano J, Koga M, Shimizu A, Kimura S. Factors contributing to the resistance of chickens to infection with Japanese *Fasciola* sp. *J Vet Med Sci.* 1992 Feb;54(1):75-9.

[64] Vaughan JL, Charles JA, Boray JC. *Fasciola hepatica* infection in farmed emus (*Dromaius novaehollandiae*). *Aust Vet J.* 1997 Nov;75(11):811-3.

[65] Soares MP, da Silva SS, Nizoli LQ, Felix SR, Schild AL. Chronic fascioliasis in farmed and wild greater rheas (Rhea americana). *Vet Parasitol.* 2007 Apr 10;145(1-2):168-71.

[66] Spithill TW, Smooker PM, Copeman DB. *Fasciola gigantica*: Epidemiology, control, immunology and molecular biology. In: Dalton JP, editor. *Fasciolosis.* Wallingford: CABI Publishing; 1999. p. 465-526.

[67] Fox NJ, White PC, McClean CJ, Marion G, Evans A, Hutchings MR. Predicting impacts of climate change on *Fasciola hepatica* risk. *PLoS One.* 2011;6(1):e16126.

[68] Kenyon F, Sargison ND, Skuce PJ, Jackson F. Sheep helminth parasitic disease in south eastern Scotland arising as a possible consequence of climate change. *Vet Parasitol.* 2009 Aug 26;163(4):293-7.

[69] Esteban JG, Bargues MD, Mas-Coma S. Geographical distribution, diagnosis and treatment of human fascioliasis: a review. *Res Rev Parasitol.* 1998;58:13-48.

[70] Rim HJ, Farag HF, Sornmani S, Cross JH. Food-borne trematodes: Ignored or emerging? *Parasitol Today.* 1994;10(6):207-9.

[71] Hopkins DR. Homing in on helminths. *Am J Trop Med Hyg.* 1992;46(6):626-34.

[72] Keiser J, Utzinger J. Emerging foodborne trematodiasis. *Emerg Infect Dis.* 2005 Oct;11(10):1507-14.

[73] Mas-Coma MS, Esteban JG, Bargues MD. Epidemiology of human fascioliasis: a review and proposed new classification. *Bull World Health Organ.* 1999;77(4):340-6.

[74] Bogitsh BJ, Carter CE, Oeltmann TN. *Human Parasitology.* 3rd ed. London: Elsevier Academic Press; 2005.

[75] Mas-Coma S, Bargues MD, Esteban JG. Human Fasciolosis. In: Dalton JP, editor. Fasciolosis. Wallingford: CABI Publishing; 1999. p. 411-34.

[76] Mas-Coma S. Epidemiology of fascioliasis in human endemic areas. *J Helminthol.* 2005 Sep;79(3):207-16.

[77] Mas-Coma S, Angles R, Esteban JG, Bargues MD, Buchon P, Franken M, et al. The Northern Bolivian Altiplano: a region highly endemic for human fascioliasis. *Trop Med Int Health.* 1999 Jun;4(6):454-67.

[78] Hillyer GV, Soler de Galanes M, Rodriguez-Perez J, Bjorland J, Silva de Lagrava M, Ramirez Guzman S, et al. Use of the Falcon assay screening test--enzyme-linked immunosorbent assay (FAST-ELISA) and the enzyme-linked immunoelectrotransfer blot (EITB) to determine the prevalence of human fascioliasis in the Bolivian Altiplano. *Am J Trop Med Hyg*. 1992 May;46(5):603-9.

[79] Esteban JG, Flores A, Aguirre C, Strauss W, Angles R, Mas-Coma S. Presence of very high prevalence and intensity of infection with *Fasciola hepatic*a among Aymara children from the Northern Bolivian Altiplano. *Acta Trop.* 1997 Jun 24;66(1):1-14.

[80] Esteban JG, Flores A, Angles R, Strauss W, Aguirre C, Mas-Coma S. A population-based coprological study of human fascioliasis in a hyperendemic area of the Bolivian Altiplano. *Trop Med Int Health.* 1997 Jul;2(7):695-9.

[81] Esteban JG, Flores A, Angles R, Mas-Coma S. High endemicity of human fascioliasis between Lake Titicaca and La Paz valley, Bolivia. *Trans R Soc Trop Med Hyg*. 1999 Mar-Apr;93(2):151-6.

[82] O'Neill SM, Parkinson M, Strauss W, Angles R, Dalton JP. Immunodiagnosis of *Fasciola hepatica* infection (fascioliasis) in a human population in the Bolivian Altiplano using purified cathepsin L cysteine proteinase. *Am J Trop Med Hyg.* 1998 Apr;58(4):417-23.

[83] Bjorland J, Bryan RT, Strauss W, Hillyer GV, McAuley JB. An outbreak of acute fascioliasis among Aymara Indians in the Bolivian Altiplano. *Clin Infect Dis.* 1995 Nov;21(5):1228-33.

[84] Fried B, Abruzzi A. Food-borne trematode infections of humans in the United States of America. *Parasitol Res.* 2010 May;106(6):1263-80.

[85] WHO. *Control of foodborne trematode infections.* Report of a WHO study group. Geneva: World Health Organisation.1995.

[86] Mailles A, Capek I, Ajana F, Schepens C, Ilef D, Vaillant V. Commercial watercress as an emerging source of fascioliasis in Northern France in 2002: results from an outbreak investigation. *Epidemiol Infect.* 2006 Oct;134(5):942-5.

[87] Sanchez-Andrade A, Suarez JL, Arias M, Francisco I, Diez C, Cortinas J, et al. Relationships between eosinophilia, anti-Fasciola IgG, and IgM rheumatoid factors, in urban and rural areas of north-western Spain. *Ann Trop Med Parasitol.* 2008 Sep;102(6):489-98.

[88] Rokni MB, Massoud J, O'Neill SM, Parkinson M, Dalton JP. Diagnosis of human fasciolosis in the Gilan province of Northern Iran: application of cathepsin L-ELISA. *Diagn Microbiol Infect Dis.* 2002 Oct;44(2):175-9.

[89] Moghaddam AS, Massoud J, Mahmoodi M, Mahvi AH, Periago MV, Artigas P, et al. Human and animal fascioliasis in Mazandaran province, northern Iran. *Parasitol Res.* 2004 Sep;94(1):61-9.

[90] Kaya S, Demirci M, Demirel R, Aridogan BC, Ozturk M, Korkmaz M. Seroprevalence of fasciolosis and the difference of fasciolosis between rural area and city center in Isparta, Turkey. *Saudi Med J.* 2006 Aug;27(8):1152-6.

[91] Yilmaz H, Godekmerdan A. Human fasciolosis in Van province, Turkey. *Acta Trop.* 2004 Oct;92(2):161-2.

[92] Kaya M, Bestas R, Cetin S. Clinical presentation and management of Fasciola hepatica infection: Single-center experience. *World J Gastroenterol. 2011* Nov 28;17(44):4899-904.

[93] Ozturhan H, Emekdas G, Sezgin O, Korkmaz M, Altintas E. Seroepidemiology of *Fasciola hepatica* in Mersin province and surrounding towns and the role of family history of the Fascioliasis in the transmission of the parasite. *Turk J Gastroenterol.* 2009 Sep;20(3):198-203.

[94] Millan JC, Mull R, Freise S, Richter J. The efficacy and tolerability of triclabendazole in Cuban patients with latent and chronic *Fasciola hepatica* infection. *Am J Trop Med Hyg. 2000* Nov-Dec;63(5-6):264-9.

[95] Agnamey P, Fortes-Lopes E, Raccurt CP, Boncy J, Totet A. Cross-sectional serological survey of human fascioliasis in haiti. *J Parasitol Res.* 2012;2012:Article ID 751951.

[96] Rojas L, Vazquez A, Domenech I, Robertson LJ. Fascioliasis: can Cuba conquer this emerging parasitosis? *Trends Parasitol.* 2010 Jan;26(1):26-34.

[97] Tran VH, Tran TK, Nguye HC, Pham HD, Pham TH. Fascioliasis in Vietnam. *Southeast Asian J Trop Med Public Health.* 2001;32 Suppl 2:48-50.

[98] De NV, Murrell KD, Cong le D, Cam PD, Chau le V, Toan ND, et al. The food-borne trematode zoonoses of Vietnam. *Southeast Asian J Trop Med Public Health.* 2003;34 Suppl 1:12-34.

[99] Esteban JG, Gonzalez C, Curtale F, Munoz-Antoli C, Valero MA, Bargues MD, et al. Hyperendemic fascioliasis associated with schistosomiasis in villages in the Nile Delta of Egypt. *Am J Trop Med Hyg.* 2003 Oct;69(4):429-37.

[100] Mas-Coma S, Valero MA, Bargues MD. Chapter 2. *Fasciola*, lymnaeids and human fascioliasis, with a global overview on disease transmission, epidemiology, evolutionary genetics, molecular epidemiology and control. *Adv Parasitol.* 2009;69:41-146.

[101] Mas-Coma S. Human fascioliasis. In: Cotruvo JA, Dufour A, Rees G, Bartram J, Carr R, Cliver DO, et al., editors. *Waterborne Zoonoses: Identification, Causes and Control.* London: IWA Publishing; 2004. p. 305-22.

[102] Mas-Coma S, Bargues MD, Valero MA. Plant-Borne Trematode Zoonoses: Fascioliasis and Fasciolopsiasis. In: Murrell KD, Fried B, editors. *Food-Borne Parasitic Zoonoses:* Springer US; 2007. p. 293-334.

[103] Esteban JG, Gonzalez C, Bargues MD, Angles R, Sanchez C, Naquira C, et al. High fascioliasis infection in children linked to a man-made irrigation zone in Peru. *Trop Med Int Health.* 2002 Apr;7(4):339-48.

[104] Taira N, Yoshifuji H, Boray JC. Zoonotic potential of infection with Fasciola spp. by consumption of freshly prepared raw liver containing immature flukes. *Int J Parasitol.* 1997 Jul;27(7):775-9.

[105] Itagaki T, Tsutsumi K. Triploid form of Fasciola in Japan: genetic relationships between Fasciola hepatica and Fasciola gigantica determined by ITS-2 sequence of nuclear rDNA. *Int J Parasitol.* 1998 May;28(5):777-81.

[106] Itagaki T, Kikawa M, Sakaguchi K, Shimo J, Terasaki K, Shibahara T, et al. Genetic characterization of parthenogenic *Fasciola* sp. in Japan on the basis of the sequences of ribosomal and mitochondrial DNA. *Parasitology.* 2005 Nov;131(Pt 5):679-85.

[107] Itagaki T, Kikawa M, Terasaki K, Shibahara T, Fukuda K. Molecular characterization of parthenogenic *Fasciola* sp. in Korea on the basis of DNA sequences of ribosomal ITS1 and mitochondrial NDI gene. *J Vet Med Sci.* 2005 Nov;67(11):1115-8.

[108] Ashrafi K, Valero MA, Panova M, Periago MV, Massoud J, Mas-Coma S. Phenotypic analysis of adults of *Fasciola hepatica*, *Fasciola gigantica* and intermediate forms from the endemic region of Gilan, Iran. *Parasitol Int.* 2006 Dec;55(4):249-60.

[109] Le TH, De NV, Agatsuma T, Thi Nguyen TG, Nguyen QD, McManus DP, et al. Human fascioliasis and the presence of hybrid/introgressed forms of *Fasciola hepatica* and *Fasciola gigantica* in Vietnam. *Int J Parasitol.* 2008;38(6):725-30.

[110] Valero MA, Panova M, Mas-Coma S. Phenotypic analysis of adults and eggs of *Fasciola hepatica* by computer image analysis system. *J Helminthol.* 2005 Sep;79(3):217-25.

[111] Periago MV, Valero MA, El Sayed M, Ashrafi K, El Wakeel A, Mohamed MY, et al. First phenotypic description of Fasciola hepatica/Fasciola gigantica intermediate forms from the human endemic area of the Nile Delta, Egypt. *Infect Genet Evol.* 2008 Jan;8(1):51-8.

[112] Periago MV, Valero MA, Panova M, Mas-Coma S. Phenotypic comparison of allopatric populations of *Fasciola hepatica* and *Fasciola gigantica* from European and African bovines using a computer image analysis system (CIAS). *Parasitol Res.* 2006 Sep;99(4):368-78.

[113] Itagaki T, Ichinomiya M, Fukuda K, Fusyuku S, Carmona C. Hybridization experiments indicate incomplete reproductive isolating mechanism between *Fasciola hepatica* and *Fasciola gigantica*. *Parasitology.* 2011 Sep;138(10):1278-84.

[114] Amor N, Halajian A, Farjallah S, Merella P, Said K, Ben Slimane B. Molecular characterization of *Fasciola* spp. from the endemic area of northern Iran based on nuclear ribosomal DNA sequences. *Exp Parasitol.* 2011 Jul;128(3):196-204.

[115] Ichikawa M, Iwata N, Itagaki T. DNA types of aspermic *Fasciola species* in Japan. *J Vet Med Sci.* 2010 Oct;72(10):1371-4.

[116] Agatsuma T, Arakawa Y, Iwagami M, Honzako Y, Cahyaningsih U, Kang SY, et al. Molecular evidence of natural hybridization between *Fasciola hepatica* and *F. gigantica*. *Parasitol Int.* 2000 Sep;49(3):231-8.

[117] Itagaki T, Tsutsumi KI, Ito K, Tsutsumi Y. Taxonomic status of the Japanese triploid forms of *Fasciola*: comparison of mitochondrial ND1 and COI sequences with *F. hepatica* and *F. gigantica*. *J Parasitol.* 1998 Apr;84(2):445-8.

[118] Lin RQ, Dong SJ, Nie K, Wang CR, Song HQ, Li AX, et al. Sequence analysis of the first internal transcribed spacer of rDNA supports the existence of the intermediate *Fasciola* between *F. hepatica* and *F. gigantica* in mainland China. *Parasitol Res.* 2007 Aug;101(3):813-7.

[119] Huang WY, He B, Wang CR, Zhu XQ. Characterisation of Fasciola species from Mainland China by ITS-2 ribosomal DNA sequence. *Vet Parasitol.* 2004 Feb 26;120(1-2):75-83.

[120] Alasaad S, Huang CQ, Li QY, Granados JE, Garcia-Romero C, Perez JM, et al. Characterization of *Fasciola* samples from different host species and geographical localities in Spain by sequences of internal transcribed spacers of rDNA. *Parasitol Res.* 2007 Oct;101(5):1245-50.

[121] Walker SM, Prodohl PA, Fletcher HL, Hanna RE, Kantzoura V, Hoey EM, et al. Evidence for multiple mitochondrial lineages of *Fasciola hepatica* (liver fluke) within infrapopulations from cattle and sheep. *Parasitol Res.* 2007 Jun;101(1):117-25.

[122] Walker SM, Johnston C, Hoey EM, Fairweather I, Borgsteede F, Gaasenbeek C, et al. Population dynamics of the liver fluke, *Fasciola hepatica*: the effect of time and spatial separation on the genetic diversity of fluke populations in the Netherlands. *Parasitology.* 2011 Feb;138(2):215-23.

[123] Walker SM, Johnston C, Hoey EM, Fairweather I, Borgsteede FH, Gaasenbeek CP, et al. Potential role of hares in the spread of liver fluke in the Netherlands. *Vet Parasitol.* 2011 Apr 19;177(1-2):179-81.

[124] Moazeni M, Sharifiyazdi H, Izadpanah A. Characterization of *Fasciola hepatica* genotypes from cattle and sheep in Iran using cytochrome C oxidase gene (CO1). *Parasitol Res.* 2011 Dec 21.

[125] Vilas R, Vazquez-Prieto S, Paniagua E. Contrasting patterns of population genetic structure of *Fasciola hepatica* from cattle and sheep: Implications for the evolution of anthelmintic resistance. *Infect Genet Evol.* 2012 Jan;12(1):45-52.

[126] Ai L, Weng YB, Elsheikha HM, Zhao GH, Alasaad S, Chen JX, et al. Genetic diversity and relatedness of Fasciola spp. isolates from different hosts and geographic regions revealed by analysis of mitochondrial DNA sequences. *Vet Parasitol.* 2011 Sep 27;181(2-4):329-34.

[127] Hurtrez-Boussès S, Durand P, Jabbour-Zahab R, GuÉGan JF, Meunier C, Bargues MD, et al. Isolation and characterization of microsatellite markers in the liver fluke (*Fasciola hepatica*). *Mol Ecol Notes.* 2004;4(4):689-90.

[128] Gunasekar KR, Tewari AK, Sreekumar C, Gupta SC, Rao JR. Elucidation of genetic variability among different isolates of *Fasciola gigantica* (giant liver fluke) using random-amplified polymorphic DNA polymerase chain reaction. *Parasitol Res.* 2008 Oct;103(5):1075-81.

[129] Marcilla A, Bargues MD, Mas-Coma S. A PCR-RFLP assay for the distinction between *Fasciola hepatica* and *Fasciola gigantica*. *Mol Cell Probes.* 2002 Oct;16(5):327-33.

[130] Ali H, Ai L, Song HQ, Ali S, Lin RQ, Seyni B, et al. Genetic characterisation of *Fasciola* samples from different host species and geographical localities revealed the existence of *F. hepatica* and *F. gigantica* in Niger. *Parasitol Res.* 2008 Apr;102(5):1021-4.

[131] Alasaad S, Li QY, Lin RQ, Martin-Atance P, Granados JE, Diez-Banos P, et al. Genetic variability among *Fasciola hepatica* samples from different host species and geographical localities in Spain revealed by the novel SRAP marker. *Parasitol Res.* 2008 Jun;103(1):181-6.

[132] Vara-Del Rio MP, Villa H, Martinez-Valladares M, Rojo-Vazquez FA. Genetic heterogeneity of *Fasciola hepatica* isolates in the northwest of Spain. *Parasitol Res.* 2007 Sep;101(4):1003-6.

[133] Dosay-Akbulut M, Trudgett A, Stanhope M. *Understanding genetic diversity of the liver fluke Fasciola hepatica.* Z Naturforsch C. 2005 Sep-Oct;60(9-10):774-8.

[134] Li QY, Dong SJ, Zhang WY, Lin RQ, Wang CR, Qian DX, et al. Sequence-related amplified polymorphism, an effective molecular approach for studying genetic variation in Fasciola spp. of human and animal health significance. *Electrophoresis.* 2009 Jan;30(2):403-9.

[135] Nguyen TG, Van De N, Vercruysse J, Dorny P, Le TH. Genotypic characterization and species identification of *Fasciola* spp. with implications regarding the isolates infecting goats in Vietnam. *Exp Parasitol.* 2009 Dec;123(4):354-61.

[136] Ai L, Chen MX, Alasaad S, Elsheikha HM, Li J, Li HL, et al. Genetic characterization, species differentiation and detection of *Fasciola* spp. by molecular approaches. *Parasit Vectors.* 2011;4:101.

[137] Dar Y, Amer S, Courtioux B, Dreyfuss G. Microsatellite analysis of Fasciola spp. in Egypt. *Parasitology Research.* 2011 Dec;109(6):1741-4.

[138] Semyenova SK, Morozova EV, Chrisanfova GG, Gorokhov VV, Arkhipov IA, Moskvin AS, et al. Genetic differentiation in eastern European and western Asian populations of the liver fluke, *Fasciola hepatica*, as revealed by mitochondrial nad1 and cox1 genes. *J Parasitol.* 2006 Jun;92(3):525-30.

[139] Teofanova D, Kantzoura V, Walker S, Radoslavov G, Hristov P, Theodoropoulos G, et al. Genetic diversity of liver flukes (Fasciola hepatica) from Eastern Europe. *Infect Genet Evol.* 2011 Jan;11(1):109-15.

[140] Lotfy WM, Brant SV, DeJong RJ, Le TH, Demiaszkiewicz A, Rajapakse RP, et al. Evolutionary origins, diversification, and biogeography of liver flukes (Digenea, Fasciolidae). *Am J Trop Med Hyg.* 2008 Aug;79(2):248-55.

[141] Ichikawa M, Bawn S, Maw NN, Htun LL, Thein M, Gyi A, et al. Characterization of *Fasciola* spp. in Myanmar on the basis of spermatogenesis status and nuclear and mitochondrial DNA markers. *Parasitol Int.* 2011 Dec;60(4):474-9.

[142] Semyenova SK, Morozova EV, Chrisanfova GG, Asatrian A, Movsessian S, Ryskov AP. RAPD variability and genetic diversity in two populations of liver fluke, *Fasciola hepatica*. *Acta Parasitol.* 2003;48(2):125-30

[143] Kaplan RM, Dame JB, Reddy GR, Courtney CH. The prevalence of Fasciola hepatica in its snail intermediate host determined by DNA probe assay. *Int J Parasitol.* 1997 Dec;27(12):1585-93.

[144] Kozak M, Wedrychowicz H. The performance of a PCR assay for field studies on the prevalence of Fasciola hepatica infection in Galba truncatula intermediate host snails. *Vet Parasitol.* 2010 Feb 26;168(1-2):25-30.

[145] Charlier J, Bennema SC, Caron Y, Counotte M, Ducheyne E, Hendrickx G, et al. Towards assessing fine-scale indicators for the spatial transmission risk of Fasciola hepatica in cattle. *Geospat health.* 2011 May;5(2):239-45.

[146] Relf V, Good B, Hanrahan JP, McCarthy E, Forbes AB, deWaal T. Temporal studies on Fasciola hepatica in Galba truncatula in the west of Ireland. *Vet Parasitol.* 2011 Feb 10;175(3-4):287-92.

[147] Valero MA, Marcos MD, Comes AM, Sendra M, Mas-Coma S. Comparison of adult liver flukes from highland and lowland populations of Bolivian and Spanish sheep. *J Helminthol.* 1999 Dec;73(4):341-5.

[148] Valero MA, Panova M, Mas-Coma S. Developmental differences in the uterus of *Fasciola hepatica* between livestock liver fluke populations from Bolivian highlands and European lowlands. *Parasitol Res.* 2001 Apr;87(4):337-42.

[149] Walker SM, McKinstry B, Boray JC, Brennan GP, Trudgett A, Hoey EM, et al. Response of two isolates of *Fasciola hepatica* to treatment with triclabendazole in vivo and in vitro. *Parasitol Res.* 2004 Dec;94(6):427-38.

[150] Walker SM, Hoey E, Fletcher H, Brennan G, Fairweather I, Trudgett A. Stage-specific differences in fecundity over the life-cycle of two characterized isolates of the liver fluke, *Fasciola hepatica*. *Parasitology.* 2006 Aug;133(Pt 2):209-16.

[151] McConville M, Brennan GP, Flanagan A, Edgar HW, Hanna RE, McCoy M, et al. An evaluation of the efficacy of compound alpha and triclabendazole against two isolates of Fasciola hepatica. *Vet Parasitol.* 2009 May 26;162(1-2):75-88.

[152] Behm CA, Sangster NC. Pathology, pathophysiology and clinical aspects. In: Dalton JP, editor. *Fasciolosis.* Wallingford: CABI Publishing; 1999. p. 185-224.

[153] Mulcahy G, Joyce P, Dalton JP. Immunology of *Fasciola hepatica* infection. In: Dalton JP, editor. *Fasciolosis.* Wallingford: CABI Publishing; 1999. p. 341-76.

[154] Meeusen EN, Piedrafita D. Exploiting natural immunity to helminth parasites for the development of veterinary vaccines. *Int J Parasitol.* 2003;33(11):1285-90.

[155] Piedrafita D, Raadsma HW, Prowse R, Spithill TW. Immunology of the host–parasite relationship in fasciolosis (*Fasciola hepatica* and *Fasciola gigantica*). *Can J Zool.* 2004;82(2):233-50.

[156] Piedrafita D, Raadsma HW, Gonzalez J, Meeusen E. Increased production through parasite control: can ancient breeds of sheep teach us new lessons? *Trends Parasitol.* 2010 Dec;26(12):568-73.

[157] Pleasance J, Raadsma HW, Estuningsih SE, Widjajanti S, Meeusen E, Piedrafita D. Innate and adaptive resistance of Indonesian Thin Tail sheep to liver fluke: A comparative analysis of Fasciola gigantica and Fasciola hepatica infection. *Vet Parasitol.* 2011 Jan 26.

[158] Lee CG, Zimmerman GL, Bishop JK. Host influence on the banding profiles of whole-body protein and excretory-secretory product of Fasciola hepatica (Trematoda) by isoelectric focusing. *Vet Parasitol.* 1992 Feb;41(1-2):57-68.

[159] Miller CM, Howell MJ, Boray JC. Host effects on glutathione S-transferase activity in Fasciola hepatica. *Int J Parasitol.* 1993 Dec;23(8):1073-6.

[160] Caseby RH, Harriott M, Fairweather I. Ionic composition of the liver fluke *Fasciola hepatica* from different mammalian hosts and comparison with host bile. *Parasitol Res.* 1995;81(5):394-7.

[161] Brady MT, O'Neill SM, Dalton JP, Mills KH. Fasciola hepatica suppresses a protective Th1 response against Bordetella pertussis. *Infect Immun.* 1999 Oct;67(10):5372-8.

[162] O'Neill SM, Mills KH, Dalton JP. *Fasciola hepatica* cathepsin L cysteine proteinase suppresses *Bordetella pertussis*-specific interferon-gamma production in vivo. *Parasite Immunol.* 2001 Oct;23(10):541-7.

[163] Chen MG, Mott KE. Progress in assessment of morbidity due to *Fasciola hepatica* infection: a review of recent literature. *Trop Dis Bull.* 1990;87(4): R1-R38.

[164] Kaplan RM. Fasciola hepatica: a review of the economic impact in cattle and considerations for control. *Vet Ther.* 2001 Winter;2(1):40-50.

[165] Roseby FB. The effect of fasciolosis on the wool production of merino sheep. *Aust Vet J.* 1970 Aug;46(8):361-5.

[166] Hawkins CD, Morris RS. Depression of productivity in sheep infected with *Fasciola hepatica*. *Vet Parasitol.* 1978;4(4):341-251.

[167] Genicot B, Mouligneau F, Lekeux P. Economic and production consequences of liver fluke disease in double-muscled fattening cattle. *Zentralbl Veterinarmed B.* 1991 May;38(3):203-8.

[168] Wamae LW, Hammond JA, Harrison LJ, Onyango-Abuje JA. Comparison of production losses caused by chronic *Fasciola gigantica* infection in yearling Friesian and Boran cattle. *Trop Anim Health Prod.* 1998 Feb;30(1):23-30.

[169] Elitok B, Elitok OM, Kabu M. Field trial on comparative efficacy of four fasciolicides against natural liver fluke infection in cattle. *Vet Parasitol.* 2006 Feb 18;135(3-4):279-85.

[170] Mehra UR, Verma AK, Dass RS, Sharma RL, Yadav SC. Effects of Fasciola gigantica infection on growth and nutrient utilisation of buffalo calves. *Vet Rec.* 1999 Dec 11;145(24):699-702.

[171] Raadsma HW, Kingsford NM, Suharyanta, Spithill TW, Piedrafita D. Host responses during experimental infection with Fasciola gigantica or Fasciola hepatica in Merino sheep I. Comparative immunological and plasma biochemical changes during early infection. *Vet Parasitol.* 2007 Feb 28;143(3-4):275-86.

[172] Charlier J, Duchateau L, Claerebout E, Williams D, Vercruysse J. Associations between anti-*Fasciola hepatica* antibody levels in bulk-tank milk samples and production parameters in dairy herds. *Prev Vet Med.* 2007 Jan 16;78(1):57-66.

[173] Khan MK, Sajid MS, Khan MN, Iqbal Z, Arshad M, Hussain A. Point prevalence of bovine fascioliasis and the influence of chemotherapy on the milk yield in a lactating bovine population from the district of Toba Tek Singh, Pakistan. *J Helminthol.* 2010 Nov 10:1-5.

[174] Lopez-Diaz MC, Carro MC, Cadorniga C, Diez-Banos P, Mezo M. Puberty and serum concentrations of ovarian steroids during prepuberal period in Friesian heifers artificially infected with *Fasciola hepatica*. *Theriogenology.* 1998 Sep;50(4):587-93.

[175] Hammam AM, Khateeb RME, Amer HA, Abou-El-Dobal SKA, Shahat KHE, Scott W. Response of *Fasciola* free and infected buffaloes to CIDR OvSynch treatment during summer season with emphasis on sex hormone and biochemical changes. *J Am Sci.* 2011;7(9):810-20.

[176] Piedrafita D, Spithill TW, Smith RE, Raadsma HW. Improving animal and human health through understanding liver fluke immunology. *Parasite Immunol.* 2010 Aug;32(8):572-81.

[177] Aitken MM, Jones PW, Hall GA, Hughes DL, Collis KA. Effects of experimental *Salmonella dublin* infection in cattle given *Fasciola hepatica* thirteen weeks previously. *J Comp Pathol.* 1978 Jan;88(1):75-84.

[178] Cox FE. Concomitant infections, parasites and immune responses. *Parasitology.* 2001;122 Suppl:S23-38.

[179] Ezenwa VO, Jolles AE. From host immunity to pathogen invasion: the effects of helminth coinfection on the dynamics of microparasites. *Integr Comp Biol.* 2011 Oct;51(4):540-51.

[180] Cervi L, Rubinstein H, Masih DT. Involvement of excretion-secretion products from Fasciola hepatica inducing suppression of the cellular immune responses. *Vet Parasitol.* 1996 Jan;61(1-2):97-111.

[181] O'Neill SM, Brady MT, Callanan JJ, Mulcahy G, Joyce P, Mills KH, et al. *Fasciola hepatica* infection downregulates Th1 responses in mice. *Parasite Immunol.* 2000 Mar;22(3):147-55.

[182] Dowling DJ, Hamilton CM, Donnelly S, La Course J, Brophy PM, Dalton J, et al. Major secretory antigens of the helminth Fasciola hepatica activate a suppressive

dendritic cell phenotype that attenuates Th17 cells but fails to activate Th2 immune responses. *Infection and Immunity.* 2010 Feb;78(2):793-801.

[183] Flynn RJ, Irwin JA, Olivier M, Sekiya M, Dalton JP, Mulcahy G. Alternative activation of ruminant macrophages by *Fasciola hepatica. Vet Immunol Immunopathol.* 2007 Nov 15;120(1-2):31-40.

[184] Pleasance J, Wiedosari E, Raadsma HW, Meeusen E, Piedrafita D. Resistance to liver fluke infection in the natural sheep host is correlated with a type-1 cytokine response. *Parasite Immunol.* 2011 Sep;33(9):495-505.

[185] Zhang W, Moreau E, Huang W, Chauvin A. Comparison of humoral response in sheep to *Fasciola hepatica* and *Fasciola gigantica* experimental infection. *Parasite.* 2004 Jun;11(2):153-9.

[186] Zhang WY, Moreau E, Hope JC, Howard CJ, Huang WY, Chauvin A. *Fasciola hepatica* and *Fasciola gigantica*: Comparison of cellular response to experimental infection in sheep. *Exp Parasitol.* 2005;111(3):154-9.

[187] Chauvin A, Bouvet G, Boulard C. Humoral and cellular immune responses to Fasciola hepatica experimental primary and secondary infection in sheep. *Int J Parasitol.* 1995 Oct;25(10):1227-41.

[188] Hansen DS, Clery DG, Estuningsih SE, Widjajanti S, Partoutomo S, Spithill TW. Immune responses in Indonesian thin tail and Merino sheep during a primary infection with *Fasciola gigantica*: lack of a specific IgG2 antibody response is associated with increased resistance to infection in Indonesian sheep. *Int J Parasitol.* 1999;29(7):1027-35.

[189] Spithill TW, Piedrafita D, Smooker PM. Immunological approaches for the control of fasciolosis. *Int J Parasitol.* 1997 Oct;27(10):1221-35.

[190] Mulcahy G, O'Connor F, McGonigle S, Dowd A, Clery DG, Andrews SJ, et al. Correlation of specific antibody titre and avidity with protection in cattle immunized against *Fasciola hepatica. Vaccine.* 1998 May-Jun;16(9-10):932-9.

[191] Molina EC, Gonzaga EA, Sinolinding EO, Lumbao LA, Peralta AA, Barraca AP. Differences in susceptibility between cattle and swamp buffaloes to infection with *Fasciola gigantica. Trop Anim Health Prod.* 2005 Nov;37(8):611-6.

[192] Molina EC, Skerratt LF. Cellular and humoral responses in liver of cattle and buffaloes infected with a single dose of Fasciola gigantica. *Vet Parasitol.* 2005 Jul 15;131(1-2):157-63.

[193] Piedrafita D, Estuningsih E, Pleasance J, Prowse R, Raadsma HW, Meeusen EN, et al. Peritoneal lavage cells of Indonesian thin-tail sheep mediate antibody-dependent superoxide radical cytotoxicity in vitro against newly excysted juvenile Fasciola gigantica but not juvenile Fasciola hepatica. *Infect Immun.* 2007 Apr;75(4):1954-63.

[194] Spithill TW, Dalton JP. Progress in development of liver fluke vaccines. *Parasitol Today.* 1998 Dec;14(6):224-8.

[195] Martinez-Moreno A, Martinez-Moreno FJ, Acosta I, Gutierrez PN, Becerra C, Hernandez S. Humoral and cellular immune responses to experimental Fasciola hepatica infections in goats. *Parasitol Res.* 1997;83(7):680-6.

[196] Cervi L, Serradell MC, Guasconi L, Masih DT. New Insights into the Modulation of Immune Response by *Fasciola hepatica* Excretory-Secretory Products *Curr Immunol Rev.* 2009 November 2009;5(4):277-84.

[197] Falcon C, Carranza F, Martinez FF, Knubel CP, Masih DT, Motran CC, et al. Excretory-secretory products (ESP) from *Fasciola hepatica* induce tolerogenic properties in myeloid dendritic cells. *Vet Immunol Immunopathol*. 2010 Sep 15;137(1-2):36-46.

[198] Zhang W, Moreau E, Peigne F, Huang W, Chauvin A. Comparison of modulation of sheep, mouse and buffalo lymphocyte responses by *Fasciola hepatica* and *Fasciola gigantica* excretory-secretory products. *Parasitol Res*. 2005 Mar;95(5):333-8.

[199] Roberts JA, Estuningsih E, Widjayanti S, Wiedosari E, Partoutomo S, Spithill TW. Resistance of Indonesian thin tail sheep against Fasciola gigantica and F. hepatica. V*et Parasitol*. 1997 Jan;68(1-2):69-78.

[200] Baeza E, Poitou I, Boulard C. Influence of pro-inflammatory treatments on experimental infection of rats with *Fasciola hepatica*: changes in serum levels of inflammatory markers during the early stages of fasciolosis. *Res Vet Sci*. 1994 Sep;57(2):180-7.

[201] Baeza E, Poitou I, Delers F, Boulard C. Influence of anti-inflammatory treatments on experimental infection of rats with *Fasciola hepatica*: changes in serum levels of inflammatory markers during the early stages of fasciolosis. *Res Vet Sci*. 1994 Sep;57(2):172-9.

[202] van Milligen FJ, Cornelissen JB, Bokhout BA. Location of induction and expression of protective immunity against *Fasciola hepatica* at the gut level: a study using an ex vivo infection model with ligated gut segments. *J Parasitol*. 1998 Aug;84(4):771-7.

[203] van Milligen FJ, Cornelissen JB, Hendriks IM, Gaasenbeek CP, Bokhout BA. Protection of *Fasciola hepatica* in the gut mucosa of immune rats is associated with infiltrates of eosinophils, IgG1 and IgG2a antibodies around the parasites. *Parasite Immunol*. 1998 Jun;20(6):285-92.

[204] van Milligen FJ, Cornelissen JB, Gaasenbeek CP, Bokhout BA. A novel ex vivo rat infection model to study protective immunity against *Fasciola hepatica* at the gut level. *J Immunol Methods*. 1998 Apr 15;213(2):183-90.

[205] Jedlina L, Kozak-Ljunggren M, Wedrychowicz H. *In vivo* studies of the early, peritoneal, cellular and free radical response in rats infected with *Fasciola hepatica* by flow cytometric analysis. *Exp Parasitol*. 2011 Jul;128(3):291-7.

[206] Burden DJ, Bland AP, Hammet NC, Hughes DL. Fasciola hepatica: migration of newly excysted juveniles in resistant rats. *Exp Parasitol*. 1983 Oct;56(2):277-88.

[207] van Milligen FJ, Cornelissen JB, Bokhout BA. Protection against *Fasciola hepatica* in the intestine is highly correlated with eosinophil and immunoglobulin G1 responses against newly excysted juveniles. *Parasite Immunol*. 1999 May;21(5):243-51.

[208] Tliba O, Sibille P, Boulard C, Chauvin A. Local hepatic immune response in rats during primary infection with *Fasciola hepatica*. *Parasite*. 2000 Mar;7(1):9-18.

[209] Piedrafita D, Parsons JC, Sandeman RM, Wood PR, Estuningsih SE, Partoutomo S, et al. Antibody-dependent cell-mediated cytotoxicity to newly excysted juvenile *Fasciola hepatica* in vitro is mediated by reactive nitrogen intermediates. *Parasite Immunol*. 2001 Sep;23(9):473-82.

[210] Sibille P, Tliba O, Boulard C. Early and transient cytotoxic response of peritoneal cells from Fasciola hepatica-infected rats. *Vet Res*. 2004 Sep-Oct;35(5):573-84.

[211] Piedrafita D, Spithill TW, Dalton JP, Brindley PJ, Sandeman MR, Wood PR, et al. Juvenile *Fasciola hepatica* are resistant to killing in vitro by free radicals compared with larvae of *Schistosoma mansoni*. *Parasite Immunol.* 2000 Jun;22(6):287-95.

[212] Roberts JA, Estuningsih E, Wiedosari E, Spithill TW. Acquisition of resistance against Fasciola gigantica by Indonesian thin tail sheep. *Vet Parasitol.* 1997 Dec 31;73(3-4):215-24.

[213] Mulcahy G, O'Connor F, Clery D, Hogan SF, Dowd AJ, Andrews SJ, et al. Immune responses of cattle to experimental anti-*Fasciola hepatica* vaccines. *Res Vet Sci.* 1999 Aug;67(1):27-33.

[214] Tliba O, Moire N, Le Vern Y, Boulard C, Chauvin A, Sibille P. Early hepatic immune response in rats infected with Fasciola hepatica. *Vet Res.* 2002 May-Jun;33(3):261-70.

[215] Donnelly S, O'Neill SM, Sekiya M, Mulcahy G, Dalton JP. Thioredoxin peroxidase secreted by *Fasciola hepatica* induces the alternative activation of macrophages. *Infect Immun.* 2005;73(1):166-73.

[216] Walsh KP, Brady MT, Finlay CM, Boon L, Mills KH. Infection with a helminth parasite attenuates autoimmunity through TGF-beta-mediated suppression of Th17 and Th1 responses. *J Immunol.* 2009 Aug 1;183(3):1577-86.

[217] Donnelly S, Stack CM, O'Neill SM, Sayed AA, Williams DL, Dalton JP. Helminth 2-Cys peroxiredoxin drives Th2 responses through a mechanism involving alternatively activated macrophages. *FASEB J.* 2008 Nov;22(11):4022-32.

[218] Girones N, Valero MA, Garcia-Bodelon MA, Chico-Calero I, Punzon C, Fresno M, et al. Immune suppression in advanced chronic fascioliasis: an experimental study in a rat model. *J Infect Dis.* 2007 May 15;195(10):1504-12.

[219] Flynn RJ, Mannion C, Golden O, Hacariz O, Mulcahy G. Experimental Fasciola hepatica infection alters responses to tests used for diagnosis of bovine tuberculosis. *Infect Immun.* 2007 Mar;75(3):1373-81.

[220] Flynn RJ, Mulcahy G, Welsh M, Cassidy JP, Corbett D, Milligan C, et al. Co-Infection of cattle with *Fasciola hepatica* and *Mycobacterium bovis*- immunological consequences. *Transbound Emerg Dis.* 2009 Aug;56(6-7):269-74.

[221] Reddington JJ, Leid RW, Wescott RB. A review of the antigens of *Fasciola hepatica*. *Vet Parasitol.* 1984;14(3-4):209-29.

[222] Sobhon P, Anantavara S, Dangprasert T, Viyanant V, Krailas D, Upatham ES, et al. *Fasciola gigantica*: studies of the tegument as a basis for the developments of immunodiagnosis and vaccine. *Southeast Asian J Trop Med Public Health.* 1998;29(2):387-400.

[223] Hanna RE. Fasciola hepatica: glycocalyx replacement in the juvenile as a possible mechanism for protection against host immunity. *Exp Parasitol.* 1980 Aug;50(1):103-14.

[224] Keegan PS, Trudgett A. *Fasciola hepatica* in the rat: immune responses associated with the development of resistance to infection. *Parasite Immunol.* 1992;14(6):657-69.

[225] Hughes DL. Trematodes, excluding schistosomes with special emphasis on *Fasciola*. *Curr Top Microbiol Immunol.* 1985;120:241-60.

[226] Halton DW. Microscopy and the helminth parasite. *Micron.* 2004;35(5):361-90.

[227] Hamilton CM, Dowling DJ, Loscher CE, Morphew RM, Brophy PM, O'Neill SM. The Fasciola hepatica tegumental antigen suppresses dendritic cell maturation and function. *Infection and Immunity.* 2009 Jun;77(6):2488-98.

[228] Hacariz O, Sayers G, Mulcahy G. A preliminary study to understand the effect of *Fasciola hepatica* tegument on naive macrophages and humoral responses in an ovine model. *Vet Immunol Immunopathol.* 2011 Feb 15;139(2-4):245-9.

[229] Chapman CB, Mitchell GF. Proteolytic cleavage of immunoglobulin by enzymes released by *Fasciola hepatica*. *Vet Parasitol.* 1982;11(2-3):165-78.

[230] Goose J. Possible role of excretory/secretory products in evasion of host defences by *Fasciola hepatica*. *Nature.* 1978;275(5677):216-7.

[231] Serradell MC, Guasconi L, Masih DT. Involvement of a mitochondrial pathway and key role of hydrogen peroxide during eosinophil apoptosis induced by excretory-secretory products from *Fasciola hepatica*. *Mol Biochem Parasitol.* 2009 Feb;163(2):95-106.

[232] Cervi L, Masih DT. Inhibition of spleen cell proliferative response to mitogens by excretory-secretory antigens of Fasciola hepatica. *Int J Parasitol.* 1997 May;27(5):573-9.

[233] Cervi L, Rossi G, Cejas H, Masih DT. *Fasciola hepatica*-induced immune suppression of spleen mononuclear cell proliferation: role of nitric oxide. *Clin Immunol Immunopathol.* 1998 May;87(2):145-54.

[234] Moreau E, Herve S, Yu ZW, Alain C. Modulation of sheep lymphocyte responses by Fasciola hepatica excretory-secretory products. *Vet Parasitol.* 2002 Sep 26;108(3):207-15.

[235] Prowse RK, Chaplin P, Robinson HC, Spithill TW. *Fasciola hepatica* cathepsin L suppresses sheep lymphocyte proliferation in vitro and modulates surface CD4 expression on human and ovine T cells. *Parasite Immunol.* 2002;24(2):57-66.

[236] Masih DT, Cervi L, Casado JM. Modification of accessory activity of peritoneal cells from *Fasciola hepatica* infected rats. *Vet Immunol Immunopathol.* 1996 Oct;53(3-4):257-68.

[237] Jefferies JR, Barrett J, Turner RJ. Immunomodulation of sheep and human lymphocytes by *Fasciola hepatica* excretory-secretory products. *Int J Parasitol.* 1996;26(10):1119-21.

[238] Jefferies JR, Turner RJ, Barrett J. Effect of *Fasciola hepatica* excretory-secretory products on the metabolic burst of sheep and human neutrophils. *Int J Parasitol.* 1997;27(9):1025-9.

[239] Baeza E, Poitou I, Boulard C. In vitro effects of Fasciola hepatica on the main functions of polymorphonuclear leukocytes: chemotaxis and free radical generation induced by phagocytosis. *Int J Parasitol.* 1993 Dec;23(8):1077-81.

[240] El-Ghaysh A, Turner RJ, Brophy PM, Barrett J. Effect of Fasciola gigantica somatic extracts and excretory/secretory products on superoxide production by activated neutrophils. *Vet Parasitol.* 1999 Jul;84(1-2):91-100.

[241] Flynn RJ, Mulcahy G. Possible role for Toll-like receptors in interaction of Fasciola hepatica excretory/secretory products with bovine macrophages. *Infect Immun.* 2008 Feb;76(2):678-84.

[242] Smith AM, Dowd AJ, Heffernan M, Robertson CD, Dalton JP. *Fasciola hepatica*: a secreted cathepsin L-like proteinase cleaves host immunoglobulin. *Int J Parasitol.* 1993;23(8):977-83.

[243] Berasain P, Carmona C, Frangione B, Dalton JP, Goni F. Fasciola hepatica: parasite-secreted proteinases degrade all human IgG subclasses: determination of the specific

cleavage sites and identification of the immunoglobulin fragments produced. *Exp Parasitol.* 2000 Feb;94(2):99-110.

[244] Norbury LJ, Beckham S, Pike RN, Grams R, Spithill TW, Fecondo JV, et al. Adult and juvenile *Fasciola* cathepsin L proteases: Different enzymes for different roles. *Biochimie.* 2011 Mar;93(3):604-11.

[245] Carmona C, Dowd AJ, Smith AM, Dalton JP. Cathepsin L proteinase secreted by *Fasciola hepatica* in vitro prevents antibody-mediated eosinophil attachment to newly excysted juveniles. *Mol Biochem Parasitol.* 1993;62(1):9-17.

[246] Carmona C, Smith A, Dowd A, Dalton JP. A *Fasciola hepatica* cathepsin L proteinase prevents the adherence of eosinophils to newly excysted juveniles. *Biochem Soc Trans.* 1992 Feb;20(1):85S.

[247] Dowd AJ, McGonigle S, Dalton JP. *Fasciola hepatica* cathepsin L proteinase cleaves fibrinogen and produces a novel type of fibrin clot. *Eur J Biochem.* 1995;232(1):241-6.

[248] Dowd AJ, Tort J, Roche L, Ryan T, Dalton JP. Isolation of a cDNA encoding *Fasciola hepatica* cathepsin L2 and functional expression in *Saccharomyces cerevisiae*. *Mol Biochem Parasitol.* 1997;88(1-2):163-74.

[249] Donnelly S, O'Neill SM, Stack CM, Robinson MW, Turnbull L, Whitchurch C, et al. Helminth cysteine proteases inhibit TRIF-dependent activation of macrophages via degradation of TLR3. *J Biol Chem.* 2010 Jan 29;285(5):3383-92.

[250] Donnelly S, Dalton JP, Robinson MW. How Pathogen-Derived Cysteine Proteases Modulate Host Immune Responses In: Robinson MW, Dalton JP, editors. *Cysteine Proteases of Pathogenic Organisms.* 1st ed. New York: Springer Science+Business Media, LLC; 2011. p. 192-207.

[251] Meemon K, Grams R, Vichasri-Grams S, Hofmann A, Korge G, Viyanant V, et al. Molecular cloning and analysis of stage and tissue-specific expression of cathepsin B encoding genes from *Fasciola gigantica*. *Mol Biochem Parasitol.* 2004;136(1):1-10.

[252] Jefferies JR, Campbell AM, van Rossum AJ, Barrett J, Brophy PM. Proteomic analysis of *Fasciola hepatica* excretory-secretory products. *Proteomics.* 2001 Sep;1(9):1128-32.

[253] Robinson MW, Menon R, Donnelly SM, Dalton JP, Ranganathan S. An integrated transcriptomics and proteomics analysis of the secretome of the helminth pathogen *Fasciola hepatica*: proteins associated with invasion and infection of the mammalian host. *Mol Cell Proteomics.* 2009 Aug;8(8):1891-907.

[254] Cervi L, Rossi G, Masih DT. Potential role for excretory-secretory forms of glutathione-S-transferase (GST) in *Fasciola hepatica*. *Parasitology.* 1999 Dec;119 (Pt 6):627-33.

[255] Robinson MW, Hutchinson AT, Dalton JP, Donnelly S. Peroxiredoxin: a central player in immune modulation. *Parasite Immunol.* 2010 May;32(5):305-13.

[256] Robinson MW, Donnelly S, Hutchinson AT, To J, Taylor NL, Norton RS, et al. A family of helminth molecules that modulate innate cell responses via molecular mimicry of host antimicrobial peptides. *PLoS Pathog.* 2011 May;7(5):e1002042.

[257] Katz N, Chaves A, Pellegrino J. A simple device for quantitative stool thick-smear technique in Schistosomiasis mansoni. *Rev Inst Med Trop Sao Paulo.* 1972 Nov-Dec;14(6):397-400.

[258] el-Morshedy H, Shehab AY, Zaki A, Farag HF. Intra-specimen and day-to-day variations of Fasciola egg counts in human stools. *East Mediterr Health J.* 2002 Jul-Sep;8(4-5):619-25.

[259] Duthaler U, Rinaldi L, Maurelli MP, Vargas M, Utzinger J, Cringoli G, et al. Fasciola hepatica: comparison of the sedimentation and FLOTAC techniques for the detection and quantification of faecal egg counts in rats. *Exp Parasitol.* 2010 Oct;126(2):161-6.

[260] Keiser J, Duthaler U, Utzinger J. Update on the diagnosis and treatment of food-borne trematode infections. *Curr Opin Infect Dis.* 2010 Oct;23(5):513-20.

[261] Awad WS, Ibrahim AK, Salib FA. Using indirect ELISA to assess different antigens for the serodiagnosis of *Fasciola gigantica* infection in cattle, sheep and donkeys. *Res Vet Sci.* 2008;21:21.

[262] Raina OK, Sriveny D, Yadav SC. Humoral immune response against Fasciola gigantica fatty acid binding protein. *Vet Parasitol.* 2004 Sep 20;124(1-2):65-72.

[263] Arias M, Hillyer GV, Sanchez-Andrade R, Suarez JL, Pedreira J, Lomba C, et al. A 2.9 kDa Fasciola hepatica-recombinant protein based ELISA test for the detection of current-ovine fasciolosis trickle infected. *Vet Parasitol.* 2006 Apr 15;137(1-2):67-73.

[264] Arias M, Morrondo P, Hillyer GV, Sanchez-Andrade R, Suarez JL, Lomba C, et al. Immunodiagnosis of current fasciolosis in sheep naturally exposed to *Fasciola hepatica* by using a 2.9 kDa recombinant protein. *Vet Parasitol.* 2007 May 15;146(1-2):46-9.

[265] Zhang W, Rogniaux H, Huang W, Chauvin A, Moreau E. Analysis of thioredoxin peroxidase as a promising antigen for diagnosis of *Fasciola gigantica* infection: a preliminary study. *Parasitol Int.* 2011 Jun;60(2):206-8.

[266] Figueroa-Santiago O, Delgado B, Espino AM. *Fasciola hepatica* saposin-like protein-2-based ELISA for the serodiagnosis of chronic human fascioliasis. *Diagn Microbiol Infect Dis.* 2011 Jul;70(3):355-61.

[267] Silva E, Castro A, Lopes A, Rodrigues A, Dias C, Conceicao A, et al. A recombinant antigen recognized by *Fasciola hepatica*-infected hosts. *J Parasitol.* 2004 Aug;90(4):746-51.

[268] Velusamy R, Singh BP, Ghosh S, Chandra D, Raina OK, Gupta SC, et al. Prepatent detection of Fasciola gigantica infection in bovine calves using metacercarial antigen. *Indian J Exp Biol.* 2006 Sep;44(9):749-53.

[269] Morphew RM, Wright HA, LaCourse EJ, Woods DJ, Brophy PM. Comparative proteomics of excretory-secretory proteins released by the liver fluke *Fasciola hepatica* in sheep host bile and during in vitro culture ex host. *Mol Cell Proteomics.* 2007 Jun;6(6):963-72.

[270] Cornelissen JB, Gaasenbeek CP, Borgsteede FH, Holland WG, Harmsen MM, Boersma WJ. Early immunodiagnosis of fasciolosis in ruminants using recombinant Fasciola hepatica cathepsin L-like protease. *Int J Parasitol.* 2001 May 15;31(7):728-37.

[271] Neyra V, Chavarry E, Espinoza JR. Cysteine proteinases Fas1 and Fas2 are diagnostic markers for Fasciola hepatica infection in alpacas (Lama pacos). *Vet Parasitol.* 2002 Apr 19;105(1):21-32.

[272] Cordova M, Herrera P, Nopo L, Bellatin J, Naquira C, Guerra H, et al. *Fasciola hepatica* cysteine proteinases: immunodominant antigens in human fascioliasis. *Am J Trop Med Hyg.* 1997 Dec;57(6):660-6.

[273] Rokni MB, Massoud J, Hanilo A. Comparison of adult somatic and cysteine proteinase antigens of *Fasciola gigantica* in enzyme linked immunosorbent assay for serodiagnosis of human fasciolosis. *Acta Trop.* 2003 Sep;88(1):69-75.

[274] Intapan PM, Maleewong W, Nateeworanart S, Wongkham C, Pipitgool V, Sukolapong V, et al. Immunodiagnosis of human fascioliasis using an antigen of *Fasciola gigantica* adult worm with the molecular mass of 27 kDa by a dot-ELISA. *Southeast Asian J Trop Med Public Health.* 2003 Dec;34(4):713-7.

[275] Cordova M, Reategui L, Espinoza JR. Immunodiagnosis of human fascioliasis with Fasciola hepatica cysteine proteinases. *Trans R Soc Trop Med Hyg.* 1999 Jan-Feb;93(1):54-7.

[276] Espinoza JR, Timoteo O, Herrera-Velit P. Fas2-ELISA in the detection of human infection by *Fasciola hepatica. J Helminthol.* 2005 Sep;79(3):235-40.

[277] Espinoza JR, Maco V, Marcos L, Saez S, Neyra V, Terashima A, et al. Evaluation of Fas2-ELISA for the serological detection of *Fasciola hepatica* infection in humans. *Am J Trop Med Hyg.* 2007;76(5):977-82.

[278] Marcos LA, Bussalleu A, Terashima A, Espinoza JR. Detection of antibodies against *Fasciola hepatica* in cirrhotic patients from Peru. *J Helminthol.* 2009 Mar;83(1):23-6.

[279] Varghese A, Raina OK, Nagar G, Garg R, Banerjee PS, Maharana BR, et al. Development of cathepsin-L cysteine proteinase based Dot-enzyme-linked immunosorbent assay for the diagnosis of Fasciola gigantica infection in buffaloes. *Vet Parasitol.* 2012 Feb 10;183(3-4):382-5.

[280] Gupta A, Dixit AK, Dixit P, Mahajan C, Sharma RL. Evaluation of dipstick-ELISA using 28 kDa Fasciola gigantica cathepsin l cysteine proteinase (FgCL3) for serodiagnosis of fasciolosis in naturally infected goats. *Vet Parasitol.* 2011 Mar 10;176(2-3):165-9.

[281] Raina OK, Yadav SC, Sriveny D, Gupta SC. Immuno-diagnosis of bubaline fasciolosis with *Fasciola gigantica* cathepsin-L and recombinant cathepsin L 1-D proteases. *Acta Trop.* 2006 May;98(2):145-51.

[282] Sriveny D, Raina OK, Yadav SC, Chandra D, Jayraw AK, Singh M, et al. Cathepsin L cysteine proteinase in the diagnosis of bovine *Fasciola gigantica* infection. *Vet Parasitol.* 2006;135(1):25-31.

[283] Yadav SC, Saini M, Raina OK, Nambi PA, Jadav K, Sriveny D. *Fasciola gigantica* cathepsin-L cysteine proteinase in the detection of early experimental fasciolosis in ruminants. *Parasitol Res.* 2005 Dec;97(6):527-34.

[284] Dixit AK, Yadav SC, Sharma RL. Experimental bubalian fasciolosis: kinetics of antibody response using 28 kDa *Fasciola gigantica* cysteine proteinase as antigen. *Trop Anim Health Prod.* 2004 Jan;36(1):49-54.

[285] Dixit AK, Yadav SC, Sharma RL. 28 kDa Fasciola gigantica cysteine proteinase in the diagnosis of prepatent ovine fasciolosis. *Vet Parasitol.* 2002 Nov 11;109(3-4):233-47.

[286] Ruiz A, Molina JM, Gonzalez J, Martinez-Moreno FJ, Gutierrez PN, Martinez-Moreno A. Humoral response (IgG) of goats experimentally infected with *Fasciola hepatica* against cysteine proteinases of adult fluke. *Vet Res.* 2003;34(4):435-43.

[287] Cornelissen JB, Gaasenbeek CP, Boersma W, Borgsteede FH, van Milligen FJ. Use of a pre-selected epitope of cathepsin-L1 in a highly specific peptide-based immunoassay for the diagnosis of Fasciola hepatica infections in cattle. *Int J Parasitol.* 1999 May;29(5):685-96.

[288] Timoteo O, Maco V, Jr., Maco V, Neyra V, Yi PJ, Leguia G, et al. Characterization of the humoral immune response in alpacas (Lama pacos) experimentally infected with

Fasciola hepatica against cysteine proteinases Fas1 and Fas2 and histopathological findings. *Vet Immunol Immunopathol.* 2005 Jun 15;106(1-2):77-86.

[289] Jezek J, El Ridi R, Salah M, Wagih A, Aziz HW, Tallima H, et al. *Fasciola gigantica* cathepsin L proteinase-based synthetic peptide for immunodiagnosis and prevention of sheep fasciolosis. *Biopolymers.* 2008;90(3):349-57.

[290] Carnevale S, Rodriguez MI, Guarnera EA, Carmona C, Tanos T, Angel SO. Immunodiagnosis of fasciolosis using recombinant procathepsin L cystein proteinase. *Diagn Microbiol Infect Dis.* 2001 Sep-Oct;41(1-2):43-9.

[291] O'Neill SM, Parkinson M, Dowd AJ, Strauss W, Angles R, Dalton JP. Short report: Immunodiagnosis of human fascioliasis using recombinant *Fasciola hepatica* cathepsin L1 cysteine proteinase. *Am J Trop Med Hyg.* 1999 May;60(5):749-51.

[292] Intapan PM, Tantrawatpan C, Maleewong W, Wongkham S, Wongkham C, Nakashima K. Potent epitopes derived from *Fasciola gigantica* cathepsin L1 in peptide-based immunoassay for the serodiagnosis of human fascioliasis. *Diagn Microbiol Infect Dis.* 2005;53(2):125-9.

[293] Wongkham C, Tantrawatpan C, Intapan PM, Maleewong W, Wongkham S, Nakashima K. Evaluation of immunoglobulin G subclass antibodies against recombinant *Fasciola gigantica* cathepsin L1 in an enzyme-linked immunosorbent assay for serodiagnosis of human fasciolosis. *Clin Diagn Lab Immunol.* 2005 Oct;12(10):1152-6.

[294] Tantrawatpan C, Maleewong W, Wongkham C, Wongkham S, Intapan PM, Nakashima K. Serodiagnosis of human fascioliasis by a cystatin capture enzyme-linked immunosorbent assay with recombinant *Fasciola gigantica* cathepsin L antigen. *Am J Trop Med Hyg.* 2005 Jan;72(1):82-6.

[295] Tantrawatpan C, Maleewong W, Wongkham C, Wongkham S, Intapan PM, Nakashima K. Evaluation of immunoglobulin G4 subclass antibody in a peptide-based enzyme-linked immunosorbent assay for the serodiagnosis of human fascioliasis. *Parasitology.* 2007;134(14):2021-6.

[296] Strauss W, O'Neill SM, Parkinson M, Angles R, Dalton JP. Short report: Diagnosis of human fascioliasis: detection of anti-cathepsin L antibodies in blood samples collected on filter paper. *Am J Trop Med Hyg.* 1999 May;60(5):746-8.

[297] Rapsch C, Schweizer G, Grimm F, Kohler L, Bauer C, Deplazes P, et al. Estimating the true prevalence of Fasciola hepatica in cattle slaughtered in Switzerland in the absence of an absolute diagnostic test. *Int J Parasitol.* 2006 Sep;36(10-11):1153-8.

[298] Muino L, Perteguer MJ, Garate T, Martinez-Sernandez V, Beltran A, Romaris F, et al. Molecular and immunological characterization of *Fasciola* antigens recognized by the MM3 monoclonal antibody. *Mol Biochem Parasitol.* 2011 Oct;179(2):80-90.

[299] Mezo M, Gonzalez-Warleta M, Ubeira FM. The use of MM3 monoclonal antibodies for the early immunodiagnosis of ovine fascioliasis. *J Parasitol.* 2007 Feb;93(1):65-72.

[300] Valero MA, Ubeira FM, Khoubbane M, Artigas P, Muino L, Mezo M, et al. MM3-ELISA evaluation of coproantigen release and serum antibody production in sheep experimentally infected with Fasciola hepatica and F. gigantica. *Vet Parasitol.* 2009 Jan 22;159(1):77-81.

[301] Mezo M, Gonzalez-Warleta M, Castro-Hermida JA, Carro C, Ubeira FM. Kinetics of anti-Fasciola IgG antibodies in serum and milk from dairy cows during lactation, and in serum from calves after feeding colostrum from infected dams. *Vet Parasitol.* 2010 Feb 26;168(1-2):36-44.

[302] Mezo M, Gonzalez-Warleta M, Castro-Hermida JA, Muino L, Ubeira FM. Field evaluation of the MM3-SERO ELISA for detection of anti-*Fasciola* IgG antibodies in milk samples from individual cows and bulk milk tanks. *Parasitol Int.* 2010 Dec;59(4):610-5.

[303] Castro E, Freyre A, Hernandez Z. Serological responses of cattle after treatment and during natural re-infection with Fasciola hepatica, as measured with a dot-ELISA system. *Vet Parasitol.* 2000 Jun 27;90(3):201-8.

[304] Reichel MP, Vanhoff K, Baxter B. Performance characteristics of an enzyme-linked immunosorbent assay performed in milk for the detection of liver fluke (Fasciola hepatica) infection in cattle. *Vet Parasitol.* 2005 Apr 20;129(1-2):61-6.

[305] Salimi-Bejestani MR, Daniel RG, Felstead SM, Cripps PJ, Mahmoody H, Williams DJ. Prevalence of Fasciola hepatica in dairy herds in England and Wales measured with an ELISA applied to bulk-tank milk. *Vet Rec.* 2005 Jun 4;156(23):729-31.

[306] Salimi-Bejestani MR, Daniel R, Cripps P, Felstead S, Williams DJ. Evaluation of an enzyme-linked immunosorbent assay for detection of antibodies to *Fasciola hepatica* in milk. *Vet Parasitol.* 2007;149(3-4):290-3.

[307] Bennema S, Vercruysse J, Claerebout E, Schnieder T, Strube C, Ducheyne E, et al. The use of bulk-tank milk ELISAs to assess the spatial distribution of Fasciola hepatica, Ostertagia ostertagi and Dictyocaulus viviparus in dairy cattle in Flanders (Belgium). *Vet Parasitol.* 2009 Oct 28;165(1-2):51-7.

[308] McCann CM, Baylis M, Williams DJ. Seroprevalence and spatial distribution of Fasciola hepatica-infected dairy herds in England and Wales. *Vet Rec.* 2010 May 15;166(20):612-7.

[309] Duscher R, Duscher G, Hofer J, Tichy A, Prosl H, Joachim A. Fasciola hepatica - Monitoring the milky way? The use of tank milk for liver fluke monitoring in dairy herds as base for treatment strategies. *Vet Parasitol.* 2011 Jan 26.

[310] Molloy JB, Anderson GR, Fletcher TI, Landmann J, Knight BC. Evaluation of a commercially available enzyme-linked immunosorbent assay for detecting antibodies to Fasciola hepatica and Fasciola gigantica in cattle, sheep and buffaloes in Australia. *Vet Parasitol.* 2005 Jun 30;130(3-4):207-12.

[311] Molloy JB, Anderson GR. The distribution of *Fasciola hepatica* in Queensland, Australia, and the potential impact of introduced snail intermediate hosts. *Vet Parasitol.* 2006;137(1-2):62-6.

[312] Hoglund J, Dahlstrom F, Engstrom A, Hessle A, Jakubek EB, Schnieder T, et al. Antibodies to major pasture borne helminth infections in bulk-tank milk samples from organic and nearby conventional dairy herds in south-central Sweden. *Vet Parasitol.* 2010 Aug 4;171(3-4):293-9.

[313] Bennema SC, Ducheyne E, Vercruysse J, Claerebout E, Hendrickx G, Charlier J. Relative importance of management, meteorological and environmental factors in the spatial distribution of Fasciola hepatica in dairy cattle in a temperate climate zone. *Int J Parasitol.* 2011 Feb;41(2):225-33.

[314] Dumenigo BE, Espino AM, Finlay CM, Mezo M. Kinetics of antibody-based antigen detection in serum and faeces of sheep experimentally infected with *Fasciola hepatica*. *Vet Parasitol.* 2000;89(1-2):153-61.

[315] Paz-Silva A, Pedreira J, Sanchez-Andrade R, Suarez JL, Diaz P, Panadero R, et al. Time-course analysis of coproantigens in rats infected and challenged with *Fasciola hepatica*. *Parasitol Res.* 2002 Jun;88(6):568-73.

[316] Estuningsih SE, Widjayanti S, Adiwinata G, Piedrafita D. Detection of coproantigens by sandwich ELISA in sheep experimentally infected with *Fasciola gigantica*. *Trop Biomed.* 2004;21(2):51-6.

[317] Estuningsih E, Spithill T, Raadsma H, Law R, Adiwinata G, Meeusen E, et al. Development and application of a fecal antigen diagnostic sandwich ELISA for estimating prevalence of *Fasciola gigantica* in cattle in central Java, Indonesia. *J Parasitol.* 2009 Apr;95(2):450-5.

[318] Ubeira FM, Muino L, Valero MA, Periago MV, Perez-Crespo I, Mezo M, et al. MM3-ELISA detection of *Fasciola hepatica* coproantigens in preserved human stool samples. *Am J Trop Med Hyg.* 2009 Jul;81(1):156-62.

[319] Anuracpreeda P, Wanichanon C, Chawengkirtikul R, Chaithirayanon K, Sobhon P. Fasciola gigantica: immunodiagnosis of fasciolosis by detection of circulating 28.5 kDa tegumental antigen. *Exp Parasitol.* 2009 Dec;123(4):334-40.

[320] Allam G, Bauomy IR, Hemyeda ZM, Sakran TF. Evaluation of a 14.5 kDa-Fasciola gigantica fatty acid binding protein as a diagnostic antigen for human fascioliasis. *Parasitol Res.* 2011 Nov 24.

[321] Espino AM, Diaz A, Perez A, Finlay CM. Dynamics of antigenemia and coproantigens during a human *Fasciola hepatica* outbreak. *J Clin Microbiol.* 1998 Sep;36(9):2723-6.

[322] Demerdash ZA, Diab TM, Aly IR, Mohamed SH, Mahmoud FS, Zoheiry MK, et al. Diagnostic efficacy of monoclonal antibody based sandwich enzyme linked immunosorbent assay (ELISA) for detection of *Fasciola gigantica* excretory/secretory antigens in both serum and stool. *Parasit Vectors.* 2011;4:176.

[323] Almazan C, Avila G, Quiroz H, Ibarra F, Ochoa P. Effect of parasite burden on the detection of *Fasciola hepatica* antigens in sera and feces of experimentally infected sheep. *Vet Parasitol.* 2001 May 22;97(2):101-12.

[324] Viyanant V, Krailas D, Sobhon P, Upatham ES, Kusamran T, Chompoochan T, et al. Diagnosis of cattle fasciolosis by the detection of a circulating antigen using a monoclonal antibody. *Asian Pac J Allergy Immunol.* 1997 Sep;15(3):153-9.

[325] Velusamy R, Singh BP, Sharma RL, Chandra D. Detection of circulating 54 kDa antigen in sera of bovine calves experimentally infected with F. gigantica. *Vet Parasitol.* 2004 Jan 30;119(2-3):187-95.

[326] Espino AM, Finlay CM. Sandwich enzyme-linked immunosorbent assay for detection of excretory secretory antigens in humans with fascioliasis. *J Clin Microbiol.* 1994 Jan;32(1):190-3.

[327] Dumenigo BE, Espino AM, Finlay CM. Detection of *Fasciola hepatica* antigen in cattle faeces by a monoclonal antibody-based sandwich immunoassay. *Res Vet Sci.* 1996;60(3):278-9.

[328] Dumenigo BE, Mezo M. Monoclonal antibody sandwich immunoassay detection of coproantigen to evaluate the efficacy of treatment in natural ovine fasciolosis. *Res Vet Sci.* 1999 Apr;66(2):165-7.

[329] Abdel-Rahman SM, O'Reilly KL, Malone JB. Evaluation of a diagnostic monoclonal antibody-based capture enzyme-linked immunosorbent assay for detection of a 26- to 28-kd *Fasciola hepatica* coproantigen in cattle. *Am J Vet Res.* 1998 May;59(5):533-7.

[330] Mezo M, Gonzalez-Warleta M, Carro C, Ubeira FM. An ultrasensitive capture ELISA for detection of *Fasciola hepatica* coproantigens in sheep and cattle using a new monoclonal antibody (MM3). *J Parasitol.* 2004;90(4):845-52.

[331] Mezo M, Gonzalez-Warleta M, Castro-Hermida JA, Ubeira FM. Evaluation of the flukicide treatment policy for dairy cattle in Galicia (NW Spain). *Vet Parasitol.* 2008 Nov 7;157(3-4):235-43.

[332] Espino AM, Marcet R, Finlay CM. Fasciola hepatica: detection of antigenemia and coproantigens in experimentally infected rats. *Exp Parasitol.* 1997 Feb;85(2):117-20.

[333] Charlier J, De Meulemeester L, Claerebout E, Williams D, Vercruysse J. Qualitative and quantitative evaluation of coprological and serological techniques for the diagnosis of fasciolosis in cattle. *Vet Parasitol.* 2008 May 6;153(1-2):44-51.

[334] Flanagan AM, Edgar HW, Forster F, Gordon A, Hanna RE, McCoy M, et al. Standardisation of a coproantigen reduction test (CRT) protocol for the diagnosis of resistance to triclabendazole in Fasciola hepatica. *Vet Parasitol.* 2011 Feb 28;176(1):34-42.

[335] McGarry JW, Ortiz PL, Hodgkinson JE, Goreish I, Williams DJ. PCR-based differentiation of *Fasciola* species (Trematoda: Fasciolidae), using primers based on RAPD-derived sequences. *Ann Trop Med Parasitol.* 2007;101(5):415-21.

[336] Rokni MB, Mirhendi H, Mizani A, Mohebali M, Sharbatkhori M, Kia EB, et al. Identification and differentiation of *Fasciola hepatica* and *Fasciola gigantica* using a simple PCR-restriction enzyme method. *Exp Parasitol.* 2010 Feb;124(2):209-13.

[337] Ichikawa M, Itagaki T. Discrimination of the ITS1 types of Fasciola spp. based on a PCR-RFLP method. *Parasitol Res.* 2010 Feb;106(3):757-61.

[338] Barger IA. Prospects for integration of novel parasite control options into grazing systems. *Int J Parasitol.* 1996;26(8-9):1001-7.

[339] Roberts JA, Suhardono. Approaches to the control of fasciolosis in ruminants. *Int J Parasitol.* 1996 Aug-Sep;26(8-9):971-81.

[340] Fuentes MV. Remote sensing and climate data as a key for understanding fasciolosis transmission in the Andes: review and update of an ongoing interdisciplinary project. *Geospat Health.* 2006 Nov;1(1):59-70.

[341] Dutra LH, Molento MB, Naumann CR, Biondo AW, Fortes FS, Savio D, et al. Mapping risk of bovine fasciolosis in the south of Brazil using Geographic Information Systems. *Vet Parasitol.* 2010 Apr 19;169(1-2):76-81.

[342] McCann CM, Baylis M, Williams DJ. The development of linear regression models using environmental variables to explain the spatial distribution of Fasciola hepatica infection in dairy herds in England and Wales. *Int J Parasitol.* 2010 Aug 1;40(9):1021-8.

[343] Fairweather I, Boray JC. Mechanisms of fasciolicide action and drug resistance in *Fasciola hepatica*. In: Dalton JP, editor. *Fasciolosis.* Wallingford: CABI Publishing; 1999. p. 225-76.

[344] Fairweather I. Triclabendazole: new skills to unravel an old(ish) enigma. *J Helminthol.* 2005 Sep;79(3):227-34.

[345] Wolff K, Eckert J, Schneiter G, Lutz H. Efficacy of triclabendazole against Fasciola hepatica in sheep and goats. *Vet Parasitol.* 1983 Sep;13(2):145-50.

[346] Richards RJ, Bowen FL, Essenwein F, Steiger RF, Buscher G. The efficacy of triclabendazole and other anthelmintics against Fasciola hepatica in controlled studies in cattle. *Vet Rec.* 1990 Mar 3;126(9):213-6.

[347] Sanyal PK, Gupta SC. Efficacy and pharmacokinetics of triclabendazole in buffalo with induced fasciolosis. *Vet Parasitol.* 1996 May;63(1-2):75-82.

[348] Apt W, Aguilera X, Vega F, Miranda C, Zulantay I, Perez C, et al. Treatment of human chronic fascioliasis with triclabendazole: drug efficacy and serologic response. *Am J Trop Med Hyg.* 1995 Jun;52(6):532-5.

[349] Jave OJ, Alban OM, Sagastegui PC, Soriano YS. [Treatment of Human Hepatic Fasciolasis with Triclabendazole]. *Rev Gastroenterol Peru.* 1999 Jul-Sep;19(3):216-20.

[350] el-Morshedy H, Farghaly A, Sharaf S, Abou-Basha L, Barakat R. Triclabendazole in the treatment of human fascioliasis: a community-based study. *East Mediterr Health J.* 1999 Sep;5(5):888-94.

[351] Keiser J, Engels D, Buscher G, Utzinger J. Triclabendazole for the treatment of fascioliasis and paragonimiasis. *Expert Opin Investig Drugs.* 2005 Dec;14(12):1513-26.

[352] Stitt AW, Fairweather I. The effect of the sulphoxide metabolite of triclabendazole ('Fasinex') on the tegument of mature and immature stages of the liver fluke, *Fasciola hepatica. Parasitology.* 1994;108(5):555-67.

[353] Stitt AW, Fairweather I, Mackender RO. The effect of triclabendazole ("Fasinex") on protein synthesis by the liver fluke, Fasciola hepatica. *Int J Parasitol.* 1995 Apr;25(4):421-9.

[354] Stitt AW, Fairweather I. *Fasciola hepatica*: disruption of the vitelline cells in vitro by the sulphoxide metabolite of triclabendazole. *Parasitol Res.* 1996;82(4):333-9.

[355] Fairweather I. Triclabendazole progress report, 2005-2009: an advancement of learning? *J Helminthol.* 2009 Jun;83(2):139-50.

[356] Fairweather I. Reducing the future threat from (liver) fluke: realistic prospect or quixotic fantasy? *Vet Parasitol.* 2011 Aug 4;180(1-2):133-43.

[357] Halferty L, Brennan GP, Hanna RE, Edgar HW, Meaney MM, McConville M, et al. Tegumental surface changes in juvenile Fasciola hepatica in response to treatment in vivo with triclabendazole. *Vet Parasitol.* 2008 Aug 1;155(1-2):49-58.

[358] Toner E, Brennan GP, Hanna RE, Edgar HW, Fairweather I. Time-dependent changes to the tegumental system and gastrodermis of adult Fasciola hepatica following treatment in vivo with triclabendazole in the sheep host. *Vet Parasitol.* 2010 Dec 15;174(3-4):218-27.

[359] Toner E, Brennan GP, Hanna RE, Edgar HW, Fairweather I. Tegumental surface changes in adult Fasciola hepatica in response to treatment in vivo with triclabendazole in the sheep host. *Vet Parasitol.* 2010 Sep 20;172(3-4):238-48.

[360] Toner E, Brennan GP, Hanna RE, Edgar HW, Fairweather I. *Fasciola hepatica*: time-dependent disruption of spermatogenesis following in vivo treatment with triclabendazole. *Parasitol Res.* 2011 Oct;109(4):1035-43.

[361] Toner E, Brennan GP, Hanna RE, Edgar HW, Fairweather I. Disruption of egg formation by Fasciola hepatica following treatment in vivo with triclabendazole in the sheep host. *Vet Parasitol.* 2011 Apr 19;177(1-2):79-89.

[362] Hanna RE, Scarcella S, Solana H, McConnell S, Fairweather I. Early onset of changes to the reproductive system of Fasciola hepatica following in vivo treatment with triclabendazole. *Vet Parasitol.* 2011 Aug 19.
[363] Coles GC. Anthelmintic resistance--looking to the future: a UK perspective. *Res Vet Sci.* 2005 Apr;78(2):99-108.
[364] Overend DJ, Bowen FL. Resistance of *Fasciola hepatica* to triclabendazole. *Aust Vet J.* 1995;72(7):275-6.
[365] Moll L, Gaasenbeek CP, Vellema P, Borgsteede FH. Resistance of *Fasciola hepatica* against triclabendazole in cattle and sheep in The netherlands. *Vet Parasitol.* 2000;91(1-2):153-8.
[366] Alvarez-Sanchez MA, Mainar-Jaime RC, Perez-Garcia J, Rojo-Vazquez FA. Resistance of *Fasciola hepatica* to triclabendazole and albendazole in sheep in Spain. *Vet Rec.* 2006;159(13):424-5.
[367] Mitchell GB, Maris L, Bonniwell MA. Triclabendazole-resistant liver fluke in Scottish sheep. *Vet Rec.* 1998 Oct 3;143(14):399.
[368] Sargison ND, Wilson DJ, Penny CD, Bartley DJ. Unexpected production loss caused by helminth parasites in weaned beef calves. *Vet Rec.* 2010 Nov 6;167(19):752-4.
[369] Thomas I, Coles GC, Duffus K. Triclabendazole-resistant *Fasciola hepatica* in southwest Wales. *Vet Rec.* 2000;146(7):200.
[370] Lane G. Anthelmintic resistance. Vet Rec. 1998;143:332.
[371] O'Brien DJ. Fasciolosis: a threat to livestock. *Irish Vet J.* 1998;51(10):539-41.
[372] Mooney L, Good B, Hanrahan JP, Mulcahy G, de Waal T. The comparative efficacy of four anthelmintics against a natural acquired Fasciola hepatica infection in hill sheep flock in the west of Ireland. *Vet Parasitol.* 2009 Oct 14;164(2-4):201-5.
[373] Olaechea F, Lovera V, Larroza M, Raffo F, Cabrera R. Resistance of Fasciola hepatica against triclabendazole in cattle in Patagonia (Argentina). *Vet Parasitol.* 2011 Jan 11.
[374] Mottier L, Alvarez L, Fairweather I, Lanusse C. Resistance-induced changes in triclabendazole transport in *Fasciola hepatica*: ivermectin reversal effect. *J Parasitol.* 2006 Dec;92(6):1355-60.
[375] Robinson MW, Lawson J, Trudgett A, Hoey EM, Fairweather I. The comparative metabolism of triclabendazole sulphoxide by triclabendazole-susceptible and triclabendazole-resistant *Fasciola hepatica*. *Parasitol Res.* 2004 Feb;92(3):205-10.
[376] Alvarez LI, Solana HD, Mottier ML, Virkel GL, Fairweather I, Lanusse CE. Altered drug influx/efflux and enhanced metabolic activity in triclabendazole-resistant liver flukes. *Parasitology.* 2005;131(4):501-10.
[377] Devine C, Brennan GP, Lanusse CE, Alvarez LI, Trudgett A, Hoey E, et al. Effect of the metabolic inhibitor, methimazole on the drug susceptibility of a triclabendazole-resistant isolate of *Fasciola hepatica*. *Parasitology.* 2009 Feb;136(2):183-92.
[378] Devine C, Brennan GP, Lanusse CE, Alvarez LI, Trudgett A, Hoey E, et al. Enhancement of the drug susceptibility of a triclabendazole-resistant isolate of *Fasciola hepatica* using the metabolic inhibitor ketoconazole. *Parasitol Res.* 2010 Jul;107(2):337-53.
[379] Devine C, Brennan GP, Lanusse CE, Alvarez LI, Trudgett A, Hoey E, et al. Potentiation of triclabendazole sulphoxide-induced tegumental disruption by methimazole in a triclabendazole-resistant isolate of *Fasciola hepatica*. *Parasitol Res.* 2010 May;106(6):1351-63.

[380] Devine C, Brennan GP, Lanusse CE, Alvarez LI, Trudgett A, Hoey E, et al. Inhibition of cytochrome P450-mediated metabolism enhances ex vivo susceptibility of *Fasciola hepatica* to triclabendazole. *Parasitology.* 2010 Apr;137(5):871-80.

[381] Devine C, Brennan GP, Lanusse CE, Alvarez LI, Trudgett A, Hoey E, et al. Piperonyl butoxide enhances triclabendazole action against triclabendazole-resistant *Fasciola hepatica. Parasitology.* 2011 Feb;138(2):224-36.

[382] Devine C, Brennan GP, Lanusse CE, Alvarez LI, Trudgett A, Hoey E, et al. Erratum to: inhibition of triclabendazole metabolism in vitro by ketoconazole increases disruption to the tegument of a triclabendazole-resistant isolate of *Fasciola hepatica. Parasitol Res.* 2011 Oct;109(4):1209-23.

[383] Devine C, Brennan GP, Lanusse CE, Alvarez LI, Trudgett A, Hoey E, et al. Inhibition of triclabendazole metabolism in vitro by ketoconazole increases disruption to the tegument of a triclabendazole-resistant isolate of *Fasciola hepatica. Parasitol Res.* 2011 Oct;109(4):981-95.

[384] Devine C, Brennan GP, Lanusse CE, Alvarez LI, Trudgett A, Hoey E, et al. Enhancement of triclabendazole action in vivo against a triclabendazole-resistant isolate of Fasciola hepatica by co-treatment with ketoconazole. *Vet Parasitol.* 2011 May 11;177(3-4):305-15.

[385] Devine C, Brennan GP, Lanusse CE, Alvarez LI, Trudgett A, Hoey E, et al. Potentiation of triclabendazole action in vivo against a triclabendazole-resistant isolate of Fasciola hepatica following its co-administration with the metabolic inhibitor, ketoconazole. *Vet Parasitol.* 2012 Feb 28;184(1):37-47.

[386] Brennan GP, Fairweather I, Trudgett A, Hoey E, McCoy, McConville M, et al. Understanding triclabendazole resistance. *Exp Mol Pathol.* 2007 Apr;82(2):104-9.

[387] Coles GC, Rhodes AC, Stafford KA. Activity of closantel against adult triclabendazole-resistant Fasciola hepatica. *Vet Rec.* 2000 Apr 22;146(17):504.

[388] Coles GC, Stafford KA. Activity of oxyclozanide, nitroxynil, clorsulon and albendazole against adult triclabendazole-resistant Fasciola hepatica. *Vet Rec.* 2001 Jun 9;148(23):723-4.

[389] Martinez-Valladares M, del Rosario Famularo M, Fernandez-Pato N, Castanon-Ordonez L, Cordero-Perez C, Rojo-Vazquez FA. Efficacy of nitroxynil against *Fasciola hepatica* resistant to triclabendazole in a naturally infected sheep flock. *Parasitol Res.* 2010 Oct;107(5):1205-11.

[390] Mansour-Ghanaei F, Shafaghi A, Fallah M. The effect of metronidazole in treating human fascioliasis. *Med Sci Monit.* 2003 Oct;9(10):PI127-30.

[391] Saba R, Korkmaz M, Inan D, Mamikoglu L, Turhan O, Gunseren F, et al. Human fascioliasis. *Clin Microbiol Infect.* 2004 May;10(5):385-7.

[392] Hernandez-Campos A, Ibarra-Velarde F, Vera-Montenegro Y, Rivera-Fernandez N, Castillo R. Synthesis and fasciolicidal activity of 5-chloro-2-methylthio-6-(1-naphthyloxy)-1H-benzimidazole. *Chem Pharm Bull (Tokyo).* 2002 May;50(5):649-52.

[393] Ibarra F, Vera Y, Quiroz H, Canto J, Castillo R, Hernandez A, et al. Determination of the effective dose of an experimental fasciolicide in naturally and experimentally infected cattle. *Vet Parasitol.* 2004 Feb 26;120(1-2):65-74.

[394] Vera Montenegro Y, Ibarra Velarde F, Liebano Hernandez E, Quiroz Romero H, Castillo Bocanegra R, Hernandez Campos A, et al. Efficacy of an experimental

fasciolicide against immature and mature *Fasciola hepatica* in artificially infected calves. *Parasitol Res.* 2004 Feb;92(3):211-4.

[395] Keiser J, Shu-Hua X, Tanner M, Utzinger J. Artesunate and artemether are effective fasciolicides in the rat model and in vitro. *J Antimicrob Chemother.* 2006 Jun;57(6):1139-45.

[396] Keiser J, Utzinger J, Vennerstrom JL, Dong Y, Brennan G, Fairweather I. Activity of artemether and OZ78 against triclabendazole-resistant *Fasciola hepatica. Trans R Soc Trop Med Hyg.* 2007 Dec;101(12):1219-22.

[397] Keiser J, Morson G. Fasciola hepatica: tegumental alterations in adult flukes following in vitro and in vivo administration of artesunate and artemether. *Exp Parasitol.* 2008 Feb;118(2):228-37.

[398] Keiser J, Rinaldi L, Veneziano V, Mezzino L, Tanner M, Utzinger J, et al. Efficacy and safety of artemether against a natural *Fasciola hepatica* infection in sheep. *Parasitol Res.* 2008 Aug;103(3):517-22.

[399] Keiser J, Gruyer MS, Perrottet N, Zanolari B, Mercier T, Decosterd L. Pharmacokinetic parameters of artesunate and dihydroartemisinin in rats infected with Fasciola hepatica. *J Antimicrob Chemother.* 2009 Mar;63(3):543-9.

[400] Keiser J, Veneziano V, Rinaldi L, Mezzino L, Duthaler U, Cringoli G. Anthelmintic activity of artesunate against *Fasciola hepatica* in naturally infected sheep. *Res Vet Sci.* 2010 Feb;88(1):107-10.

[401] O'Neill JF, Johnston RC, Halferty L, Brennan GP, Keiser J, Fairweather I. Adult triclabendazole-resistant *Fasciola hepatica*: morphological changes in the tegument and gut following in vivo treatment with artemether in the rat model. *J Helminthol.* 2009 Jun;83(2):151-63.

[402] Shalaby HA, El Namaky AH, Kamel RO. In vitro effect of artemether and triclabendazole on adult Fasciola gigantica. *Vet Parasitol.* 2009 Mar 9;160(1-2):76-82.

[403] Hien TT, Truong NT, Minh NH, Dat HD, Dung NT, Hue NT, et al. A randomized controlled pilot study of artesunate versus triclabendazole for human fascioliasis in central Vietnam. *Am J Trop Med Hyg.* 2008 Mar;78(3):388-92.

[404] Keiser J, Sayed H, el-Ghanam M, Sabry H, Anani S, el-Wakeel A, et al. Efficacy and safety of artemether in the treatment of chronic fascioliasis in Egypt: exploratory phase-2 trials. *PLoS Negl Trop Dis.* 2011 Sep;5(9):e1285.

[405] Duthaler U, Smith TA, Keiser J. In vivo and in vitro sensitivity of *Fasciola hepatica* to triclabendazole combined with artesunate, artemether, or OZ78. *Antimicrob Agents Chemother.* 2010 Nov;54(11):4596-604.

[406] Rossignol JF, Abaza H, Friedman H. Successful treatment of human fascioliasis with nitazoxanide. *Trans R Soc Trop Med Hyg.* 1998 Jan-Feb;92(1):103-4.

[407] Kabil SM, Ashry EE, Ashraf NK. An open-label clinical study of nitazoxanide in the treatment of human fascioliasis. *Curr Therap Res.* 2000;61(6):339-45.

[408] Favennec L, Jave Ortiz J, Gargala G, Lopez Chegne N, Ayoub A, Rossignol JF. Double-blind, randomized, placebo-controlled study of nitazoxanide in the treatment of fascioliasis in adults and children from northern Peru. *Aliment Pharmacol Ther.* 2003 Jan;17(2):265-70.

[409] Massoud A, El Sisi S, Salama O. Preliminary study of therapeutic efficacy of a new fasciolicidal drug derived from Commiphora molmol (myrrh). *Am J Trop Med Hyg.* 2001 Aug;65(2):96-9.

[410] Soliman OE, El-Arman M, Abdul-Samie ER, El-Nemr HI, Massoud A. Evaluation of myrrh (Mirazid) therapy in fascioliasis and intestinal schistosomiasis in children: immunological and parasitological study. *J Egypt Soc Parasitol.* 2004 Dec;34(3):941-66.

[411] Abo-Madyan AA, Morsy TA, Motawea SM, Morsy AT. Clinical trial of Mirazid in treatment of human fascioliasis, Ezbet El-Bakly (Tamyia Center) Al-Fayoum Governorate. *J Egypt Soc Parasitol.* 2004 Dec;34(3):807-18.

[412] El-Mathal EM, Fouad MA. Human fascioliasis among immigrant workers in Saudi Arabia. *J Egypt Soc Parasitol.* 2005 Dec;35(3 Suppl):1199-207.

[413] Abdul-Ghani RA, Loutfy N, Hassan A. Myrrh and trematodoses in Egypt: an overview of safety, efficacy and effectiveness profiles. *Parasitol Int.* 2009 Sep;58(3):210-4.

[414] Massoud AM, El-Sherbini ET, Mos N, Saleh NM, Abouel-Nour MF, Morsy AT. Mirazid in treatment of three zoonotic trematodes in Beni-Sweif and Dakhalia Governorates. *J Egypt Soc Parasitol.* 2010 Apr;40(1):119-34.

[415] Botros SS, El-Lakkany NM, Badawy AA, Mahmoud SS, Ebeid FA, Fenwick A. Mirazid shows insignificant activity against ovine fascioliasis. *Ann Trop Med Parasitol.* 2009 Oct;103(7):605-16.

[416] Osman MM, El-Taweel HA, Shehab AY, Farag HF. Ineffectiveness of myrrh-derivative Mirazid against schistosomiasis and fascioliasis in humans. *East Mediterr Health J.* 2010 Sep;16(9):932-6.

[417] Stevenson CR, Mahoney RH, Fisara P, Strehlau G, Reichel MP. The efficacy of formulations of triclabendazole and ivermectin in combination against liver fluke (*Fasciola hepatica*) and gastro-intestinal nematodes in cattle and sheep and sucking lice species in cattle. *Aust Vet J.* 2002 Nov;80(11):698-701.

[418] Meaney M, Allister J, McKinstry B, McLaughlin K, Brennan GP, Forbes AB, et al. *Fasciola hepatica*: morphological effects of a combination of triclabendazole and clorsulon against mature fluke. *Parasitol Res.* 2006;99(5):609-21.

[419] Meaney M, Allister J, McKinstry B, McLaughlin K, Brennan GP, Forbes AB, et al. *Fasciola hepatica*: ultrastructural effects of a combination of triclabendazole and clorsulon against mature fluke. *Parasitol Res.* 2007 Apr;100(5):1091-104.

[420] Hutchinson GW, Dawson K, Fitzgibbon CC, Martin PJ. Efficacy of an injectable combination anthelmintic (nitroxynil+clorsulon+ivermectin) against early immature Fasciola hepatica compared to triclabendazole combination flukicides given orally or topically to cattle. *Vet Parasitol.* 2009 Jun 10;162(3-4):278-84.

[421] Fairweather I, Boray JC. Fasciolicides: efficacy, actions, resistance and its management. *Vet J.* 1999 Sep;158(2):81-112.

[422] Virkel G, Lifschitz A, Sallovitz J, Ballent M, Scarcella S, Lanusse C. Inhibition of cytochrome P450 activity enhances the systemic availability of triclabendazole metabolites in sheep. *J Vet Pharmacol Ther.* 2009 Feb;32(1):79-86.

[423] Lifschitz A, Virkel G, Ballent M, Sallovitz J, Lanusse C. Combined use of ivermectin and triclabendazole in sheep: in vitro and in vivo characterisation of their pharmacological interaction. *Veterinary Journal.* 2009 Nov;182(2):261-8.

[424] de Vasconcellos MC, de Amorim A. Activity of *Euphorbia splendens* var. *hislopii* N.E.B. (Euphorbiaceae) latex against *Lymnaea columella* (Say, 1817) (*Pulmonata: Lymnaeidae*), intermediate host of *Fasciola hepatica*, Linnaeus, 1758 (Trematoda: Fasciolidae). 2: limited field-testing. *Mem Inst Oswaldo Cruz.* 2003 Oct;98(7):981-5.

[425] Kumar P, Singh DK. Molluscicidal activity of Ferula asafoetida, Syzygium aromaticum and Carum carvi and their active components against the snail Lymnaea acuminata. *Chemosphere.* 2006 Jun;63(9):1568-74.
[426] Hammami H, Ayadi A. Molluscicidal and antiparasitic activity of Solanum nigrum villosum against *Galba truncatula* infected or uninfected with *Fasciola hepatica. J Helminthol.* 2008 Sep;82(3):235-9.
[427] Kumar P, Singh VK, Singh DK. Combination of molluscicides with attractant carbohydrates and amino acids in bait formulation against the snail *Lymnaea acuminata. Eur Rev Med Pharmacol Sci.* 2011 May;15(5):550-5.
[428] Singh SK, Yadav RP, Singh A. Molluscicides from some common medicinal plants of eastern Uttar Pradesh, India. *J Appl Toxicol.* 2010 Jan;30(1):1-7.
[429] Sunita K, Singh DK. Fascioliasis control: in vivo and in vitro phytotherapy of vector snail to kill *fasciola* larva. *J Parasitol Res.* 2011;2011:240807.
[430] Waller PJ. Anthelmintic resistance. *Vet Parasitol.* 1997;72(3-4):391-405.
[431] Suhardono, Roberts JA, Copeman DB. Biological control of Fasciola gigantica with Echinostoma revolutum. *Vet Parasitol.* 2006 Aug 31;140(1-2):166-70.
[432] Calienes AF, Fraga J, Pointier JP, Yong M, Sanchez J, Coustau C, et al. Detection and genetic distance of resistant populations of *Pseudosuccinea columella* (Mollusca: Lymnaeidae) to *Fasciola hepatica* (Trematoda: Digenea) using RAPD markers. *Acta Trop.* 2004 Sep;92(1):83-7.
[433] Bargues MD, Artigas P, Khoubbane M, Flores R, Gloer P, Rojas-Garcia R, et al. Lymnaea schirazensis, an overlooked snail distorting fascioliasis data: genotype, phenotype, ecology, worldwide spread, susceptibility, applicability. *PLoS One.* 2011;6(9):e24567.
[434] Roberts JA, Widjayanti S, Estuningsih E, Hetzel DJ. Evidence for a major gene determining the resistance of Indonesian Thin Tail sheep against Fasciola gigantica. *Vet Parasitol.* 1997 Mar;68(4):309-14.
[435] Graczyk TK, Fried B. Human Waterborne Trematode and Protozoan Infections. *Adv Parasitol.* 2007;64:111-60.
[436] Curtale F, Hassanein YA, Savioli L. Control of human fascioliasis by selective chemotherapy: design, cost and effect of the first public health, school-based intervention implemented in endemic areas of the Nile Delta, Egypt. *Trans R Soc Trop Med Hyg.* 2005 Aug;99(8):599-609.
[437] Wijffels GL, Salvatore L, Dosen M, Waddington J, Wilson L, Thompson C, et al. Vaccination of sheep with purified cysteine proteinases of *Fasciola hepatica* decreases worm fecundity. *Exp Parasitol.* 1994;78(2):132-48.
[438] Dalton JP, McGonigle S, Rolph TP, Andrews SJ. Induction of protective immunity in cattle against infection with *Fasciola hepatica* by vaccination with cathepsin L proteinases and with hemoglobin. *Infect Immun.* 1996;64(12):5066-74.
[439] Piacenza L, Acosta D, Basmadjian I, Dalton JP, Carmona C. Vaccination with cathepsin L proteinases and with leucine aminopeptidase induces high levels of protection against fascioliasis in sheep. *Infect Immun.* 1999 Apr;67(4):1954-61.
[440] van Milligen FJ, Cornelissen JB, Bokhout BA. Fasciola hepatica: an antigen fraction derived from newly excysted juveniles, containing an immunoreactive 32-kDa protein, induces strong protective immunity in rats. *Exp Parasitol.* 2000 Mar;94(3):163-71.

[441] Golden O, Flynn RJ, Read C, Sekiya M, Donnelly SM, Stack C, et al. Protection of cattle against a natural infection of *Fasciola hepatica* by vaccination with recombinant cathepsin L1 (rFhCL1). *Vaccine.* 2010 Aug 2;28(34):5551-7.

[442] Jayaraj R, Piedrafita D, Dynon K, Grams R, Spithill TW, Smooker PM. Vaccination against fasciolosis by a multivalent vaccine of stage-specific antigens. *Vet Parasitol.* 2009 Mar 23;160(3-4):230-6.

[443] Reszka N, Cornelissen JB, Harmsen MM, Bienkowska-Szewczyk K, de Bree J, Boersma WJ, et al. *Fasciola hepatica* procathepsin L3 protein expressed by a baculovirus recombinant can partly protect rats against fasciolosis. *Vaccine.* 2005 Apr 27;23(23):2987-93.

[444] Wedrychowicz H, Kesik M, Kaliniak M, Kozak-Cieszczyk M, Jedlina-Panasiuk L, Jaros S, et al. Vaccine potential of inclusion bodies containing cysteine proteinase of Fasciola hepatica in calves and lambs experimentally challenged with metacercariae of the fluke. *Vet Parasitol.* 2007 Jun 20;147(1-2):77-88.

[445] Dabrowska M, Kaliniak M, Wedrychowicz H. [Influence of vaccination of calves with recombinant cysteine proteinase of Fasciola hepatica on development and infectivity of miracidia]. *Wiad Parazytol.* 2006;52(4):305-9.

[446] Wedrychowicz H, Lamparska M, Kesik M, Kotomski G, Mieszczanek J, Jedlina-Panasiuk L, et al. The immune response of rats to vaccination with the cDNA or protein forms of the cysteine proteinase of *Fasciola hepatica. Vet Immunol Immunopathol.* 2003 Jul 15;94(1-2):83-93.

[447] Kesik M, Jedlina-Panasiuk L, Kozak-Cieszczyk M, Plucienniczak A, Wedrychowicz H. Enteral vaccination of rats against *Fasciola hepatica* using recombinant cysteine proteinase (cathepsin L1). *Vaccine.* 2007 May 4;25(18):3619-28.

[448] Kofta W, Mieszczanek J, Plucienniczak G, Wedrychowicz H. Successful DNA immunisation of rats against fasciolosis. *Vaccine.* 2000 Jul 1;18(26):2985-90.

[449] Harmsen MM, Cornelissen JB, Buijs HE, Boersma WJ, Jeurissen SH, van Milligen FJ. Identification of a novel Fasciola hepatica cathepsin L protease containing protective epitopes within the propeptide. *Int J Parasitol. 2004* May;34(6):675-82.

[450] Villa-Mancera A, Quiroz-Romero H, Correa D, Ibarra F, Reyes-Perez M, Reyes-Vivas H, et al. Induction of immunity in sheep to *Fasciola hepatica* with mimotopes of cathepsin L selected from a phage display library. *Parasitology.* 2008 Oct;135(12):1437-45.

[451] Morphew RM, Wright HA, Lacourse EJ, Porter J, Barrett J, Woods DJ, et al. Towards delineating functions within the fasciola secreted cathepsin l protease family by integrating in vivo based sub-proteomics and phylogenetics. *PLoS Negl Trop Dis.* 2011;5(1):e937.

[452] Hillyer GV, Haroun ET, Hernandez A, de Galanes MS. Acquired resistance to *Fasciola hepatica* in cattle using a purified adult worm antigen. *Am J Trop Med Hyg.* 1987 Sep;37(2):363-9.

[453] Hillyer GV. Induction of immunity in mice to *Fasciola hepatica* with a *Fasciola/Schistosoma* cross-reactive defined immunity antigen. *Am J Trop Med Hyg.* 1985 Nov;34(6):1127-31.

[454] Muro A, Ramajo V, Lopez J, Simon F, Hillyer GV. Fasciola hepatica: vaccination of rabbits with native and recombinant antigens related to fatty acid binding proteins. *Vet Parasitol.* 1997 May;69(3-4):219-29.

[455] Ramajo V, Oleaga A, Casanueva P, Hillyer GV, Muro A. Vaccination of sheep against Fasciola hepatica with homologous fatty acid binding proteins. *Vet Parasitol.* 2001 May 9;97(1):35-46.

[456] Casanueva R, Hillyer GV, Ramajo V, Oleaga A, Espinoza EY, Muro A. Immunoprophylaxis against *Fasciola hepatica* in rabbits using a recombinant Fh15 fatty acid-binding protein. *J Parasitol.* 2001 Jun;87(3):697-700.

[457] Lopez-Aban J, Casanueva P, Nogal J, Arias M, Morrondo P, Diez-Banos P, et al. Progress in the development of Fasciola hepatica vaccine using recombinant fatty acid binding protein with the adjuvant adaptation system ADAD. *Vet Parasitol.* 2007 Apr 30;145(3-4):287-96.

[458] Muro A, Casanueva P, Lopez-Aban J, Ramajo V, Martinez-Fernandez AR, Hillyer GV. Identification of Fasciola hepatica recombinant 15-kDa fatty acid-binding protein T-cell epitopes that protect against experimental fascioliasis in rabbits and mice. *J Parasitol.* 2007 Aug;93(4):817-23.

[459] Lopez-Aban J, Esteban A, Vicente B, Rojas-Caraballo J, Del Olmo E, Martinez-Fernandez AR, et al. Adaptive Immune Stimulation Is Required to Obtain High Protection with Fatty Acid Binding Protein Vaccine Candidate against *Fasciola hepatica* in Balb/C Mice. *J Parasitol.* 2011 Dec 22.

[460] Hillyer GV, Garcia Rosa MI, Alicea H, Hernandez A. Successful vaccination against murine *Schistosoma mansoni* infection with a purified 12 Kd *Fasciola hepatica* cross-reactive antigen. *Am J Trop Med Hyg.* 1988 Jan;38(1):103-10.

[461] Aban JL, Ramajo V, Arellano JL, Oleaga A, Hillyer GV, Muro A. A fatty acid binding protein from *Fasciola hepatica* induced protection in C57/BL mice from challenge infection with *Schistosoma bovis*. *Vet Parasitol.* 1999 Jun 15;83(2):107-21.

[462] Abane JL, Oleaga A, Ramajo V, Casanueva P, Arellano JL, Hillyer GV, et al. Vaccination of mice against schistosoma bovis with a recombinant fatty acid binding protein from *Fasciola hepatica*. *Vet Parasitol.* 2000 Jul 24;91(1-2):33-42.

[463] Tendler M, Vilar MM, Brito CA, Freire NM, Katz N, Simpson A. Vaccination against schistosomiasis and fascioliasis with the new recombinant antigen Sm14: potential basis of a multi-valent anti-helminth vaccine? *Mem Inst Oswaldo Cruz.* 1995 Mar-Apr;90(2):255-6.

[464] Tendler M, Brito CA, Vilar MM, Serra-Freire N, Diogo CM, Almeida MS, et al. A Schistosoma mansoni fatty acid-binding protein, Sm14, is the potential basis of a dual-purpose anti-helminth vaccine. *Proc Natl Acad Sci U S A.* 1996 Jan 9;93(1):269-73.

[465] Vilar MM, Barrientos F, Almeida M, Thaumaturgo N, Simpson A, Garratt R, et al. An experimental bivalent peptide vaccine against schistosomiasis and fascioliasis. *Vaccine.* 2003 Dec 8;22(1):137-44.

[466] Almeida MS, Torloni H, Lee-Ho P, Vilar MM, Thaumaturgo N, Simpson AJ, et al. Vaccination against *Fasciola hepatica* infection using a *Schistosoma mansoni* defined recombinant antigen, Sm14. *Parasite Immunol.* 2003 Mar;25(3):135-7.

[467] Sexton JL, Wilce MC, Colin T, Wijffels GL, Salvatore L, Feil S, et al. Vaccination of sheep against *Fasciola hepatica* with glutathione S-transferase. Identification and mapping of antibody epitopes on a three-dimensional model of the antigen. *J Immunol.* 1994 Feb 15;152(4):1861-72.

[468] Morrison CA, Colin T, Sexton JL, Bowen F, Wicker J, Friedel T, et al. Protection of cattle against *Fasciola hepatica* infection by vaccination with glutathione S-transferase. *Vaccine.* 1996 Dec;14(17-18):1603-12.

[469] Spithill TW, Smooker PM, Sexton JL, Bozas E, Morrison CA, Creaney J, et al. Development of vaccines against *Fasciola hepatica*. In: Dalton JP, editor. *Fasciolosis.* Wallingford: CABI Publishing; 1999. p. 377-410.

[470] De Bont J, Claerebout E, Riveau G, Schacht AM, Smets K, Conder G, et al. Failure of a recombinant Schistosoma bovis-derived glutathione S-transferase to protect cattle against experimental Fasciola hepatica infection. *Vet Parasitol.* 2003 Apr 18;113(2):135-44.

[471] Preyavichyapugdee N, Sahaphong S, Riengrojpitak S, Grams R, Viyanant V, Sobhon P. Fasciola gigantica and Schistosoma mansoni: vaccine potential of recombinant glutathione S-transferase (rFgGST26) against infections in mice. *Exp Parasitol.* 2008 Jun;119(2):229-37.

[472] Estuningsih SE, Smooker PM, Wiedosari E, Widjajanti S, Vaiano S, Partoutomo S, et al. Evaluation of antigens of *Fasciola gigantica* as vaccines against tropical fasciolosis in cattle. *Int J Parasitol.* 1997;27(11):1419-28.

[473] Jaros S, Jaros D, Wesolowska A, Zygner W, Wedrychowicz H. Blocking Fasciola hepatica's energy metabolism - a pilot study of vaccine potential of a novel gene - phosphoglycerate kinase. *Vet Parasitol.* 2010 Sep 20;172(3-4):229-37.

[474] Dewilde S, Ioanitescu AI, Kiger L, Gilany K, Marden MC, Van Doorslaer S, et al. The hemoglobins of the trematodes *Fasciola hepatica* and *Paramphistomum epiclitum*: a molecular biological, physico-chemical, kinetic, and vaccination study. *Protein Sci.* 2008 Oct;17(10):1653-62.

[475] Espino AM, Hillyer GV. A novel *Fasciola hepatica* saposinlike recombinant protein with immunoprophylactic potential. *J Parasitol.* 2004 Aug;90(4):876-9.

[476] Espino AM, Morales A, Delgado B, Rivera FM, Figueroa O, Suarez E. Partial immunity to Fasciola hepatica in mice after vaccination with FhSAP2 delivered as recombinant protein or DNA construct. *Ethn Dis.* 2010 Winter;20(1 Suppl 1):S1-17-23.

[477] Espino AM, Osuna A, Gil R, Hillyer GV. Fasciola hepatica: humoral and cytokine responses to a member of the saposin-like protein family following delivery as a DNA vaccine in mice. *Exp Parasitol.* 2005 Aug;110(4):374-83.

[478] Espino AM, Rivera F. Quantitation of cytokine mRNA by real-time RT-PCR during a vaccination trial in a rabbit model of fascioliasis. *Vet Parasitol.* 2010 Apr 19;169(1-2):82-92.

[479] Mendes RE, Perez-Ecija RA, Zafra R, Buffoni L, Martinez-Moreno A, Dalton JP, et al. Evaluation of hepatic changes and local and systemic immune responses in goats immunized with recombinant Peroxiredoxin (Prx) and challenged with *Fasciola hepatica*. *Vaccine.* 2010 Apr 1;28(16):2832-40.

[480] Buffoni L, Martinez-Moreno FJ, Zafra R, Mendes RE, Perez-Ecija A, Sekiya M, et al. Humoral immune response in goats immunised with cathepsin L1, peroxiredoxin and Sm14 antigen and experimentally challenged with Fasciola hepatica. *Vet Parasitol.* 2011 Sep 28.

[481] Maggioli G, Silveira F, Martin-Alonso JM, Salinas G, Carmona C, Parra F. A recombinant thioredoxin-glutathione reductase from *Fasciola hepatica* induces a protective response in rabbits. *Exp Parasitol.* 2011 Dec;129(4):323-30.

[482] Acosta D, Cancela M, Piacenza L, Roche L, Carmona C, Tort JF. *Fasciola hepatica* leucine aminopeptidase, a promising candidate for vaccination against ruminant fasciolosis. *Mol Biochem Parasitol.* 2008 Mar;158(1):52-64.
[483] Maggioli G, Acosta D, Silveira F, Rossi S, Giacaman S, Basika T, et al. The recombinant gut-associated M17 leucine aminopeptidase in combination with different adjuvants confers a high level of protection against *Fasciola hepatica* infection in sheep. *Vaccine.* 2011 Nov 8;29(48):9057-63.
[484] Smooker PM, Kennedy NJ, Steeper KR, Christopoulos H, Spithill TW. Fasciola: kinetics and quality of humoral responses to fatty acid binding protein and cathepsin l following delivery as DNA vaccines in mice. *Exp Parasitol.* 2001 Mar;97(3):154-60.
[485] Smooker PM, Rainczuk A, Kennedy N, Spithill TW. DNA vaccines and their application against parasites--promise, limitations and potential solutions. *Biotechnology annual review.* 2004;10:189-236.
[486] Lopez-Aban J, Nogal-Ruiz JJ, Vicente B, Morrondo P, Diez-Banos P, Hillyer GV, et al. The addition of a new immunomodulator with the adjuvant adaptation ADAD system using fatty acid binding proteins increases the protection against Fasciola hepatica. *Vet Parasitol.* 2008 May 6;153(1-2):176-81.
[487] Martinez-Fernandez AR, Nogal-Ruiz JJ, Lopez-Aban J, Ramajo V, Oleaga A, Manga-Gonzalez Y, et al. Vaccination of mice and sheep with Fh12 FABP from Fasciola hepatica using the new adjuvant/immunomodulator system ADAD. *Vet Parasitol.* 2004 Dec 15;126(3):287-98.
[488] Legocki AB, Miedzinska K, Czaplinska M, Plucieniczak A, Wedrychowicz H. Immunoprotective properties of transgenic plants expressing E2 glycoprotein from CSFV and cysteine protease from *Fasciola hepatica*. *Vaccine.* 2005 Mar 7;23(15):1844-6.
[489] Paykari H, Dalimi A, Madani R. Immunization of sheep against Fasciola gigantica with glutathione S-transferase. *Vet Parasitol.* 2002 Apr 30;105(2):153-9.
[490] Raina OK, Nagar G, Varghese A, Prajitha G, Alex A, Maharana BR, et al. Lack of protective efficacy in buffaloes vaccinated with *Fasciola gigantica* leucine aminopeptidase and peroxiredoxin recombinant proteins. *Acta Trop.* 2011 Jun;118(3):217-22.
[491] Kumar N, Anju V, Gaurav N, Chandra D, Samanta S, Gupta SC, et al. Vaccination of buffaloes with *Fasciola gigantica* recombinant glutathione S-transferase and fatty acid binding protein. *Parasitol Res.* 2012 Jan;110(1):419-26.
[492] Raina OK, Tripathy A, Sriveny D, Gupta PK, Samanta S, Gupta SC, et al. Immune responses to polyethylenimine delivered plasmid DNA encoding a *Fasciola gigantica* fatty acid binding protein in mice and rabbits. *J Helminthol.* 2009 Sep;83(3):275-83.
[493] Smitha S, Raina OK, Singh BP, Samanta S, Velusamy R, Dangoudoubiyam S, et al. Immune responses to polyethylenimine-mannose-delivered plasmid DNA encoding a *Fasciola gigantica* fatty acid binding protein in mice. *J Helminthol.* 2010 Jun;84(2):149-55.
[494] Nambi PA, Yadav SC, Raina OK, Sriveny D, Saini M. Vaccination of buffaloes with *Fasciola gigantica* recombinant fatty acid binding protein. *Parasitol Res.* 2005 Sep;97(2):129-35.

[495] McManus DP, Dalton JP. Vaccines against the zoonotic trematodes *Schistosoma japonicum, Fasciola hepatica* and *Fasciola gigantica. Parasitology.* 2006;133 Suppl:S43-61.

[496] Hillyer GV. *Fasciola* antigens as vaccines against fascioliasis and schistosomiasis. *J Helminthol.* 2005 Sep;79(3):241-7.

[497] Robinson MW, Tort JF, Lowther J, Donnelly SM, Wong E, Xu W, et al. Proteomics and phylogenetic analysis of the cathepsin L protease family of the helminth pathogen *Fasciola hepatica*: expansion of a repertoire of virulence-associated factors. *Mol Cell Proteomics.* 2008 Jun;7(6):1111-23.

[498] Russell SL, McFerran NV, Hoey EM, Trudgett A, Timson DJ. Characterisation of two calmodulin-like proteins from the liver fluke, *Fasciola hepatica. Biol Chem.* 2007 Jun;388(6):593-9.

[499] Ryan LA, Hoey E, Trudgett A, Fairweather I, Fuchs M, Robinson MW, et al. *Fasciola hepatica* expresses multiple alpha- and beta-tubulin isotypes. *Mol Biochem Parasitol.* 2008 May;159(1):73-8.

[500] Chemale G, Morphew R, Moxon JV, Morassuti AL, Lacourse EJ, Barrett J, et al. Proteomic analysis of glutathione transferases from the liver fluke parasite, *Fasciola hepatica. Proteomics.* 2006 Dec;6(23):6263-73.

[501] Boukli NM, Delgado B, Ricaurte M, Espino AM. *Fasciola hepatica* and *Schistosoma mansoni*: identification of common proteins by comparative proteomic analysis. *J Parasitol.* 2011;97(5):852-61.

[502] De la Torre Escudero E, Manzano-Roman R, Valero L, Oleaga A, Perez-Sanchez R, Hernandez-Gonzalez A, et al. Comparative proteomic analysis of Fasciola hepatica juveniles and Schistosoma bovis schistosomula. *J Proteomics.* 2011 Aug 24;74(9):1534-44.

[503] Chemale G, Perally S, LaCourse EJ, Prescott MC, Jones LM, Ward D, et al. Comparative proteomic analysis of triclabendazole response in the liver fluke *Fasciola hepatica. J Proteome Res.* 2010 Oct 1;9(10):4940-51.

[504] Wilson RA, Wright JM, de Castro-Borges W, Parker-Manuel SJ, Dowle AA, Ashton PD, et al. Exploring the *Fasciola hepatica* tegument proteome. *Int J Parasitol.* 2011 Nov;41(13-14):1347-59.

[505] Gourbal BE, Guillou F, Mitta G, Sibille P, Theron A, Pointier JP, et al. Excretory-secretory products of larval *Fasciola hepatica* investigated using a two-dimensional proteomic approach. *Mol Biochem Parasitol.* 2008 Sep;161(1):63-6.

[506] Moxon JV, LaCourse EJ, Wright HA, Perally S, Prescott MC, Gillard JL, et al. Proteomic analysis of embryonic Fasciola hepatica: characterization and antigenic potential of a developmentally regulated heat shock protein. *Vet Parasitol.* 2010 Apr 19;169(1-2):62-75.

[507] Hernandez-Gonzalez A, Valero ML, del Pino MS, Oleaga A, Siles-Lucas M. Proteomic analysis of in vitro newly excysted juveniles from *Fasciola hepatica. Mol Biochem Parasitol.* 2010 Aug;172(2):121-8.

[508] Cancela M, Ruetalo N, Dell'Oca N, da Silva E, Smircich P, Rinaldi G, et al. Survey of transcripts expressed by the invasive juvenile stage of the liver fluke *Fasciola hepatica. BMC Genomics. 2010*;11:227.

[509] Dalton JP, Neill SO, Stack C, Collins P, Walshe A, Sekiya M, et al. Fasciola hepatica cathepsin L-like proteases: biology, function, and potential in the development of first generation liver fluke vaccines. *Int J Parasitol.* 2003 Sep 30;33(11):1173-81.

[510] Dixit AK, Dixit P, Sharma RL. Immunodiagnostic/protective role of cathepsin L cysteine proteinases secreted by Fasciola species. *Vet Parasitol.* 2008 Jul 4;154(3-4):177-84.

[511] Robinson MW, Dalton JP, Donnelly S. Helminth pathogen cathepsin proteases: it's a family affair. *Trends Biochem Sci.* 2008 Dec;33(12):601-8.

[512] Stack C, Dalton JP, Robinson MW. The phylogeny, structure and function of trematode cysteine proteases, with particular emphasis on the *Fasciola hepatica* cathepsin L family. *Adv Exp Med Biol.* 2011;712:116-35.

[513] Smooker PM, Jayaraj R, Pike RN, Spithill TW. Cathepsin B proteases of flukes: the key to facilitating parasite control? *Trends Parasitol.* 2010 Oct;26(10):506-14.

[514] Cancela M, Acosta D, Rinaldi G, Silva E, Duran R, Roche L, et al. A distinctive repertoire of cathepsins is expressed by juvenile invasive *Fasciola hepatica. Biochimie.* 2008 Oct;90(10):1461-75.

[515] Young ND, Hall RD, Jex AR, Cantacessi C, Gasser RB. Elucidating the transcriptome of *Fasciola hepatica* - a key to fundamental and biotechnological discoveries for a neglected parasite. *Biotechnol Adv.* 2010;28(2):222-31.

[516] Irving JA, Spithill TW, Pike RN, Whisstock JC, Smooker PM. The Evolution of Enzyme Specificity in *Fasciola* spp. *J Mol Evol.* 2003;57(1):1-15.

[517] Smooker PM, Whisstock JC, Irving JA, Siyaguna S, Spithill TW, Pike RN. A single amino acid substitution affects substrate specificity in cysteine proteinases from *Fasciola hepatica. Protein Sci.* 2000 Dec;9(12):2567-72.

[518] Stack CM, Caffrey CR, Donnelly SM, Seshaadri A, Lowther J, Tort JF, et al. Structural and functional relationships in the virulence-associated cathepsin L proteases of the parasitic liver fluke, *Fasciola hepatica. J Biol Chem.* 2008 Apr 11;283(15):9896-908.

[519] Berasain P, Goni F, McGonigle S, Dowd A, Dalton JP, Frangione B, et al. Proteinases secreted by *Fasciola hepatica* degrade extracellular matrix and basement membrane components. *J Parasitol.* 1997 Feb;83(1):1-5.

[520] Corvo I, Cancela M, Cappetta M, Pi-Denis N, Tort JF, Roche L. The major cathepsin L secreted by the invasive juvenile *Fasciola hepatica* prefers proline in the S2 subsite and can cleave collagen. *Mol Biochem Parasitol.* 2009;167(1):41-7.

[521] Lowther J, Robinson MW, Donnelly SM, Xu W, Stack CM, Matthews JM, et al. The importance of pH in regulating the function of the *Fasciola hepatica* cathepsin L1 cysteine protease. *PLoS Negl Trop Dis.* 2009;3(1):e369.

[522] Robinson MW, Corvo I, Jones PM, George AM, Padula MP, To J, et al. Collagenolytic activities of the major secreted cathepsin L peptidases involved in the virulence of the helminth pathogen, *Fasciola hepatica. PLoS Negl Trop Dis.* 2011;5(4):e1012.

[523] Marcos LA, Terashima A, Yi P, Andrade R, Cubero FJ, Albanis E, et al. Mechanisms of Liver Fibrosis Associated with Experimental *Fasciola hepatica* Infection: Roles of Fas2 Proteinase and Hepatic Stellate Cell Activation. *J Parasitol.* 2011 Feb;97(1):82-7.

[524] McGonigle L, Mousley A, Marks NJ, Brennan GP, Dalton JP, Spithill TW, et al. The silencing of cysteine proteases in *Fasciola hepatica* newly excysted juveniles using RNA interference reduces gut penetration. *Int J Parasitol.* 2008;38(2):149-55.

[525] Rinaldi G, Morales ME, Cancela M, Castillo E, Brindley PJ, Tort JF. Development of Functional Genomic Tools in Trematodes: RNA Interference and Luciferase Reporter Gene Activity in *Fasciola hepatica*. *PLoS Negl Trop Dis* [serial on the Internet]. 2008; 2(7).

[526] Beckham SA, Piedrafita D, Phillips CI, Samarawickrema N, Law RH, Smooker PM, et al. A major cathepsin B protease from the liver fluke Fasciola hepatica has atypical active site features and a potential role in the digestive tract of newly excysted juvenile parasites. *Int J Biochem Cell Biol.* 2009 Jul;41(7):1601-12.

[527] Young ND, Jex AR, Cantacessi C, Hall RS, Campbell BE, Spithill TW, et al. A portrait of the transcriptome of the neglected trematode, Fasciola gigantica--biological and biotechnological implications. *PLoS Negl Trop Dis.* 2011;5(2):e1004.

In: Parasites
Editor: Gilmar S. Erzinger

ISBN: 978-1-62257-692-0

Chapter 6

PROTOZOAN PARASITE DETECTION AND CLASSIFICATION TECHNIQUES OF MICROSCOPE IMAGES FOR COMPUTER-AIDED DIAGNOSIS

***Ching-Hao Lai*[1,*] *and J. Somasekar*[2]**

[1]Smart Network System Institute, Institute for Information Industry, Taipei, Taiwan
[2]Department of Computer Science and Engineering (CSE), Gopalan College of Engineering and Management, Bangalore, India

ABSTRACT

Protozoan parasites can cause human to get many diseases, such as, Malaria, Typhoid fever, Amebiasis, Sleeping sickness, Babesiosis etc., and these diseases are serious global health problems. The precise diagnosis and determination of parasitemia are necessary for the accurate medication. Recent researches [1-4] have suggested that the diagnosis of amebiasis through automatic parasite detection method using image processing techniques. In this chapter, two approaches are proposed for diagnosis of amebiasis by means: automatic parasite detection and counting in digital microsope images. The first approach detects the RBCs infected with E. histolytica parasites, and then the second approach counts the number of the infected cells to help diagnose parasite diseases by computers and get appropriate treatments. The proposed approaches achieve this goal using novel edge detection, segmentation and morphological processing techniques. In this chapter, several experiments are performed to test the efficacy of the proposed approaches using statistical measure correlation coefficients, and the results show a perfect positive correlation between automatic and manual count parasitemia values. This work is useful in medical applications and can help to identify amebiasis and construct a computer-aided diagnosis system in rural areas where health care manpower is limited.

* E-mail address: chinghaolai@gmail.com, somasekar_nitk@yahoo.com.

INTRODUCTION

A parasite is an organism that lives on or in a host organism and gets its food from or at the expense of its host. There are three main classes of parasites that can cause disease in humans: protozoa, helminthes, and ectoparasites. Protozoa are microscopic, one-celled organisms that can be free-living or parasitic in nature. *E. histolytica* is a protozoan and a microscope is necessary to view this parasite. Helminths are large, multicellular organisms that are generally visible to the naked eye in their adult stages. Ectoparasites can broadly include blood-sucking arthropods such as mosquitoes.

The clinical symptoms of Amebiasis caused by the protozoan parasites *E. histolytica*. Amebiasis is one of the world's most prevalent and fatal infectious diseases. Primarily it is a problem of the developing world; around 500 million people are infected worldwide while 75,000 die of the disease annually. Behind malaria and *schistosomiasis* it ranks third on the list of parasitic causes of death worldwide [5, 6]. In 1993, Diamond and Clark presented a host of data to confirm the existence of *E. histolytica* and *E. dispar* and to honor the hypothesis put forward by Brumpt in 1925. Therefore, *E. histolytica* and *E. dispar* are currently recognized as distinct species, mostly based on genetic, biochemical and immunological studies and 90% of the infections were due to a second species, *E. dispar*. - E. *histolytica* is a potential pathogen that can cause serious damage to the large intestine and other extra-intestinal organs, mainly the liver, whereas *E. dispar* is a non pathogenic commensal species that is morphologically identical to *E. histolytica.*

Mammals such as dogs and cats can become infected transiently, but are not thought to contribute significantly to transmission. The actual incidence of amebiasis remains unknown in the worldwide. Survey indicates that the infection rates vary from 0.2 to 50%, being directly correlated with sanitary conditions. An epidemiologic study reported that more than 20% of the population was infected with amebiasis in the 5-years to 10-years period preceding the study (see Fig.1). Various factors, such as poor education, poverty, overcrowding, contaminated water supply, and unsanitary conditions, contribute to the fecal-oral transmission.

E. histolytica exists in two forms: a motile and invasive trophozoite and an infective Cyst. The trophozoite varies in size; from 12 to 30 μm and the cyst initially from 10 to 16 μm in diameter. The cyst has a single nucleus but finally has 4 nuclei. It is then known as a mature cyst. There are four stages in the lifecycle of *E. histolytica* namely the trophozoite, precyst, cyst and metacyst. Colonization with the morphologically identical parasite, E. dispar is three times more common in developing countries and at least 10 times more common in developed countries.

The best initial diagnostic approach at this time is detection of *E. histolytica* antigen in stool [10]. Laboratory diagnosis of amebiasis is usually based on microscopy and serological methods including ELISA, IHA and latex agglutination. During the last decade, molecular biology-based diagnostic procedures are used to detect *E. histolytica* to the point where today they are the preferred approach. Accurate diagnosis is important not just for patients with dysentery but also for the 90% of *E. histolytica* infections that are asymptomatic because infection may easily be transmitted from person to person, especially in developing countries which have poor hygienic conditions and inadequate water treatment.

Computer is a useful tool in many applications in clinical and technical medical activities particularly in the last two decades. Different diagnosis methods available for amebiasis but these are time consuming, need technical expertise and high cost. Detection is the most important task whereas the parasite count is necessary for an appropriate treatment. The aim of our study is to develop an automatic image processing system for the rapid and accurate diagnosis of amebiasis through the parasite detection and counting.

Lai *et al.* [1] proposed a protozoan parasite extraction technique for the parasite microscope images to solve and reduce the problem that the protozoan parasite image with poor quality and complex background. Lai *et al.* first used a proposed *gamma equalization* method to solve illumination variance problem when parasite image are took from different experiment conditions and environments. Then *two-classes edge enhancement* and *two-means clustering* techniques are used to enhance the boundaries of parasites and smooth the background to take parasite objects in each parasite image. Besides, Lai *et al.* [3] also further proposed modified *connected component* method and *circle mask scoring* technique to more accurately extract multiple nuclei from each protozoan parasite.

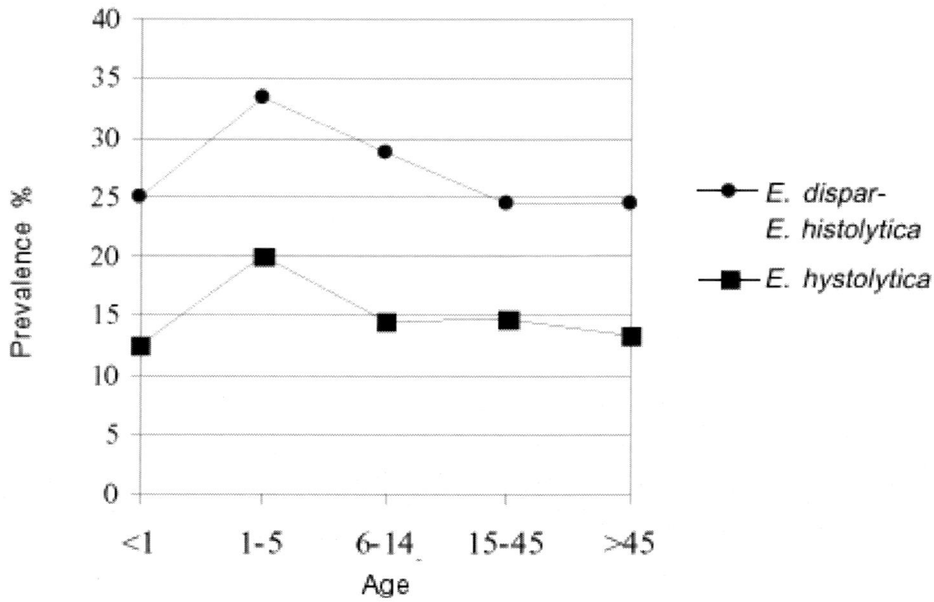

Figure 1. Age related rates and stool colonization with *E. histolytica* and *E. histolytica - E. dispar*: solid squares indicated individuals colonized with E. histolytica and with solid circles individuals colonized with the *E. dipar - E. histolytica* complex.

1. Materials and Methods

The proposed parasite detection and parasite count method for diagnosis of amebiasis is shown in Figure 2. The details of each stage are presented in the following subsections.

2. Image Acquisition

This is an essential task, which required interface between the camera, the microscope and the computer. The ideal system should be designed to cope with images acquired under various uncontrolled conditions and perform the same processing on them. Digital images were randomly captured with the aid of a microscope and stored. Stored images were used as a raw data for the detection and parasite count (below) for accurate amebiasis diagnosis.

Image Pre-Processing

Pre-processing step includes noise reduction and smoothening for images. In this chapter the median filter is used to smooth images to reduce most noises. The median filter is a non-linear digital filtering technique to reduce noise from images without losing the important information of cells and parasites. In median filtering, each pixel is replaced by the median of its neighboring pixel values in a corresponding sub-image window. The median filter is given by

$$g(x, y) = \text{median}(f(s, t)),$$
$$(s, t) \in S_{xy}$$

Where $g(x, y)$ is the restored image at point (x, y) of the original image $f(x, y)$ and S_{xy} represents the set of coordinates in a rectangular sub-image window. The input image after median filter is given in Figure 3A-3. The pre-processed image is used to the following edge detection processing.

3. Edge Detection

Image edge detection is an important work for object recognition of the human organs, and it also is an important preprocessing step in medical image segmentation. In this chapter, we used a better edge detection method [7] based on mathematical morphology which can detect the edges of parasite infected cells accurately when compared to early algorithms such as gradient-based and template-based algorithm. The main reason for choosing this algorithm that it can help obtain better results for accurate parasite counting. We summarized this algorithm as follows.

Let $M(x, y)$ denote a gray-scale two-dimensional image and A denote structuring element. Dilation and Erosion of a gray scale image $M(x, y)$ by a gray-scale structuring element $A(i, j)$, are defined as follows,

$$(M \circledast A)(x, y) = \max\{M(x-i, y-j)+A(i, j)\},$$

$$(M * A)(x, y) = \min\{M(x-i, y-j)-A(i, j)\}.$$

The edge of image M, which is denoted by $E_d(M)$, is defined as the difference set of the dilation domain of M and the domain of M, such as,

$$E_d(M) = (M \circledcirc A) - M.$$

The perfect edges of infected RBCs will be got by performing the difference between the processed image by the above process and the image before dilation. This edge detection algorithm [7] is,

$$(B \bullet A) \circledcirc A - B \bullet A,$$

Where, $B = (M \bullet A) \circ A.$

$M \circ A$ and $M \bullet A$ are opening and closing operations of gray-scale image M by gray-scale structure element A and defined $(M * A) \circledcirc B$ and $(M \circledcirc A) * A$, respectively. The edge detection image of input image (Figure 3A-1) is shown as Figure 3A-4.

Based on this edge detection method, we extract the components infected parasite in binary image form. Now we applied morphological operation namely image erosion on previous binary image. The binary mask of infected RBCs of input image result as shown in Figure 3A-5. Let E be a Euclidean space or an integer grid, and A is a binary image in E. The erosion of the binary image A by the structuring element B is defined by:

$$A * B = \{z | E / B_z \subseteq A\}.$$

Where B_z is the translation of B by the vector Z. i.e., $B_z = \{b + z / b \subseteq B\}, \forall z \in E.$

After morphological operations, the proposed approach labels the objects infected with parasites and the labeled image is shown in Figure 3A-6. Finally, the proposed approach counts the number of cells infected with parasites and highlights the parasite infected components which is shown as Figure 3A-7. The number of cells infected with E. histolytica parasite is shown by a message window as Figure 3B in our system.

4. Experimental Results

The proposed method is applied to several microscopic images from different blood smears. All the programs are implemented using the MATLAB tool. The various stages of processing of microscopic blood images are graphically illustrated in Figure 3(A). Figure 3A-1 and Figure 3A-4 shows a raw image and image after edge detection, respectively. The final detection result as shown in Figure 3A-7, and the automatic parasite count values are shown as Figure 3(B).

In order to evaluate the performance of our proposed method, we apply it to some of parasite uninfected images. Figure 4 shows the result of testing uninfected blood cell image.

Our proposed method indicating that no cell is infected with E. histolytica parasite. Also the edge detection not showed any edge in resultant image (Figure 4-b) due to no cell is infected with parasites. We applied other well known edge detection methods [8, 9] for the same image and final resultant image yields some edges. Hence the presented method is more accurate for parasite count for amebiasis diagnosis.

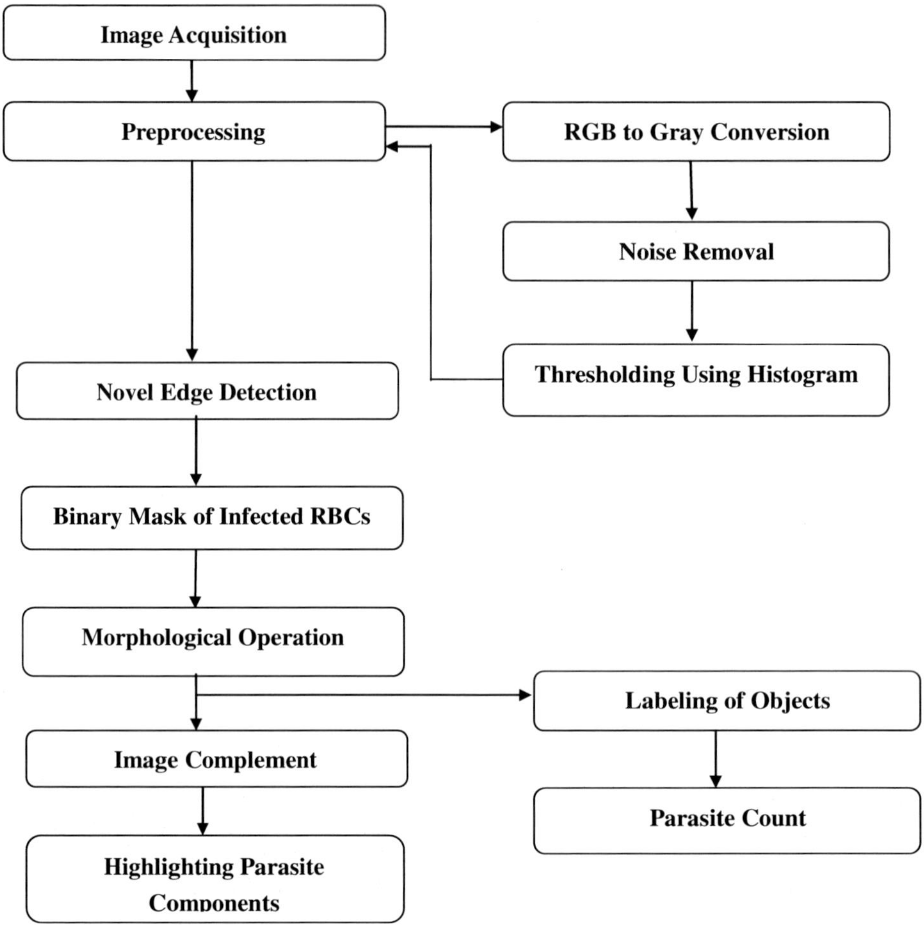

Figure 2. Flow chart of the proposed method.

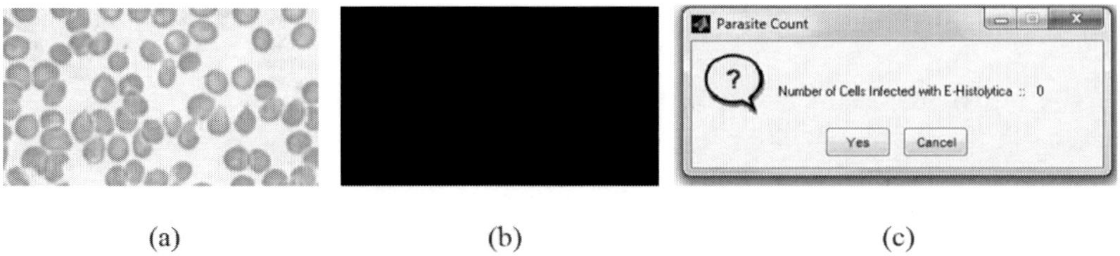

Figure 4. Parasite uninfected RBC: (a) Original image, (b) new edge detection image, and (c) Parasite count.

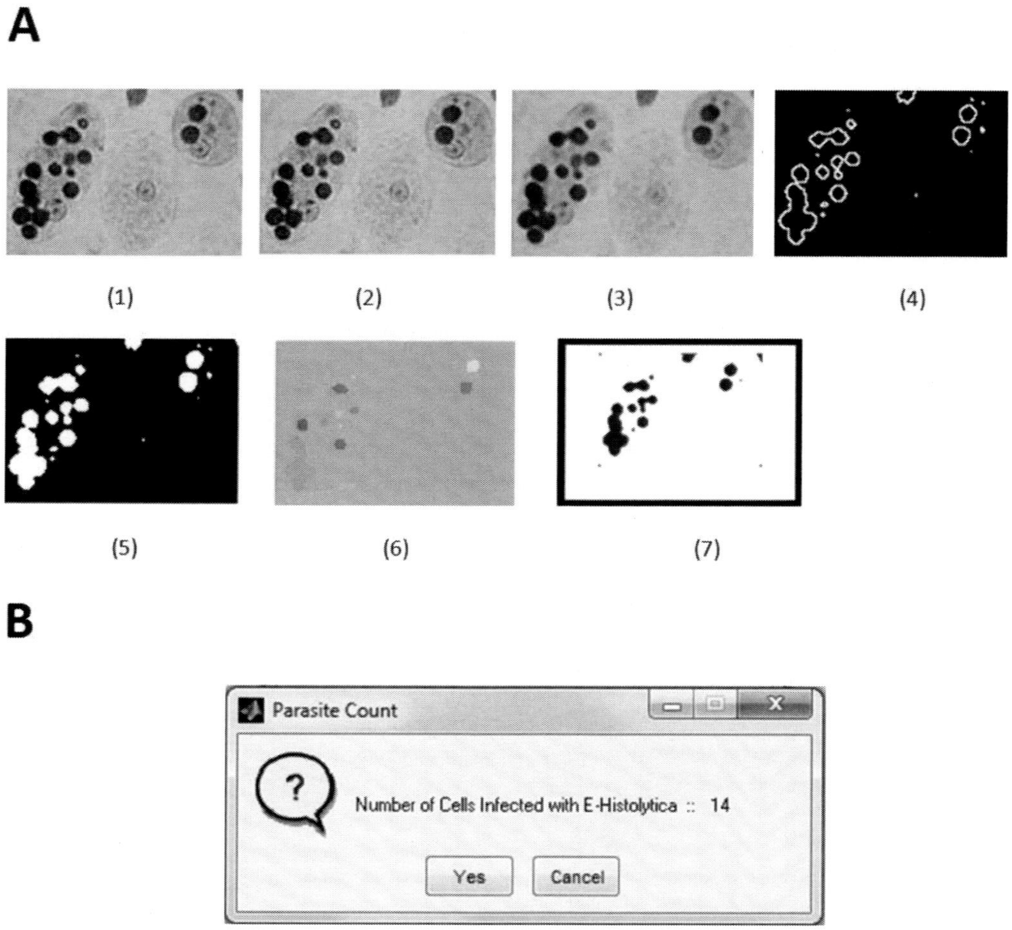

Figure 3. The various stages of processing of microscopic blood images are graphically illustrated in (A) in the following sequence: (1) Original color image, (2) RGB to Gray conversion, (3) Gray level image after median filtering, (4) parasitic edges detected, (5) binary mask of infected RBCs after edge detection, (6) label 8-connected components after morphological operation, (7) highlighting the parasite identification, and the parasite count is shown in (B).

5. Manual Parasite Counting vs. Automatic Parasite Counting

Images processed through the proposed method counts the total number of cells infected with E. histolytica parasite and manual count of parasite as shown in Table 1. However, there was significant variation between manual count and automatic count for parasitemia in three images (image no. 3, 5 and 10 from Table 1). When analyzed further, it was observed that this blood smear contained predominantly poorly stained and overlapping of more cells. 10 images processed through automatic system. Figure 5(b) shows the graphical comparison of manual and automatic parasitemia counting. Investigations into person-to-person variability in parasitemia determination by manual counting in comparison to automatic parasite counting reproducibility revealed that there was marked person-to-person variation. In contrast, automatic parasite count gave consistently identical results for all images processed.

Table 1. Summarizes result of manual and automatic count of infected cells

Image No.	1	2	3	4	5	6	7	8	9	10
Manual Parasite Count	3	5	10	6	10	1	2	5	14	5
Automatic Parasite Count	3	5	12	6	9	1	2	5	14	6

6. Statistical Analysis

A statistical measure namely coefficient of correlation used to judge the quality of the proposed method. Correlation coefficient is a measure of intensity or degree of linear relationship between two variables. We calculated correlation coefficient between manual and automatic count of parasite values of 10 microscopic blood images (Table 1) and value is, $r = 0.9918 \approx 1$. This shows that there is an *app. Perfect* positive correlation between automatic and manual parasite values and graphically as shown in Figure 5(a). Also to test the reliability, we calculated P.E(r) and given by,

$$\text{P.E}(r) = 0.6745 \times \text{S.E}(r) < 0.9918.$$

Therefore, the reliability of an observed correlation coefficient is significant.

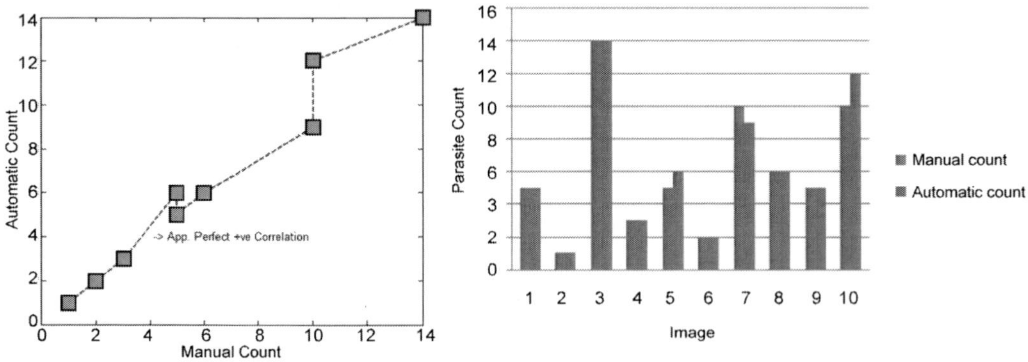

(a) (b)

Figure 5. (a) Correlation between manual and automatic parasite count; (b) Histogram of parasite values obtained by manual count and automatic count of parasites.

CONCLUSION AND FUTURE WORK

This chapter investigates the possibility of detecting E.histolytica parasites and counting parasite infected RBCs for accurate diagnosis of amebiasis using image processing techniques. Furthermore, we compared manual parasite count and automatic parasite count and results showed that the proposed method yields a well correlation between manual and automatic count. We also test this software for both infected and uninfected images. Our system excels over the manual system of detection because extremely fast detection of the parasite within seconds, highly accurate in true detection and no detection, simple to operate and serves as a reliable second opinion to pathologists and microbiologists. It is clear from the experimental results that the proposed technique has high diagnosis ability for amebiasis. Future researches should aim at detection and differentiation of E. histolytica and E. dispar and identify the exact stage of the life cycle and the degree of E. histolytica parasites.

REFERENCES

[1] Ching-Hao Lai, Shyr-Shen Yu, Hsiao-Yun Tseng, Meng-Hsiun Tsai, "A protozoan parasite extraction scheme for digital microscopic images," *Computerized Medical Imaging and Graphics* 34 (2010), pp. 122–130.

[2] Hsiao-Yu Wang, Hung-Yuan Chung, "A Computer Vision Approach to Micro-Nucleus Automatic Detection for Protozoan Parasites Segmentation," *Lecture Notes in Engineering and Computer Science 2181* (2010), pp. 1446-1452.

[3] Ching-Hao Lai, Hsiao-Yu Wang, Yi-Chuan Tsai and Shyr-Shen Yu," An Iterative Based Novel Multi-Nucleus Detection Scheme For Protozoan Parasite Microscopic Images," *International Journal of Innovative Computing, Information and Control* 5 (2009), pp. 3875-3886.

[4] J. Somasekar, A. Rama Mohan Reddy, and L. Sreenivasulu Reddy, *"An Efficient Algorithm for Automatic Malaria Detection in Microscopic Blood Images,"* ObCom 2011, Part II, CCIS 270 (2012), pp. 431–440, Springer-Verlag Berlin Heidelberg 2012.

[5] J. A. Walsh, "Problems in recognition and diagnosis of amebiasis: estimation of the global magnitude of morbidity and mortality," *Rev Infect Dis* 8 (1986), pp. 228-38.

[6] E. Li, & S. L. Stanley "protozoa: Amebiasis," *Gastroenterol clin. Noerh Am* 25 (1996), pp. 471-92.

[7] Y. Q. Zhao, W. H. Gui, Z. C. Chen, J. T. Tang, L. Y. Li,"Medical Images Edge Detection Based on Mathematical Morphology," *Proceedings of the 2005 IEEE Engineering in Medicine and Biology 27th Annual Conference Shanghai,* China, September 1-4, 2005.

[8] R. C. Gonzalez, R. E. Woods, and S. L. Eddins, *Digital Image Processing using MATLAB.*Prentice Hall, 2004.

[9] R. C. Gonzalez and R. E. Woods, *Digital Image Processing.* 2nd ed. Prentice Hall; 2002.

[10] W. A. Petri Jr., U. Singh "Diagnosis and management of amebiasis," *Clin Infect Dis* 29 (1999), pp. 1117–1125.

In: Parasites
Editor: Gilmar S. Erzinger

ISBN: 978-1-62257-692-0

Chapter 7

LACTOFERRIN: A PROTEIN OF THE INNATE IMMUNE SYSTEM CAPABLE OF KILLING PARASITIC PROTOZOA

***Cynthia Ordaz-Pichardo*[1,*], *Nidia León-Sicairos*[2,*], *Adrián Canizalez-Román*[2], *Miguel A. Cornejo-Cortés*[3], *Guillermo Ortiz-Estrada*[4] *and Mireya de la Garza*[4,†]**

[1]Laboratorio de Biología Celular y Productos Naturales, Escuela Nacional de Medicina y Homeopatía, Instituto Politécnico Nacional, Ticomán, México
[2]Unidad de Investigación de la Facultad de Medicina, Universidad Autónoma de Sinaloa, Culiacán, Sinaloa, México
[3]Departamento de Ciencias Biológicas, Facultad de Estudios Superiores Cuautitlán, Universidad Nacional Autónoma de México
Carretera Cuautitlán, Cuautitlán Izcalli, Estado de México, México
[4]Departamento de Biología Celular, Centro de Investigación y de Estudios Avanzados del IPN. México, México

ABSTRACT

Lactoferrin (LF) is a cationic iron-binding glycoprotein found in mucosal secretions and fluids of mammals such as those from respiratory, vaginal, digestive and intestinal tract, tears and saliva and is particularly abundant in colostrum and milk. LF is also produced by the secondary granules of neutrophils and secreted by these cells in blood and sites of infection. This glycoprotein of about 80 kDa is a component of the innate immune system devoted to capture ferric iron in order to be unavailable for pathogens growth when they try to colonize and invade a host. LF is truly a multifunctional molecule since is immunomodulator, anti-carcinogenic and microbicidal. The iron-free form of LF (apoLF) is active against numerous infective species of bacteria, fungi, virus, and protozoa. LF is digested by the gastric enzyme pepsin, producing N-terminal peptides named Lactoferricins (LFcins), which often are more potent antimicrobials than the native protein. Milk-derived bovine LF as well as the recombinant human and bovine LF has been utilized to treat microbial infections in animal models and in trials with patients, resulting in an excellent option to treat several diseases. The research about LF in the treatment of parasitic diseases is limited; however, it is an exciting field of investigation, given the high morbidity and mortality rate produced by parasites, especially in developing countries, and also by the high number of resistance strains to commonly used drugs. In this chapter, we review the use of apoLF and LFcins to combat diseases caused by parasites and the known mechanisms involved in the action of LF on the parasitic cell and the immune status of the host.

INTRODUCTION

Parasitic pathogenic protozoa are broadly spread pathogens responsible for important and often fatal diseases worldwide. Parasites and their corresponding hosts have co-evolved during their life on Earth. Parasites generally obtain all of the nutrients from the host in order to survive and colonize. Some parasites spend one part of their life cycle in an invertebrate vector and the other part in a vertebrate host in which express the pathogenicity determinants [1-6]. Parasites can live in intracellular niches [7,9-10] but some species can live extracellularly in fluids and secretions [11]; in both sites, anti-parasitic drugs require a rapid and efficient killing effect in order to combat the disease. There are several reasons by which some drugs have limited use against parasitic protozoa, for example, their toxicity or low specificity, and the emergence of resistant strains. In many cases, the discovering or design and trials of new drugs in animals employ more time than that used by parasites to develop resistance to them. These reasons have encouraged the seeking of new products that fulfill the parasiticidal requirements without causing side effects to patients. Additionally, in some infections a combination of antimicrobial drugs of clinical use with peptides and proteins of the innate immune system have been explored with the aim to reduce the drug dose and then diminish the toxic effects.

1. IRON IS A CRUCIAL ELEMENT FOR ALL LIVING BEINGS, COMPRISING PARASITES AND THEIR HOSTS

Iron (Fe) is a crucial element for the cellular metabolism of all living organisms, with the exception of some *Lactobacillus* species. In nature, iron is mainly found in two oxidation forms (Fe^{2+}, Fe^{+3}). However, iron is toxic because free Fe^{2+}, in the presence of oxygen, produces highly reactive oxygen species (ROS) by the Fenton reaction, damaging all types of macromolecules [11,12] . One manner in which cells and tissues avoid iron toxicity is binding this metal to proteins, thus all biological systems possess diverse mechanisms to maintain the iron homeostasis in the body, through the evolutionary design of diverse iron-containing proteins [13-15].

In mammals, intracellular iron can be found as a component of the heme group of electron-transporter proteins. This element is also a co-factor in redox processes catalyzed by enzymes and in the Fe-S center of several proteins [16-18]. Extracellular iron is primarily bound by the iron-binding proteins transferrin (TF) that is an iron-transporter that provides iron to all cells, mainly the erythrocyte precursors for the synthesis of hemoglobin; and in lactoferrin (LF), a protein that sequesters iron in mucosae and infection sites. Excess of iron in the cell is deposited in the intracellular protein ferritin that mainly stores iron in the liver.

As an innate antimicrobial defense strategy, mammals have developed an elaborate non-specific iron-withholding system that remarkably reduces the amount of iron available to invaders. Under normal conditions, the free iron concentration in body fluids is approximately 10^{-18} M, far too small an amount to support pathogen growth, which is in the micromolar range [16,17-21]. This iron-withholding mechanism is mainly carried out by TF circulating in serum, lymph and cerebrospinal fluid, and LF in mucosae [20,22]. LF is a critical component of the first line of defense of the immune system against microbes. Indeed, LF is an amazing multifunctional protein that possesses a broad scope of activities, such as modulation of humoral and cellular components involved in inflammatory and immune responses, and it has anti-tumor properties [22-23]. All of these properties have noticeably a range of implications and potential clinical applications; in this chapter, we will review the anti-parasitic usage.

2. Lactoferrin, an Anti-Microbial Iron-Binding Glycoprotein

Lactoferrin is a non-heme monomeric glycoprotein of approximately 80 kDa that belongs to the transferrin´s family, the members of which have the property of chelating iron. It can reversibly bind up to two ferric ions with greater affinity (~10^{-23} M) than the other proteins of the family, and is highly cationic showing an isoelectric point of 8.5-9.4. LF loses iron in the acidic range of pH (2.5-4.0) [24]. Tertiary structure of LF consists in two main N and C lobes, in turn, organized in N1, N2 and C1 and C2 domains. Both N and C lobes are linked at N1 and C1 domains by a three-turn alpha chain [25]. Between N1/N2 and C1/C2, a cleft is formed by R-amino acid groups of one aspartic, two tyrosines and one histidine where iron (Fe^{3+}) along with a carbonate (CO_3^{2-}) ion are bound reversibly to LF [22,25-32]. Therefore, LF can be in nature as hololactoferrin (holoLF), with one or two ferric-iron ions bound, and as an iron-free form known as apolactoferrin (apoLF). Both types of LF interact with pathogenic microorganisms in a contradictory manner: apoLF can be microbicidal whereas holoLF can be used by bacteria, fungi and protozoa as an iron source for growth [33-34].

LF is a protein secreted in its iron-free form in colostrum and milk from the majority of mammalian species [35]. Evolutionarily, LF has a highly conserved structure; the amino acid sequence of LF shows a homology of approximately 70% between humans and mice and >70% between humans and cattle [36]. LF and other milk components form part of the protective effect of breast-feeding to newborns. LF is synthetized by glandular epithelial cells and thus normally found in exocrine secretions and body fluids, including tears, saliva, synovial fluid, vaginal secretion, semen, nasal and bronchial secretions, bile, gastrointestinal fluids and urine [34,37-40]. Moreover, LF is stored in secondary granules of neutrophils which normally release small amounts of this glycoprotein in serum as a consequence of their degranulation [35, 37]. This host strategy, which attempts to deprive microbes of growth-vital iron, is called "nutritional immunity" [20, 37; 41-42]. Therefore, LF is present in the secretions that protect mucosal surfaces and also in infection sites with the aim of sequestering iron to make it unavailable to microbes, avoiding the colonization and invasion.

During an infection, the host iron-withholding defense system is alerted and strengthened. Several strategies additionally decrease the iron available to microbes, and importantly, the level of LF in serum significantly increases during the inflammatory response [31,38]. Also, iron absorption and TF saturation levels diminish and ferritin synthesis increases. In general, iron is removed from sites of microbial invasion, which is facilitated by neutrophils releasing apoLF and macrophages scavenging for holoLF (with iron); in both cases, the aim is to restrict the access of iron to microbes. Moreover, depletion of intracellular iron represents a mechanism by which macrophages can kill intracellular pathogens [43]. All of these events are called "hyposiderhemia (or hypoferremia) of infection" and are part of the nutritional immunity [14, 17, 41; 44-47].

Milk is a healthy and nutritious source of energy, proteins, vitamins and minerals. In addition to the nutritional value of milk, there has been a growing research interest in its capacity to kill microbes, and this knowledge can be applied to feeding recommendations for babies (breast milk) and also to the production of functional foods for people. LF can be isolated from cow's milk by various purification methods, or it can be expressed by recombinant methods. Because of the multiple functions of both, bovine LF (bLF) and human LF (hLF), they have been used for clinical treatments and industrial applications. One of the first presentations of LF was in infant formulas. In some countries, LF is currently added to immune system-enhancing nutraceuticals, cosmetics, drinks, etc. The ability of LF to prevent nososcomial infections in infants has been assayed and the results showed lower infection levels. Additionally, several studies showed that infants fed with formula lacking LF had less intestinal iron absorption than breastfed infants. It was proposed that LF also promotes the proliferation of lactic acid bacteria in the bowel, such as *Bifidobacterium* and *Lactobacillus*, which protect the host from harmful bacteria [48].

Due to the capacity of apoLF to chelate iron, by which it avoids the growth of microbes, LF was initially considered as microbiostatic. Years later, because of its highly cationic burden, it was found that LF interacts with electronegative charges on bacterial surface lipopolysaccharide (LPS) and porins in Gram-negative bacteria and teicoic acids in Gram-positive bacteria, destabilizing the membrane and causing death, including multidrug-resistant bacteria [32,49-52]; thus, LF is now considered as microbicidal. LF also inhibits bacterial biofilm development, a major problem in clinical treatments [53]. The antimicrobial activity of LF has been demonstrated towards a great number of bacterial species such as *Streptococcus mutans, Vibrio parahaemolyticus, Vibrio cholerae*, and others [54].

The presence of stable antimicrobial peptides resulting from the proteolysis of LF was shown twenty years ago. It was demonstrated that limited acid proteolysis of bLF yielded a hydrolysate that had greater antibacterial activity than the native LF. In 1992, Bellamy and co-workers described the generation, after pepsin hydrolysis, of fragments from human and bovine origin that showed enhanced antimicrobial activity. From bLF a 25 amino acid peptide consisting of residues 17–41, lactoferricin B (LFcinB), is generated, while a 47 amino acid peptide, lactoferricin H (LFcinH), consisting of residues 1–47 can be generated from hLF. LFcin is a region devoid of iron chelating activity, characterized by its strong cationic charge due, in hLF, to four arginines; this structure seems to be better suited for making contact with pathogen membranes [30,55-57]. Both peptides are derived from the N-terminal region of the N1-lobe and have greater antibacterial activity than their parent proteins [58-59]. LFcinB is the most potent peptide of the two and has antimicrobial activity against a wide range of microorganisms, including Gram-positive and Gram-negative bacteria, viruses, protozoa, and fungi. Recently, several anti-inflammatory activities of lactoferrin-derived peptides have been described, suggesting that they have other properties in the defense against microbes in addition to their antimicrobial activity [60].

Importantly, LF exhibits a natural synergistic interaction with two proteins of the innate immune system, IgA and lysozyme against numerous bacterial species [61-62], and with other antimicrobial peptides from the innate immune system such as β-defensins of the gastrointestinal tract and from neutrophils [63]. LF also synergizes with antibiotics and drugs promoting the eradication of microorganisms that cause chronic diseases, such as those caused by *Candida albicans* [64] and *Staphylococcus aureus* [65], and against bacteria that infect the diabetic foot [52-66]. Interestingly, LF also acts against some viruses [67-68]. Furthermore, apoLF has been reported to act as a parasiticidal in several protozoan species, mainly in extracellular parasites and this is the theme of discussion in this chapter.

In patients, the ingestion of bLF as well as hLF has been reported to have beneficial effects on infections of both digestive and non-digestive tract tissues [69]. Clinical studies of LF therapy have been performed against hepatitis C virus [70-71], the gastric infection caused by *Helicobacter pylori* [72-73], and in tinea pedis caused by the fungus *Trichophyton mentagrophytes* [74]. In the digestive tract, intact and partially digested LF, which peptides can retain biological activity, bacteriostatic and bactericidal effects are exerted. In non-digestive tracts, oral administered LF or its digestion products could act initially on the intestinal immune system and then promote systemic protective immunity as a secondary effect, thus, LF action in infections is a complex phenomenon and probably due to multiple mechanisms [69,75].

In summary, LF and its derived peptides have a broad potential to treat diverse infections by both a direct interaction with pathogens and the immune system stimulation.

3. Lactoferrin, a Protein from the Innate Immune System

Depending on the organism, immunity may involve innate immunity only, as for most invertebrates, or both innate and adaptive immunity, as for mammals. Whatever the systems brought into play, they are always complex and tightly regulated, because they intend to be as harmful as possible against the infectious agents, whereas they must be as harmless as possible for the host itself. LF is definitely one of the molecules that play key roles not only against microbes but also against excessive and harmful host responses in mammals: a sword and a shield at the same time. Those two aspects of LF roles in host defense, together with its potency to modulate both innate and adaptive immunities, make LF an amazing multifunctional molecule in immunity [22]. LF is regarded a modulator molecule able to modify positively or negatively the production of humoral mediators and/or the activity of cellular components involved in specific immune responses [23]. LF displays a wide array of effects on cellular components of host adaptive defense including, for instance, its role as maturation factor for B lymphocytes with regard to their phenotype and function [76].

As we already mentioned, LF is a multifunctional glycoprotein found in milk and exocrine fluids. It is also a major component of polymorphonuclear neutrophils. LF isolated from milk and tears has previously been described as an inhibitor of the classical complement pathway. Further studies have shown that LF inhibits the formation of the classical C3 convertase, thus preventing the cleavage of C3. The complement inhibition can be reversed by addition of Fe^{3+}, but not by Ca^{2+} or Mg^{2+} [60].

The multifunctionality of LF makes difficult the understanding of its activities. The literature is indeed crammed with evidences for LF roles in the host defense but, unfortunately, most of them are not necessarily informative on the actual mechanisms of action in immunity. Indeed, whereas many *in vivo* studies report protective effects against infections or septic shock, it is not perfectly clear whether those effects are the result of a direct antimicrobial activity or of a boost of cell-mediated immunity modulation. However, recent studies undoubtedly demonstrate that bLF supports the immune and antioxidant status in healthy human volunteers [22].

In vitro approaches are very suited to comprehend the actual roles of LF in immunity, since they are conducted in homologous conditions (e.g., LF with human cells), and using reagents of the highest purity and quality. As we mentioned, LF binds the bacterial LPS with high affinity and this proinflammatory component is a common contaminant in LF preparations which can readily activate transduction pathways of the inflammatory response (usually MAP kinase and NFκB pathways). *In vitro* studies have demonstrated a role of LF in modulating activation of cells involved either in innate immunity (most particularly macrophages, neutrophils, basophils, esosinophils, mastocytes and NK cells) or in adaptive immunity (B- and T-cells and dendritic cells), in both as a proinflammatory or an anti-inflammatory agent. In most cases, the modulation of the activity of immune cells by LF results in decreased or increased expression of proinflammatory interleukins (IL-1, IL-6, IL-18, TNF-α) by immune cells or adhesion molecules by endothelial cells (E-selectin) [22].

In conclusion, many functions have been proposed, and continue being proposed, for LF, some of which are independent of metal binding. As more structural information becomes available with expression of recombinant forms of both full-length LF and its fragments, new technologies will permit a better understanding of the antimicrobial and immunomodulator properties of this truly multifunctional protein.

4. Iron-deprived Lactoferrin Can Be Microbiostatic and Microbicidal against Extra- and Intra-cellular Parasitic Protozoa

It has been found that LF shows a specific antiprotozoan activity. LF could be an immunomodulator in the polarization of the response Th2 to Th1 in infectious processes caused by extracellular parasites. In this way, LF leads to the proinflammatory cytokines production and then is the bridge in the activation of the adaptive response and destruction of these pathogens through specific mechanisms mediated by Abs, opsonins, TLRs and lectins. Additionally, LF could promote a Th1 response for resolving infections caused by intracellular parasites, for example, in the destruction of *Plasmodium* merozoites, due to the upregulation of IFN-γ, IL-12 and TNF-α. Also, it is known that LF modulates the phagocytic capacity of PMN neutrophils and monocytes/macrophages in the resolution of infections [22,31]. In summary, Table 1 and Table 2 show the parasites in which LF and its peptides have been used to control infections, respectively. Table 3 shows the known mechanisms of action of LF against parasites.

Entamoeba histolytica

Entamoeba histolytica is a protozoan that causes amoebiasis, a parasitic intestinal infection responsible for high morbidity and mortality worldwide. The parasite infects, together with the morphologically similar but non-pathogenic *Entamoeba dispar*, around 10% of the world population. Estimates place the current number of infections by *E. histolytica* at ~50 million people and accounts for 100,000 deaths each year. All people are at risk of infection, although infection/disease occurs predominantly among infants and travelers to endemic countries [77-79]. Several aspects of the host–parasite relationship such as parasite virulence and host susceptibility are not totally understood, and no effective vaccination strategies exist against amoebiasis [80].

This protozoan infects exclusively human beings and several animal models have been used to study the disease that it causes. Parasite cysts are ingested by the fecal-oral route through contaminated water and foods. *E. histolytica* cysts support the acidic pH of stomach and reach the small intestine excysting in the terminal ileum and invading the large intestine mucosa where trophozoites multiply and cause colitis and dysentery, and due to diverse secreted virulence determinants they produce abscesses and perforation of the intestinal tissue. In severe cases, amoebae travel to the liver, causing amoebic liver abscesses [81-83].

Different types of drugs have been tested for use in treating amoebiasis. According to their site of action, these drugs have been classified as luminals (for example, quinfamide) and both luminals and extra-luminals (for example, metronidazole and emetine) [84]. All of the drugs presently employed produce negative side effects in patients. Metronidazole is the drug of election to treat hepatic amoebiasis, but this drug is toxic and might be mutagenic for patients when it is used at high doses or as long-term treatment [85-86]. In addition, *E. histolytica* strains resistant to metronidazole have been found in cultures *in vitro* [87]. Thus, the necessity for an effective and safe anti-amoebic agent is growing.

5. *E. Histolytica* Trophozoites Are Killed by Lactoferrin *In Vitro*

In an axenic culture, *E. histolytica* trophozoites depend on an exogenous iron source for growth, and they can use both ferric and ferrous ions. Amoebae are capable of using several iron-containing proteins *in vitro*: hemoglobin, holotransferrin (holoTF), hololactoferrin (holoLF) and ferritin, suggesting that this parasite could use these proteins as an iron source in the diverse organs that infects [88-92]. On the contrary, iron-free LF (apoLF) can be amoebicidal.

Table 1. Parasites in which Lactoferrin has been studied *in vitro* or used to control infections

Protozoa and stage	Origin of LF	Studies *in vitro*	Animal models	Studies in patients	Reference
Extracellular					
Entamoeba histolytica (trophozoite)	apo-bLF, apo-hLF	Trophozoites cultured *in vitro*	IA in C3H/HeJ mice, ALA in hamster		[León-Sicairos *et al.*, 2006a; 2006b; 2012, in press; Ordaz-Pichardo *et al.*, 2012, in press]
Giardia lamblia (trophozoite)	apo-bLF, apo-hLF	Trophozoites cultured *in vitro*		children of 12 to 36 months of age.	[Turchany *et al.*, 1995; 1997; Ochoa *et al.*, 2008]
Trypanosoma cruzi (trypomastigote)	apo-hLF	MPM, HBM			[Lima and Kierszenbaum, 1987b]
Intracellular					
Annchaliia algerae (spore)	hLF	C2Bbe1 cells			[Leitch and Ceballos, 2009]
Babesia caballi (intra-erythrocytic form)	apo-bLF	Cultured *in vitro*			[Ikadai *et al.*, 2005]
Encephalitozoon intestinalis (spore)	h-bLF	C2Bbe1 cells			[Leitch and Ceballos, 2009]
Plasmodium falciparum (merozoite)	apo-hLF, holo-hLF	Human erythrocytes, CHO cells, C32 AMC			[Fritsch *et al.*, 1987] [Eda *et al.*, 1999]
Pneumocystis carinii (cyst)	apo-bLF, apo-hLF	HLEC			[Cirioni *et al.*, 2000]
(trophozoite)	apo-bLF, apo-hLF	HLEC			[Cirioni *et al.*, 2000]
Toxoplasma gondii tachyzoite (RH strain)	apo-bLF, holo-bLF	MPM, MEC	ICR mice		[Tanaka *et al.*, 1996; 1998]
tachyzoite (BK strain)	holo-hLF	HEC, MFC			[Dzitko *et al.*, 2007]
Trypanosoma cruzi (amastigote)	apo-hLF	MPM, HBM			[Lima and Kierszenbaum, 1987a]

C32 AMC: C32 amelanotic melanoma cells; ALA: amoebic liver abscess; bLF: bovine lactoferrin; hLF: apo-human lactoferrin; CHO cells: Chinese hamster ovary cells; HBM: human blood monocytes; HEC: human epithelial cells; HLEC: human long epithelial cells; IA: intestinal amoebiasis; MEC: mouse embryonal cells; MFC: murine fibroblast cells; MPM: mouse peritoneal macrophages.

Table 2. Lactoferrin peptides (natural and synthetic) that have been used against parasites

Protozoa and stage	Peptide	Studies *in vitro*	Animal models	Reference
Extracellular				
Entamoeba histolytica (trophozoite)	LFcinB, 4-14	Trophozoites cultured *in vitro*	IA in C3H/HeJ mice	[León-Sicairos *et al.*, 2006b]
	LFcinB, 17-30		ALA in hamster	[López-Soto *et al.*, 2010]
	Lfampin, 265-284	Trophozoites cultured *in vitro*		[López-Soto *et al.*, 2010]
	Lfchimera (fusion peptide) LfcinB 17-30 + LFampin 265-284)	Trophozoites cultured *in vitro*		[López-Soto *et al.*, 2010]
Giardia lamblia (trophozoite)	LFcinB 17-41, LFcinH 18-40	Trophozoites cultured *in vitro*		[Turchany *et al.*, 1995]
Intracellular				
Annchaliia algerae (spore)	LFcinB 4-14	C2Bbe1 cells		[Leitch and Ceballos, 2009]
Cryptosporidium parvum (sporozoite)	LFcinB, LFcinH			[Carryn *et al.*, 2011]
Eimeria stiedai (sporozoite)	LFcinB 17-41	RHC	Rabbit	[Omata *et al.*, 2001]
Encephalitozoon intestinalis (spore)	LFcinB 4-14	C2Bbe1 cells		[Leitch and Ceballos, 2009]
Plasmodium falciparum (merozoite)	LFcinH 25-37	CHO cells, C32 AMC		[Eda *et al.*, 1999]
Toxoplasma gondii cyst (Beverly strain)	LFcinB 17-41		ICR mice	[Tanaka *et al.*, 1995; Isamida *et al.*, 1998; Omata *et al.*, 2001]
sporozoite (Beverly strain)	LFcinB 17-41	MEC	Mice	[Tanaka *et al.*, 1995; Omata *et al.*, 2001]
tachyzoite (Beverly strain)	LFcinB 17-41	MEC	Mice	[Tanaka *et al.*, 1995; Omata *et al.*, 2001]

C32 AMC: C32 amelanotic melanoma cells; ALA: amoebic liver abscess; CHO cells: chinese hamster ovary cells; LFcinB: bovine lactoferricin; LFcinH: human lactoferricin; LFampin: lactoferrampin; LFchimera: lactoferrin chimera; HBM: human blood monocytes; HEC: human epithelial cells; IA: intestinal amoebiasis; MEC: mouse embryonal cells; MFC: murine fibroblast cells; MPM: mouse peritoneal macrophages; RHC: rabbit hepatobiliary cells. Numbers after peptides correspond to the residues in lactoferrin.

Table 3. Mechanism of action *in vitro* of Lactoferrin and its derivative peptides in parasites

Protozoa and stage	LF alone or in combination with other antimicrobial agents	LFcin alone or in combination with other antimicrobial agents	Mechanism of action	Reference
Extracellular				
Entamoeba histolytica (trophozoite)	apo-bLF alone apo-hLF alone apo-bLF + sIgA + Lysozyme apo-hLF + sIgA + Lysozyme (apo-bLF, apo-hLF) + MTZ		Interaction with the membrane lipids, (phospholipids and cholesterol) resulting in injury of the membrane and lysis. Synergic effect, mechanism unknown Synergistic effect, mechanism unknown	[León-Sicairos *et al* ., unpublished [León-Sicairos *et al* ., 2006a; 2006b
		LFcinB alone, LFcinB + MTZ, Lfcin, LFampin, LF chimera	LFcinB is able to penetrate amoeba cell and reach the nucleus and causing cell lysis. Synergistic effect, mechanism unknown Unknown Unknown	[León-Sicairos *et al* ., 2006b; [León-Sicairos *et al* ., 2006b; [López-Soto *et al* ., 2010] [López-Soto et al., 2010]
Giardia lamblia (trophozoite)	apo-bLF alone, apo-hLF alone	LFcinB alone LFcinH alone	Ultra-structural changes to trophozoites due to the interaction of LF and the LFcins with the cell surface. These changes affected the plasmalemma, endomembranes and cytoskeleton	[Turchany *et al* ., 1995; 1997]
Intracellular				
Anchaliia algerae (spore)	bLF alone	LFcinB alone	It is likely that the parasiticidal effect of hLF could be carried out in combination with antimicrobial peptides (defensins).	[Leitch and Ceballos, 2009]
Babesia caballi (intra-erythrocytic form)	apo-bLF alone		Inhibitory effect of apo-bLF appears to be related to inactivation or inhibition of a growth factor in the culture medium, specifically needed by *B. caballi.*	[Ikadai *et al* ., 2005]
Cryptosporidium parvum (sporozoite)	LFcinB, LFcinH		Unknown	[Carryn *et al.* , 2011]
Eimeria stiedai (sporozoite)		LFcinB alone	Inactivation of the parasite´s penetration capacity to cells after the pre-treatment with LFcinB.	[Omata *et al* ., 2001]
Enterocytozoon intestinalis (spore)	bLF alone	LFcinB alone	It is likely that the parasiticidal effect of bLF could be carried out in combination with antimicrobial peptides (defensins).	[Leitch and Ceballos, 2009]

Protozoa and stage	LF alone or in combination with other antimicrobial agents	LFcin alone or in combination with other antimicrobial agents	Mechanism of action	Reference
Enterocytozoon intestinalis (spore)	bLF alone	LFcinB alone	It is likely that the parasiticidal effect of bLF could be carried out in combination with antimicrobial peptides (defensins).	[Leitch and Ceballos, 2009]
Plasmodium falciparum (merozoite)	apo-hLF alone		Lactoferrin/iron complex generates oxygen free radicals, which may cause membrane damage of both erythrocyte and parasite In addition, LF strongly inhibited the adherence of PE to immobilized CSA.	[Fritsch *et al* ., 1987; Eda *et al* ., 1999]
		LfcinB	LF peptide inhibited CSA-, CD36-, and TSP-mediated cytoadherence of PE	[Eda *et al* ., 1999]
Pneumocystis carinii (cyst and trophozoite)	apo-bLF alone apo-hLF alone apo-bLF + clarithromycin		It is suggested that additive effect of clarithromicin may result from the cumulative inhibitory effect on different essential metabolic pathways of *P. carini* and the growth inhibitory effect of this parasite could be mediate by iron-chelating activity of apoLF.	[Cirioni *et al* ., 2000]
Toxoplasma gondii (tachyzoite)	apo-bLF		It appears that the *Toxoplasma* growth-inhibitory activity induced by native-bLF in macrophages may be mediated by an *l*-arginine-dependent effector pathway that does not involve NO production. In addition, it is suggest that effect of LF might be mediated by LF-receptors on macrophages, with signal transduction leading to tyrosine phosphorylation of a 30 kDa protein and finally to the inhibition of intracellular tachyzoites multiplication.	[Tanaka *et al* ., 1995; Isamida *et al* ., 1998; Omata *et al* ., 2001]
		LFcinB alone	Inactivation of the parasite´s capacity to penetrate cells after treatment with LFcinB, even earlier than the loss of membrane integrity.	[Tanaka *et al* ., 1995; Omata *et al* ., 2001]
Trypanosoma cruzi (amastigote)	apo-hLF		LF stimulated phagocytosis and intracellular killing capacity of HBM or MPM. The killing of internalized AMA by LF-treated MPM is mediated in part by H_2O_2- and O_2-.	[Lima and Kierszenbaum,1985; 1987a]

apo-bLF: bovine apolactoferrin; apo-hLF: human apolactoferrin; LFcinB: bovine lactoferrin; LFampin: bovine lactoferrampin;

CSA: chondroitin sulfate A; HBM: human blood monocytes; sIgA: human secretory immunoglobulin type A;

MTZ: metronidazole; MPM: mouse peritoneal macrophages; PE: *P. falciparum* infected erythrocytes;

TSP: thrombospondin

Interestingly, more than 30 years ago, observations among milk-drinking African nomads of the Maasai tribe, who consumed Cebu-cattle milk with a lower iron concentration than that typically reported for bovine milk, showed an unusual freedom from infection with *E. histolytica,* compared with similar nomads consuming a mixed diet. The authors proposed that the amoebicidal effect was due to the host's iron-binding proteins, TF and LF [93].

The capacity of human and bovine milk to kill amoebae *in vitro* was studied by our group of research [91]. In the presence of iron, the milk microbicidal effect was observed to be concentration-dependent and this effect increased in the absence of iron when trophozoites were incubated for 1 h in 5%, 10%, 15% and 20% of fresh human milk (86%, 78%, 66% and 62% cell viability, respectively). LF was purified from milk, and in the iron-free state, LF inhibited the amoeba growth at a concentration of 1 mg/mL (12.5 μM) [91]. In fact, both hLF and bLF were able to practically eliminate amoeba culture by using 100 μM LF. Furthermore, the peptide LFcinB (aa RRWQWRMKKLG) (600 μM) also inhibited the amoeba growth [91]. We found that in addition of LF, IgA and in less-grade lysozyme, were amoebicidal; it is unknown which other components of milk are amoebicidal.

Importantly, LF showed synergism with lysozyme and IgA against cultured amoebae [91]. Although the sensitivity of *E. histolytica* toward IgA, apoLF and lysozyme was assayed *in vitro*, the parasites might also be affected *in vivo* due to the presence of these proteins in the large intestine, the region where *E. histolytica* infection occurs. Thus, LF appears to be particularly important in the health and function of the intestinal tract. Also, LF greatly reduces inflammation in some bowel diseases, and LF concentration has been found to be augmented in the large intestine in patients with colitis or active ulcerative colitis, colon cancer, colon disorders and in other pathologies. These results indicate the crucial participation of LF at the intestine level.

Both LF and the peptide LFcinB were able to kill amoebae in a concentration-dependent manner. This effect was modulated by the age of the culture, since parasites in stationary phase were more susceptible than those in early logarithmic phase. Temperature of incubation, presence of ions and pH also influenced the parasiticidal activity of LF and the peptide. It is important to say that bLF, hLF, and LFcinB were able to synergize with low doses of metronidazole *in vitro* to eradicate amoeba in culture, suggesting that the combination of LF with this drug could be effective in amoebiasis treatment and, in turn, diminishing the risks to this drug [85].

We also have evaluated the effect of three bovine LF-derived peptides as anti-amoebics: LFcin (aa 17-30), lactoferrampin (LFampin) (aa 265-284), and a fusion of the two peptides (LF chimera). At 50 μM of peptide concentration, LFcin and LFampin had a moderate amoebicidal effect, since by 50% of viable trophozoites were obtained at 12 h. However, LF chimera showed the same mortality in only 3 h. Also, we observed the effect by confocal-laser scanning microscopy. The LF peptides were observed interacting with the amoebic surface at 30 s approximately, and in posterior times into cytosolic vesicles. This data suggests that the lactoferrin-peptides LFcin and LFampin posses a moderate amoebicidal effect, however, the LF chimera has a better amoebicidal activity on a time- and concentration-dependent manner [93].

Concerning the amoebicidal mechanism of LF, we have observed by confocal microscopy that the target sites could be mainly membrane phospholipids and cholesterol. After 1 h incubation with LF, this protein caused morphological changes on trophozoites, leading to cell rounding, vacuolization, membrane damage and lysis.

By electronic microscopy, we observed perforation of the cytoplasmic membrane. These changes were on surface since LF did not enter the amoeba cell. However, LFcinB penetrated the amoeba cell and reach the nucleus [91].

6. Bovine Lactoferrin Diminishes the Intestinal and Hepatic Abscesses Caused by E. histolytica Trophozoites in Animal Models

Two animal models, mouse and hamster, have been mainly used to study intestinal and hepatic amoebiasis, respectively. We have recently evaluated the therapeutic effect of LF in the control of both types of amoebiasis. Firstly, bLF was orally administered daily during a week to susceptible mice of the C3H/HeJ strain. Results showed that 20 mg LF/kg weight were able to eliminate the infection in 63% of the mice, since neither trophozoites nor evidence of epithelial damage and/or swelling were found in tissue sections of the cecum. The rest of the treated animals (37%) showed a decrease in the trophozoites number and mucus secreted to the lumen, as compared with untreated and infected mice. By immune-histochemistry, the profile of secreted cytokines into the cecum revealed that infected but untreated animals showed a mixed Th1/regulatory cytokines profile, whereas the cecum of mice treated (cured) showed a Th2 cytokine profile (IL-4) and expression of the multifunctional IL-6. In addition, cytokines and increasing cecal production of total IgA Ab were found associated with little inflammation and disease control observed in the cecum of LF-treated animals. These results suggest that oral administration of bLF could control intestinal amoebic infection probably by killing amoebae and favoring their removal and re-establishing the anti-inflammatory intestinal environment [94].

Concerning hepatic amoebiasis, we evaluated the therapeutic effect of bLF in comparison with metronidazole in a model of amoebic liver abscesses (ALA) in hamster, in which ALA were induced by intrahepatic injection of amoebae. Bovine LF was administered intragastrically during 8 days. Interestingly, bLF (2.5 mg/100 g weight) alone was able to reduce the ALA in hamsters, which showed no clinical signs of disease; in these animals, ALA significantly diminished, detecting only 0.63% of liver lesion, in comparison with 63% in non-treated animals. In hamsters where a mixture of metronidazole with LF was applied (0.5 and 2.5 mg/100 g weight, respectively), hamsters did not show clinical signs of disease and ALA was practically eradicated, detecting only 0.15% of lesion. Furthermore, liver function and blood cells trended to normal levels by the treatment with bLF, either alone or mixed with metronidazole. These results suggest that LF alone, or in combination with low doses of metronidazole, may help in the therapy of amoebiasis possibly without producing undesirable effects to patients [95].

Most probably, LF could be effective against intestinal and hepatic amoebiasis. However, it must be administered in the free-iron state (i.e., apoLF) and probably in enteric-coated formulas that permit the passage through the site infected by *E. histolytica* in the colon. In the case of hepatic amoebiasis, the peptides derived from LF generated in the stomach by the action of pepsin are absorbed in the small intestine and can act throughout the body and ultimately reach the liver.

In conclusion, the observed specific effect of physiologically occurring proteins like LF against *E. histolytica,* may open up a new approach for drug development against this parasite protozoan.

Giardia Lamblia

Giardia lamblia (syn. *Giardia duodenalis* or *Giardia intestinalis*), is one of the most common protozoan infections of the human small intestine and a leading cause of diarrheal disease in children. An estimated 200 million people have symptomatic giardiasis worldwide. Infection rates of close to 100% have been reported for some developing countries, whereas incidence rates of up to 7% giardiasis have been observed in developed ones. A rising incidence of giardiasis has been noted for children in day-care centers, which has led to the designation of giardiasis as a re-emerging infectious disease in the developed world. In addition, *Giardia* infection is a common cause of diarrhea in domestic animals (e.g. dairy calves, dogs, and cats) associated with substantial economic losses [96-97].

The parasite has a relatively simple life cycle consisting of two forms, the infectious cyst, which is resistant to many environmental factors, and the trophozoite, which colonizes the intestinal lumen and causes disease. Cysts are ingested through contaminated drinking water or occasionally food, or are acquired by person-to-person contact. Excystation from cysts to trophozoites occurs in response to luminal host signals encountered in the stomach and duodenum, i.e. low pH followed by elevated pH and proteases, whereas encystation requires an elevated pH and bile. *Giardia* mostly colonizes the proximal small intestine, although a significant proportion of patients also have ileal, colon, or stomach colonization. Trophozoites resist removal through the bulk flow in the intestinal lumen by proliferation and attachment to the mucus and intestinal epithelium, but they do not invade the mucosa. The non-invasive nature of the infection and close attachment to the epithelium (in an "off-shore" location) are characteristic features of giardiasis, and are important in the consideration of host defenses against the parasite [96]. Importantly, in animal models, the oral supplementation of vitamin A and LF has influenced growth of the ileum and colon.

7. Studies of Lactoferrin against *G. lamblia In Vitro* and in Patients with Diarrhea

Little is known about human defenses against *G. lamblia*. The clinical course of giardiasis is variable, and serum Abs do not appear to be protective. Secretory IgA Ab has a central role in the anti-giardial defense. LF is produced by epithelial cells in the small intestine in humans where it probably contributes to the protection of nursing infants from giardiasis [Eckmann, 2003]. It has been proposed that natural factors produced by intestinal tissue, transported into the intestine, or ingested in milk (ie, by breastfed babies) might promote resistance to this disease. Among these factors, a bile salt-stimulated lipase (a fatty acid esterase) may be a heat-labile giardiacidal component of the non-immune human milk [97]. It is likely that both immune and non-immune defenses influence the course and severity of the disease [96].

Bovine and human LF and two of their N-terminal peptides were shown to kill *G. lamblia* trophozoites *in vitro.* Interestingly, trophozoites harvested from very-low-density cultures were highly resistant to LF, and somewhat resistant to the peptides, in comparison with those cultured in late logarithmic or stationary phase, which were sensitive. On the other hand, ferric iron protected *Giardia* trophozoites against LF and the peptides although the latter lacking iron-binding sites [97]. Two years after, these authors demonstrated ultra-structural changes to trophozoites due to the interaction of LF and the peptides with the cell surface. These changes affected the plasmalemma, endomembranes and cytoskeleton [97].

Concerning the use of LF in giardiasis prevention, a study demonstrated a lower prevalence of colonization with Giardia and better growth among 52 Peruvian children of 12 to 36 months of age, to whom two daily doses of 500 mg each of recombinant bLF were administered for nine months [Ochoa et al., 2008]. These results suggest that LF might have therapeutic value in patients infected with Giardia spp. This fact is important given the huge number of children infected with this parasite worldwide.

Plasmodium falciparum

Malaria is a mosquito-borne disease of humans and other animals caused by protozoa of the genus *Plasmodium.* The disease results from the multiplication of parasites within red blood cells (erythrocytes), causing symptoms that typically include fever and headache, in severe cases, progressing to coma and death. It is widespread in tropical and subtropical regions, where it remains as the main cause of morbidity and mortality. There were an estimated 225 million cases of malaria worldwide in 2009, and from them 781,000 people died according to the [98]; 90% of malaria-related deaths occurred in sub-Saharan Africa, and the majority in young children. Five species of *Plasmodium* can infect humans. Severe disease is largely caused by *P. falciparum,* while those caused by *P. vivax*, *P. ovale*, and *P. malariae* are generally a milder disease and rarely fatal. On the other hand, *P. knowlesi* is a zoonosis that causes malaria in macaques but can also infect humans.

P. falciparum is a unicellular Apicomplexan parasite that spends part of its life-cycle in the mosquito (*Anopheles*), and part in humans and it has evolved a special form of motility that enables it to burrow into animal cells, a process termed "host cell invasion". The acute, life threatening, phase of malarial infection arises when the merozoite form of the parasite undergoes multiple cycles of invasion to erythrocytes and a rapid proliferation [99]. The particular virulence of *P. falciparum*, compared with other malaria species which naturally infect humans, is thought to be due to the way in which the parasite modifies the surface of the infected erythrocytes. Approximately 16 h into the asexual cycle, parasite encodes proteins which appear on the erythrocyte surface and mediate adherence to a variety of host tissues. Binding of infected erythrocytes to the vascular endothelium, a process which occurs in all of the infections, it is an important factor in the pathogenesis of severe disease where concentration of organisms occurs particularly in brain. Binding to uninfected erythrocytes to form erythrocyte rosettes, a property of some isolates, is linked to disease severity [100].

Antimalarial drugs that interfere with haem detoxification, or target other parasite-speific processes, have been effective in the treatment of malaria, but their use has been dogged by the development of resistance. Similarly, efforts to obtain an effective blood vaccine are hindered by the variability of surface-exposed antigens [101]. Currently, this parasite is resistant in most endemic areas to the widely used blood drugs schizonticide and chloroquine. Many strains are also resistant to antifols (e.g. pyrimethamine, proguanil), and some are no longer eliminated by quinine. These polyresistant parasites also have an enhanced ability to resist new drugs such as mefloquine and halofantrine. There are indications that *P. vivax* is also becoming resistant to chloroquine in some African countries where primaquine resistance of the hypnozoites also exists. New developments in antimalarials include the search for radically new compounds, for drugs that reverse chloroquine resistance and for new strategies to stop the progress of this problem. One of the strategies that have been approached is the use of molecules of the human immune response, such as the protein LF and its derived peptides.

8. INTERACTION BETWEEN *PLASMODIUM FALCIPARUM* AND LACTOFERRIN *IN VITRO* AND *IN VIVO*

The first evidence about the *in vitro* growth inhibition of *P. falciparum* by LF was reported by Kharazmi *et al.* [102]. This group of research presented evidence that neutrophils (which contain LF in their secondary granules) from chronic granulomatous disease (CGD) patients (who are defective in oxidative metabolism), have the capacity to inhibit the *in vitro* multiplication of *P. falciparum*. By using a micro-titre inhibition assay, they incubated peripheral-blood neutrophils from CGD patients and from normal individuals with *P. falciparum* isolate F32 in a culture system. There was an inhibition of parasite growth by neutrophils after 48 h of culture. At neutrophil to erythrocyte ratio of 1:50, an inhibition of parasite growth of 57% by normal neutrophils, and 39-68% by CGD patient cells, was found. In this work, when the neutrophils were stimulated by phorbol myristate acetate, both cell types enhanced inhibition of parasite growth. These findings suggest that the oxygen-independent systems of human neutrophils are involved in parasite destruction, such as acid hydrolases, LF, and cationic proteins, which could be regarded as potential mediators of parasite destruction [102].

Three years after, another group of research reported the *P. falciparum* growth inhibition *in vitro* by LF and the iron-chelators desferriferrithiocin and desferricrocin. These microbial chelators were able to inhibit the growth of *P. falciparum* in a dose-dependent way. Parasite multiplication was stopped at 25-30 μM desferriferrithiocin, whereas 60-90 μM desferricrocin was needed to exhibit the same effect. Obviously, after iron saturation, the microbial chelators were ineffective; in contrast, both iron-free and iron-saturated hLF (30 μM) inhibited *P. falciparum* growth. A 3-day incubation of host erythrocytes with iron-free and iron-saturated hLF prior to infection enhanced the effect, which was therefore attributed to LF bound iron. In this work, authors concluded that this effect could be due to LF/iron complex generates oxygen-free radicals, which may cause membrane damage to both erythrocyte and parasite and then this process could be considered to lead growth inhibition of the parasite [103].

In 1999, Eda et al. reported that LF (100 μg/mL) strongly inhibited the adherence of *P. falciparum*-infected erythrocytes (PE) to immobilized chondroitin sulfate A (CSA)-conjugated albumin, and LF also blocked the PE binding to CD36-expressing Chinese hamster ovary (CHO) cells, as well as immobilized CD36 at concentrations of 5 μg/mL and

100 μg/mL, respectively. In this work, biotinylated LF bound to CD36 in a saturable manner, and such binding was inhibited by unlabeled LF and the anti-CD36 monoclonal antibody (mAb) 8A6, suggesting specificity of binding. Furthermore, LF (100 μg/mL) inhibited PE binding to immobilized thrombospondin (TSP); specific binding of LF to TSP was confirmed using biotinylated LF. LF inhibited PE binding to C32 amelanotic melanoma cells in a dose-dependent manner. In addition, a peptide of LF, RNMRKVRGPPVSC (aa 25-37 of LF), which has been suggested to contribute to LF binding to various materials, including CSA, inhibited PE binding to immobilized CSA-conjugated albumin, immobilized CD36, CD36-expressing CHO cells, immobilized TSP, and C32 amelanotic melanoma cells, as well as LF itself. These results suggest that LF peptide may provide the basis for developing agents that are able to inhibit CSA-, CD36-, and TSP-mediated cytoadherence of PE [104].

CD36 and CSA are points of contact between the infected erythrocyte and vascular endothelial cells. One of the best characterized among these receptors is CD36, an 88 kDa integral membrane protein found on monocytes, platelets and endothelial cells; infected erythrocytes from most parasite isolates bind to CD36. However, CD36 has not been detected in endothelial cells of the cerebral blood vessels, and parasites from clinical isolates tend to adhere to both CD36 and intracellular adhesion molecule-1 (ICAM1). ICAM1 is a member of the immunoglobulin superfamily and functions in cell-cell adhesion. In addition, sequestration of infected erythrocytes and ICAM1 expression has been co-localized in the brain. Also, chondroitin sulfate A (CSA) has been implicated in the cytoadherence within the placenta and may contribute to the adverse effects of *P. falciparum* during pregnancy [101].

It is important to clear the fact that during its intra-erythrocyte period, the parasite modifies this host cell to make it a more suitable habitat. For example, the erythrocyte membrane becomes more permeable to small molecular weight metabolites, presumably reflecting the needs of an actively growing parasite. Another modification of the host cell concerns the cytoadherence of PE to endothelial cells and the resulting sequestration of mature parasites in capillaries and post-capillary venules. This sequestration likely leads to microcirculatory alteration and metabolic dysfunction which could be responsible for many of the manifestations of severe falciparum malaria. The cytoadherence to endothelial cells confers at least two advantages to the parasite, a microaerophilic environment which is better suited for its metabolism, and avoidance of the spleen and subsequent destruction in this organ [99].

Finally, Kassim et al [105] established the presence of factors in breast milk and sera against *in vitro* growth of *P. falciparum* parasite. They showed the presence of significant Ab titers to ring, trophozoite, schizont and gametocyte stages of *P. falciparum* in 144 Nigerian maternal-milk samples and also in paired maternal and infant sera. This study also demonstrates a significant *in vitro* growth inhibition of *P. falciparum* by using the maternal and infant sera, but most notably by breast milk samples and breast milk constituents, such as LF and sIgA, suggesting a protective *in vivo* role for breast milk in the possible modulation of malaria frequency, severity, and complications [105].

In conclusion, whether a molecule such as LF or its peptides is able to avoid the binding of *P. falciparum*-infected erythrocytes to some of its target or point of contacts located in the host endothelial cells, these could be potential anti-malarial compounds.

Toxoplasma gondii

Toxoplasmosis is a disease caused by *Toxoplasma gondii*, an obligate intracellular protozoan that has a worldwide distribution and is considered one of the most successful parasites on Earth [106-107]. The tissue cyst-forming coccidium of *T. gondii* has a facultative heterogeneous life cycle and can probably infect all warm-blood animals (mammals and birds), with the cat being the only definitive host. *T. gondii* is of veterinary and human importance, because it is the causative agent of a zoonosis and by the economical losses produced due to the treatment to infected cattle [106]. Infection is mainly acquired by ingestion of food or water contaminated with oocysts shed by cats or by eating undercooked or raw meat containing tissue cysts, both leading to the parasite replicative form, the tachyzoite [108]. In the feline intestinal epithelium, the parasites develop through a sexual cycle, resulting in oocyst shedding. The oocysts are resistant to the environmental stress and undergo sporogony to form sporozoites which are infective in intermediate and definitive hosts [109]. Although up to one third of the world human population is infected with *T. gondii* [108], most infections are asymptomatic since a primary infection is usually subclinical but in some patients, cervical lymphadenopathy or ocular disease can be present. Infection acquired during pregnancy may cause severe damage to the fetus. In immune-compromised patients, reactivation of latent disease can cause life-threatening encephalitis [106,108].

9. Lactoferrin and Lactoferricin B Protect against *Toxoplasma gondii*

The effect of LF in *T. gondii* has been carefully studied by Susuki´s research group. They found out that LF has no direct parasiticidal effect on the protozoan (RH strain), since the treatment with LF did not cause loss of the parasite membrane integrity or loss in its cell penetration activity [110-111]. For these experiments, bovine LF at the iron-saturation level found in milk was used native-bLF), thus although the authors did not mention, the method they used produces LF with ~10% Fe^{3+} [112 -113].

Even though bLF does not have a direct effect in *T. gondii* tachyzoites, the development of intracellular parasites (RH strain) was actually inhibited by this molecule. This was assessed in mouse peritoneal macrophages (MPM) and embryo cells incubated with tachyzoites [111]. Treatment with native-bLF showed a tendency to suppress toxoplasma growth in a dose-dependent manner, reaching over 70% suppression with 1 mg/mL bLF [111]. This observation also applies for holo-hLF, since it inhibited the *in vitro* replication of *T. gondii* (BK strain) in human epithelial and murine fibroblast cell lines at the same concentration, and it did not present a direct parasiticidal activity against tachyzoites [114]. This concentration might seem high but actually is a physiological concentration [54,114-115], so a high concentration of LF in mucosa could effectively prevent the intensive replication at the site of infection in the gut. Interestingly, apo-bLF and holo-bLF showed similar effects in *T. gondii* development of intracellular parasites (tachyzoites). These results suggest that bovine LF induces the inhibitory effect on the development of intracellular parasites in an iron-independent way, since apoLF, nativeLF, and holoLF displayed very similar anti-toxoplasma activity [111].

It has been observed that actually apo-bLF binds to tachyzoites, probably using the same binding protein than that for apo-bTF and apo-ovoTF (from duck), since both compete with apo-bLF for the binding. This unspecific binding could be through a 42 kDa apo-bLF binding protein, detected by far Western blot of tachyzoite proteins [Tanaka *et al.*, 2003]. However, the exact function of this putative apo-bLF binding protein is unknown, since it has been explored that bLF does not directly influence the viability of *T. gondii* [110-111], so further studies will be required to determine the function of this binding protein.

The mechanism by which LF inhibits the growth of *Toxoplasma* in cells has not been completely elucidated. When bLF was added after the *T. gondii* inoculation of MPM, it suppressed *T. gondii* growth, but bLF had no effect when it was added before the inoculation. The mechanism seems to be different from cell to cell since when embryo cells were incubated with bLF before the *T. gondii* inoculation there was an inhibitory effect in the parasite growth, while when incubated with bLF after the inoculation, there was no effect [111]. Actually, the mechanisms described of intracellular killing of *T. gondii* are different in these cells. In macrophages activated by INF-γ, there is oxygen-dependent killing through reactive oxygen species and reactive nitrogen intermediates derived from L-arginine [116-119], while in human fibroblasts activated by INF-γ, the suppression of growth of the intracellular parasites was due to starvation for tryptophan [120]. However, when production of O^{2-} was measured, it was diminished in cultures of macrophages supplemented with native-bLF and this effect was dose- and time-dependent. These findings suggest that the *Toxoplasma* growth-inhibitory activity induced by LF in macrophages is not mediated by O^{2-} or nitric oxide (NO) molecules [121]. A competitive inhibitor of the L-arginine-dependent effector pathway, NG-monomethyl-*l*-arginine diminished the inhibitory activity induced by native-bLF in cultures. From these findings, it appears that the *Toxoplasma* growth-inhibitory activity induced by native-bLF in macrophages may be mediated by an *l*-arginine-dependent effector pathway that does not involve NO production [121]. This is in opposition of what has been found in other models where reactive oxygen intermediates are involved in the intracellular killing of *Trypanosoma cruzi* amastigotes by LF treated macrophages [122].

By using an anti-[phosphorylated tyrosine] mAb, phosphorylation of tyrosine residues was detected in protein(s) of approximately 30 kDa in macrophages incubated with native-bLF, and the density of the band increased dose-dependently. Inhibition of the native lactoferrin-induced tyrosine phosphorylation by genistein led to a loss of the native lactoferrin-induced growth inhibitory effect against the parasites, suggesting that native-bLF induces tyrosine phosphorylation in macrophages and phosphorylation seems to be associated with the induction of the growth inhibitory activity exerted against intracellular *T. gondii* [119]. A possibility is that stimulatory effect of LF might be mediated by LF-receptors, with signal transduction leading to tyrosine phosphorylation of this 30 kDa protein and finally to the inhibition of intracellular tachyzoites multiplication; however, further research is needed to completely elucidate the anti-*Toxoplasma* mechanism. In fact, very little is known about the actual mechanisms of action of LF on mammalian cells expressing LF-receptors [22]; however, it has been observed that interactions of LF with receptors on cell surface is responsible for the activation of signaling pathways and/or endocytosis and targeting of all or part of LF in the cell nucleus where it can act as a transcriptional activator [22,38,123].

Particularly active against *T. gondii* seems to be the bovine LF-derived peptide LFcinB, but unlike the whole LF, LFcinB has a direct parasiticidal effect on *Toxoplasma*. In 1995, Tanaka et al. described that LFcinB has an anti-toxoplasma activity. The treatment of *T. gondii* (RH strain) with LFcinB caused an irreversible loss of trypan-blue dye exclusion activity [110]. It was observed inactivation of the parasite´s capacity to penetrate cells after the treatment with LFcinB, even earlier than the loss of membrane integrity. This was assessed *in vitro* using tachyzoites and sporozoites (Beverly strain) entering into mouse embryonal cells (MEC), and *in vivo* using mice inoculated with tachyzoites, sporozoites, and cysts pre-treated with LFcinB [109-110]. A LFcinB dose of 100 μg/mL or higher achieved efficacy, consistently with the effect showed to other microorganisms [109,124].

Oral LFcinB also had a protective effect against oral administration of *T. gondii* cysts (Beverly strain) in mice, and this effect was dose-dependent [125]. The protection could be caused by the direct interaction of LFcinB with *T. gondii*, consistent with Tanaka previous work [110]. Also, when mice were intraperitoneally challenged with cysts, oral LFcinB protection was incomplete, supporting the idea that direct contact is needed. This lead to suggest that LFcinB could be used as a food additive, due to its anti-toxoplasma activity [125].

Trypanosoma cruzi

Chagas or American trypanosomiasis is a tropical infection, one of the most important endemic diseases in Latin America caused by the flagellate protozoan Trypanosoma cruzi. T. cruzi is commonly transmitted to humans and other mammals by an insect vector, the blood-sucking insects of the subfamily Triatominae (family Reduviidae) [126]. The disease may also be spread through blood transfusion and organ transplantation, ingestion of food contaminated with parasites, and from a mother to her fetus [126-127]. Chagas disease affects 8-10 million people living in endemic Latin American countries, with an additional 300,000–400,000 living in non- endemic countries. An estimated 41,200 new cases occur annually in endemic countries, and 14,400 infants are born with congenital Chagas disease annually. About 20,000 deaths are attributed to this disease each year [126,128]. The disease is present in 18 countries on the American continent, ranging from the Southern United States to Northern Argentina. In the last two decades, many efforts have been made to reduce the incidence of Chagas disease in endemic countries [126,128], but exchange of population between Latin America and Europe, the United States, Australia and Japan has resulted in increased detection of T. cruzi in these countries [126].

The symptoms of Chagas disease vary over the course of an infection. In the early acute stage, symptoms are mild and usually produce no more than local swelling at the site of infection. The initial acute phase is responsive to anti-parasitic treatments, with 60–90% cure rates. After 4–8 weeks, individuals with active infections enter the chronic phase of Chagas disease that is asymptomatic for 60–80% of chronically infected individuals through their lifetime. The human disease occurs in two stages: an acute stage, which occurs shortly after an initial infection, and a chronic stage that develops over many years [122,129].

T. cruzi is an obligate intracellular parasite, their infective forms are metacyclic and bloodstream trypomastigote and amastigote. Metacyclic trypomastigotes are released in the insect feces while amastigotes and bloodstream trypomastigotes are released from the infected host cells of the vertebrate host after a complex intracellular life cycle. The recognition between parasite and mammalian host cell involves numerous molecules present in both cell types. T. cruzi predominantly invades macrophages and cardiomyocytes, leading to persistent infection [128]. After invasion, the clinical manifestations of Chagas disease are due to cell death in the target tissues that occurs during the infective cycle, by sequentially inducing an inflammatory response, cellular lesions, and fibrosis. For example, intracellular amastigotes destroy the intramural neurons of the autonomic nervous system in the intestine and heart, leading to megaintestine and heart aneurysms, respectively. If left untreated, Chagas disease can be fatal, in most cases due to heart muscle damage [126,128].

Benznidazole and nifurtimox are drugs that have been used to treat the disease. However, both drugs induce severe side effects [128-129]. The need of effective drugs, without adverse effects, has stimulated the search for new compounds with potential clinical utility; also, the participation of the human immune response has been the aim of many groups of researchers. Different studies have defined CD8+ T lymphocytes, IFN-γ and macrophages as important elements controlling parasite replication during the acute phase of infection; due the host innate immune system playing a pivotal role in the recognition of T. cruzi infection and the subsequent development of adaptive immunity [130].

Concerning works related to the research field of LF and Trypanosoma, in 1987, Lima and Kierszenbaum reported that this glycoprotein may contribute to macrophage clearance of amastigote forms (AMA) released from infected host cells. In their experiments, they co-cultured mouse peritoneal macrophages (MPM) or human blood monocytes (HBM) with intracellular AMA of T. cruzi in the presence of hLF. Results displayed a greater amount of organisms inside phagocytes and a high proportion of phagocytes (taking up AMA) in the presence of hLF. Moreover, pretreatment of either MPM or AMA with LF also enhanced cell-parasite association. LF was capable of binding to the AMA forms and phagocytes. In other assays, hLF stimulated phagocytosis and intracellular killing capacity of HBM or MPM. Thus, with the results, researchers suggested that LF may contribute to macrophage clearance of AMA released from infected host cells [131].

The same group of research reported that the presence of iron is required for the LF to stimulating intracellular killing by macrophages but not to enhancing the uptake of particles and microorganisms. In these studies, the iron-binding protein TF did not alter the capacity of MPM to either take up or kill the AMA, indicating that the noted LF effects were not shared by all iron-binding proteins [122,132]. They also reported in this work that killing of internalized AMA by LF-treated MPM previously reported to be mediated in part by H_2O_2 and O_2^-. Authors discussed about the role of AMA LF receptor its biological significance in parasite-host interaction, since mononuclear phagocytes also express a LF receptor, and treatment of these cells with LF has been shown to increase their capacities to take up and kill T. cruzi amastigotes in vitro [132].

10. Parasites of Veterinary Importance

Babesia caballi

Equine piroplasmosis is considered the most important tick-borne disease of horses and is caused by two Apicomplexan protozoan parasites, the exclusively intra-erythrocytic *B. caballi* and the intra-erythrocytic and intra-lymphocytic *Theileria equi* [133]. Piroplasmosis occurs in regions and environments (pasture) where horses are exposed to vector ticks, especially of the genera *Dermacentor*, *Hyalomma* and *Rhipicephalus* [134-136]. The disease is widely distributed in most tropical and subtropical areas of the world as well as in some temperate zones [137], and represents a serious threat to the horse-raising industry and international movement of horses [138-139]. The clinical manifestations of the disease are often variable, acute *B. caballi* infection is characterized by high fever (40 °C), anemia, anorexia, malaise, tachypnea, and dyspnea [140]. Following the acute phase, horses remain persistently infected and serve as reservoirs for transmission by tick vectors [141]. *T. equi* infections are considered more severe since they are more commonly associated with hemoglobinuria and death [142].

11. Lactoferrin Interacts in Different Manner with *Babesia* and *Theileria*

The *in vitro* intra-erythrocytic growth of *T. equi* was not inhibited in the presence of 5 mg/mL native-bLF, holo-bLF, or apo-bLF. In contrast, the growth of *B. caballi* was significantly suppressed only in the presence of apo-bLF, with an IC_{50} value of 2.7 mg/mL. *B. caballi* growth was neither inhibited in the presence of 5 mg/mL of a LF hydrolyzate produced by pepsin cleavage [143]. In an attempt to determine the apo-bLF mechanism of action in *B. caballi*, the authors mixed apo-bLF with the growth media, and then they removed the apo-bLF using a heparin (HP) column, after what they use this medium to repeat the experiments. They found the same IC_{50} for this apo-bLF-HP medium than for the apo-bLF medium, suggesting that the inhibitory effect on the growth of *B. caballi* is not due to a direct interaction with apo-bLF. They determined the iron content in the normal medium and in apo-LF-HP medium, and it was the same, this way ruling out an iron sequestering role of apo-bLF in the parasiticidal effect [143]. The inhibitory effect of apo-bLF on *B. caballi* appears to be related to inactivation or inhibition of a growth factor in the culture medium, specifically needed by *B. caballi* and not for *T. equi*. Growth inhibition of *B. caballi* did not appear to be related to general actions of LF, suggesting that this is a novel antimicrobial mechanism of LF [143].

Eimeria stiedai

Coccidioses of rabbits are ubiquitous infections caused by obligatory intracellular protozoan parasites belonging to the genus *Eimeria* are considered major causes of significant morbidity and mortality worldwide [144-145]. *E. stiedai* is the only one that invades the hepatobiliary epithelial cells and is the causative agent of hepatic lesions and obstructive jaundice due to oocyst shedding [109,146]. The consequences of this condition could result in overwhelming diseases that may coincide with economic losses in rabbit rearing industries [145].

Just like it happens with the previously mentioned coccidium *T. gondii*, with *E. stiedai,* the inactivation of the parasite´s penetration capacity to cells after the pre-treatment with bovine LFcin was observed. This was assessed *in vitro* using sporozoites entering into rabbit hepatobiliary cells and *in vivo* using rabbits inoculated with sporozoites pre-treated with LFcin. It is conceivable that a sufficient amount of LFcin provided in the rabbit diet may be effective to inactivate *E. stiedai* [109].

12. Other Parasites Less Studied (Cryptosporidium, Pneumocystis, Encephalitozoon, Nosema, Enterocytozoon)

Cryptosporidium parvum

Cryptosporidium is a protist belonging to the large Apicomplexa phylum that includes important pathogens such as *Plasmodium falciparum* and *Toxoplasma gondii* [147]. *C. parvum* is transmitted via the oocyst stage; oocysts can be ingested by contaminated water or foodstuffs. In contrast to other Apicomplexan parasites, the infection is minimally invasive as it is restricted to epithelial cells lining the gastro-intestinal tract [148,149]. *C. parvum* is a major etiological agent of calf diarrhea [150]. Point prevalence estimates for cattle <3 months of age is up to 100%, and the high numbers of oocysts shed in feces ($>5x10^6$ oocysts per gram), by even apparently healthy animals, pose a significant threat to environmental contamination and transmission to humans [150,151]. In healthy humans, gastrointestinal tract infection with *C. parvum* leads to a self-limiting diarrhea. In immune-compromised hosts such as AIDS patients, however, a severe, life threatening, cholera-like infection can be developed [152]. Dissemination to extra-intestinal sites may occur and contribute to morbidity [147, 153].

13. Bovine and Human LF-derived Peptides are Microbicidal for *C. parvum*

LF had no effect on either viability or infectivity of *C. parvum* sporozoites, however both its derivatives, LFcinH and LFcinB reduced the sporozoite viability and *in vitro* infectivity. Reduction of infectivity by these cationic peptides is due principally to its effect on sporozoite viability, and its parasiticidal activity is greatest on extracellular stages [154]. The effect of LF is similar to what happens with *T. gondii* [110], where incubation with LFcinB (100 μg/mL, 60 min) significantly reduced the parasite viability and infectivity to levels comparable to those observed in the present study, but *C. parvum* seems to need lower concentrations and a shorter incubation period (10 μg/mL, 15 min) [154].

Thus, LFcinB is apparently more effective against *C. parvum*, although only a parallel experiment with *T. gondii* would confirm this.

Pneumocystis carinii

Pneumocystis genus comprises diverse species of parasites that proliferate in the alveolus of a wide variety of mammalian hosts. *P. carinii* is an extracellular lung pathogen that causes a severe pneumonia (*Pneumocystis* pneumonia or PcP) in immunological-compromised hosts, especially those with AIDS, affecting 15-28% of these patients annually. Malignant blood diseases, premature babies and people who receive organ transplants and take drugs to suppress the immune system constitute other groups at risk of infection with this pathogen. Infective forms of *P. carinii* are acquired through inhalation. These parasites bind to type-I pneumocytes and prevent gas exchange between capillaries and alveoli, causing hypoxemia and death [155-157]. In animal models, it has been found that *P. carinii* is able to replicate in the lungs of immune-competent hosts and can be transmitted by airborne route to susceptible organisms or to a second immune-competent host [158]. Trophic forms, sporocytes and mature cysts are usually considered the three main morphological forms involved in the *Pneumocystis* life cycle. Trophic forms are the most abundant of all *Pneumocystis* life-cycle stages, representing 90-95% of the total population in the lungs of hosts with PcP [159].

The efficacy of several anti-*P. carinii* new compounds has been extensively evaluated *in vitro* and in animal models. Combinations of macrolides and clindamycin, primaquine, atovaquone, dihydrofolate-reductase inhibitors, and trimethoprim-sulphamethoxazole (TMP-SMX) have a good activity against *P. carinii in vitro,* in the rat model and in humans [160-162]. However, agents in current clinical use have limitations, thus there is a need for more potent and less toxic compounds or new drug combinations [163-166]. Importantly, it has been demonstrated that iron chelators are effective as anti-*P. carinii* agents in experimental models [160-162].

Cirioni et al. [167] investigated the *in vitro* activity of LF alone and in combination with other clinically used antimicrobial agents against *P. Carinii* [167]. They found that iron-free bLF alone was the most active agent, since it suppressed the growth of cystic and trophic forms by >60% (IC_{50} 20 mg/L). Iron-free hLF, at the same concentration, was able to suppress the growth of cystic and trophic forms by >50% (IC_{50} 20 mg/L). Interestingly and importantly, both LFs at 20 mg/L combined with clarithromycin (4 mg/L) had a high anti-*P. carinii* activity, with >90% decrease in cystic and trophic form counts.

These results suggest that bovine and human apo-LF may inhibit *P. carinii* growth in human lungs and could act synergistically with other clinically used compounds. Iron is a vital element for the growth of *Pneumocystis* in the host lung and although the authors did not mention, it is likely that growth inhibitory effect of this pathogen was mediated by iron-chelating activity of LF. The antimicrobial activity of macrolides (such as clarithromycin), results from of the ability to inhibit protein synthesis by binding to the transpeptidation site of the larger ribosomal subunit [168]. Then, it is suggested that additive effect of clarithromycin may result from the cumulative inhibitory effect on different, essential metabolic pathways of *P. carinii* and iron-chelating activity of LF. These findings provide experimental support to the use of apo-LF in combination with macrolides in the therapy of *Pneumocystis* infections [167].

Microsporidia

Microsporidia are unicellular, obligate intracellular parasites infecting a broad range of vertebrate and invertebrate hosts. The phylum Microsporidia consist of nearly 150 genera with more than 1200 species, but only seven genera [*Enterocytozoon*, *Encephalitozoon* (including *Septata*), *Pleistophora*, *Trachipleistophora*, *Vittaforma*, *Brachiola* and *Nosema*] as well as unclassified microsporidia have been described as pathogens in humans [169]. Microsporidiosis is an emerging and opportunistic infection in AIDS patients, organ transplant recipients, children, travelers, contact lens wearers and the elderly. It is associated with a wide range of clinical syndromes of microsporidiosis in humans. Additionally, these organisms have long been known to be causative agents of economically important diseases in insects (silk worms and honey bees), fish and mammals (rabbits, fur-bearing animals, and laboratory rodents) [170]. Species of microsporidia infecting humans have been identified in water sources as well as in wild, domestic and food-producing farm animals, raising concerns of water-borne, food-borne and zoonotic transmission.

Microsporidia life cycle includes a proliferative merogonic stage, followed by a sporogonic stage resulting in characteristically small, environmentally resistant, infective spores. The spores contain a long, coiled, tubular extrusion apparatus (polar tube) and has a crucial role in host cell invasion. Upon extrusion from the spore, it injects the sporoplasm into the cytoplasm of a new host cell after piercing the cell plasmalemma or the membrane of phagosomes containing the endocytosed spores. In the gastrointestinal tract, where the majority of human microsporidia infections occur, infection of differentiated mucosal epithelial cells most likely results from impalement via spores that germinate in the lumen in close proximity to the target cells [171-173]. Parasites often cause a severe, non-bloody, non-mucoid diarrhea with up ten or even more bowel movements per day, slowly progressive weight loss, and malabsorption of fat, D-xylose, and vitamin B_{12}. In addition, in humans, environmental spores infect the nasopharynx, lungs, eyes and skin [173-174].

Several human-infecting microsporidia (*Vittaforma corneae, Encephalitozoon hellem, Encephalitozoon cuniculi, Encephalitozoon intestinalis, Trachipleistophora hominis*) have been cultured in different mammalian cell lines (e.g., rabbit and monkey kidney cells, human fetal lung fibroblasts); this fact has been of enormous benefit in understanding the biologic aspects of microsporidia. In addition, animal models are necessary to study routes of transmission, immune responses, and therapeutic strategies. Different immune-deficient mice [(e.g., severe combined immunodeficient (SCID) athymic mice)], rabbits and simian immune-deficiency rhesus monkeys have been infected with several human-infecting microsporidia [169].

14. LACTOFERRIN AND LACTOFERRICIN B INHIBIT THE GROWTH OF MICROSPORIDIA

The most effective drugs for treating microsporidiosis in humans currently include albendazole, which is effective against the *Encephalitozoon* species but not against *Enterocytozoon bieneusi*, and fumagillin, which has broader anti-microsporidia activity but is toxic in mammals. Further work is necessary to identify other drugs, which can be both effective and devoid of adverse effects [175]. Currently, the inhibition of spore germination process is an attractive target for therapy of microsporidiosis [172,176]. In 2009, clinical isolates from microsporidia species were used in spore germination and enterocyte-like (C2Bbe1) cell infection assays to determine the effect of a panel of antimicrobial peptides and the glycoprotein LF. Spores were incubated with hLF and LfcinB, or with some defensins, each alone or in combination with lysozyme, prior to the germination assay [63]. They found that *Anncaliia (*before grouped in the genera *Brachiola* or *Nosema) algerae* spore germination was inhibited by all of the compounds tested. The peptide concentration used in these experiments was in the range 2 μg-2 mg/mL. The *A. algerae* minimum inhibitory concentration (MIC) for hLF was 10 μg/mL. With the smaller peptide LfcinB, the MIC for *A. algerae* germination was in the range of 5-10 μg/mL. Interestingly, hLF significantly inhibited germination in the presence or absence of Lz. Human LF and lysozyme are ubiquitous in the gastrointestinal tract, and is likely that in the human intestine LF and Lisozyme interact with microsporidia spores and could potentially inhibit their germination. In addition, the parasiticidal effect of hLF could be carried out in combination with other antimicrobial peptides such as defensins, which are found on the intestinal villus surface, in the lumen of the crypt of Lieberkuhn, and in the *lamina propria* neutrophils [177-178]. Thus, LF appears to be particularly important in maintenance and functionality of the intestinal tract.

Concerning the infection of differentiated enterocytes by the microsporidia species, Leitch and Ceballos [69] found that a combination of the neutrophils defensin with both hLF and Lysozyme resulted in a statistically significant greater decrease of C2Bbe1 cells infection by *A. algerae*. In addition, lysozyme alone only inhibited infection of differentiated enterocytes by *A. algerae*, while hLF inhibited infection by *Encephalitozoon intestinalis* (generally infects the human gastrointestinal tract) and *A. algerae*. This data suggests that intestinal antimicrobial peptides and LF can contribute to the defense of the intestine against infection by luminal microsporidia spores and may open a new approach for drug development against these parasites [69].

Conclusion

Parasitic pathogens are microorganisms that cause serious diseases to humans and other animals worldwide. In an effort to have new and safe drugs to combat parasites without producing side effects to patients, the protein of the mammalian innate-immune system Lactoferrin and several of its N-terminal derived-peptides have been studied in assays *in vitro* and *in vivo*. Results about its anti-parasitic activity and safe use in animal models and humans are very promising; in addition, Lactoferrin modulates the immune response to parasites in benefit of the host. Therefore, this protein cold being used alone and/or in combination with anti-parasitic drugs to treat different diseases in a near future.

ACKNOWLEDGMENTS

This work was supported by Consejo Nacional de Ciencia y Tecnología (CONACyT), México, grant 60102.

REFERENCES

[1] Bannister, L.H., Hopkins, J.M., Fowler, R.E., Krishna, S., Mitchell, G.H. A brief illustrated guide to the ultrastructure of *Plasmodium falciparum* asexual blood stages. *Parasitol Today.* 2000 Oct;16(10):427-33.

[2] Sacks, D.L. *Leishmania*-sand fly interactions controlling species-specific vector competence. *Cell Microbiol.* 2001 Apr;3(4):189-96.

[3] Macedo, A.M., Machado, C.R., Oliveira, R.P., Pena, S.D. *Trypanosoma cruzi*: genetic structure of populations and relevance of genetic variability to the pathogenesis of chagas disease. *Mem Inst Oswaldo Cruz.* 2004 Feb;99(1):1-12.

[4] Garcia, E.S., Ratcliffe, N.A., Whitten, M.M., Gonzalez, M.S., Azambuja, P. Exploring the role of insect host factors in the dynamics of *Trypanosoma cruzi-Rhodnius prolixus* interactions. *J Insect Physiol.* 2007 Jan;53(1):11-21.

[5] Carvalho, S., Cruz, T., Santarem, N., Castro, H., Costa, V., Tomas, A.M. Heme as a source of iron to *Leishmania infantum* amastigotes. *Acta Trop.* 2009 Feb;109(2):131-5.

[6] Lambert, H., Barragan, A. Modeling parasite dissemination: host cell subversion and immune evasion by *Toxoplasma gondii. Cell Microbiol.* 2010 Mar;12(3):292-300.

[7] Simpson, A.G., Stevens, J.R., Lukes, J. The evolution and diversity of kinetoplastid flagellates. *Trends Parasitol.* 2006 Apr;22(4):168-74.

[8] Ronnebaumer, K., Gross, U., Bohne, W. The nascent parasitophorous vacuole membrane of *Encephalitozoon cuniculi* is formed by host cell lipids and contains pores which allow nutrient uptake. *Eukaryot Cell.* 2008 Jun;7(6):1001-8.

[9] Ueno, N., Bratt, C.L., Rodriguez, N.E., Wilson, M.E. Differences in human macrophage receptor usage, lysosomal fusion kinetics and survival between logarithmic and metacyclic *Leishmania infantum chagasi* promastigotes. Cell Microbiol. 2009 Dec;11(12):1827-41.

[10] Landfear, S.M. Nutrient transport and pathogenesis in selected parasitic protozoa. *Eukaryot Cell.* 2011 Apr;10(4):483-93.

[11] Archibald, F. *Lactobacillus plantarum*, an organism not requiring iron. FEMS Microbiology Letters. 1983;19(1):29-32.

[12] Chiancone, E., Ceci, P., Ilari, A., Ribacchi, F., Stefanini, S. Iron and proteins for iron storage and detoxification. *Biometals.* 2004 Jun;17(3):197-202.

[13] Finch, C. Regulators of iron balance in humans. *Blood.* 1994 Sep 15;84(6):1697-702.

[14] Weinberg, E.D. Iron-loading and disease surveillance. *Emerg Infect Dis.* 1999 May-Jun;5(3):346-52.

[15] Sharp, P., Srai, S.K. Molecular mechanisms involved in intestinal iron absorption. *World J Gastroenterol.* 2007 Sep 21;13(35):4716-24.

[16] Weinberg, E.D. Iron and infection. *Microbiol Rev.* 1978 Mar;42(1):45-66.

[17] Griffiths, W.J., Kelly, A.L., Cox, T.M. Inherited disorders of iron storage and transport. *Mol Med Today.* 1999 Oct;5(10):431-8.

[18] Conrad, M.E, Umbreit, J.N. Disorders of iron metabolism. *N Engl J Med.* 2000 Apr 27;342(17):1293-4.

[19] Weinberg, E.D. Infection and iron metabolism. *Am J Clin Nutr.* 1977 Sep;30(9):1485-90.

[20] Bullen, J.J. The significance of iron in infection. *Rev Infect Dis.* 1981 Nov-Dec;3(6):1127-38.

[21] Dunn, L.L., Rahmanto, Y.S., Richardson, D.R. Iron uptake and metabolism in the new millennium. *Trends Cell Biol.* 2007 Feb;17(2):93-100.

[22] Legrand, D., Mazurier, J. A critical review of the roles of host lactoferrin in immunity. *Biometals.* 2010 Jun;23(3):365-76.

[23] Actor, J.K., Hwang, S.A., Kruzel, M.L. Lactoferrin as a natural immune modulator. *Curr Pharm Des.* 2009;15(17):1956-73.

[24] Mazurier, J., Spik, G. Comparative study of the iron-binding properties of human transferrins. I. Complete and sequential iron saturation and desaturation of the lactotransferrin. *Biochim Biophys Acta.* 1980 May 7;629(2):399-408.

[25] Baker, E.N., Baker, H.M. Molecular structure, binding properties and dynamics of lactoferrin. *Cell Mol Life Sci.* 2005 Nov;62(22):2531-9.

[26] Anderson, B.F., Baker, H.M., Dodson, E.J., Norris, G.E., Rumball, S.V., Waters, J.M., et al. Structure of human lactoferrin at 3.2-A resolution. *Proc Natl Acad Sci U S A.* 1987 Apr;84(7):1769-73.

[27] Moore, S.A., Anderson, B.F., Groom, C.R., Haridas, M., Baker, E.N. Three-dimensional structure of diferric bovine lactoferrin at 2.8 A resolution. *J Mol Biol.* 1997 Nov 28;274(2):222-36.

[28] Jameson, G.B., Anderson, B.F., Norris, G.E., Thomas, D.H., Baker, E.N. Structure of human apolactoferrin at 2.0A resolution. Refinement and analysis of ligand-induced conformational change. *Acta Crystallogr D Biol Crystallogr.* 1998 Nov 1;54(Pt 6 Pt 2):1319-35.

[29] Testa, U. Lactoferrin. In: Testa, U., editor. *Proteins of Iron Metabolism Boca Raton,* Florida: CRS Press; 2002. p. 71-141.

[30] Pan, Y., Rowney, M., Guo, P., Hobman, P. Biological properties of lactoferrin: an overview. *Anglais.* 2007;62(1):31-42.

[31] Legrand, D., Pierce, A., Elass, E., Carpentier, M., Mariller, C., Mazurier, J. Lactoferrin structure and functions. *Adv Exp Med Biol.* 2008;606:163-94.

[32] Jenssen, H., Hancock, R.E. Antimicrobial properties of lactoferrin. *Biochimie.* 2009 Jan;91(1):19-29.

[33] Weinberg, E.D. Iron availability and infection. *Biochim Biophys Acta.* 2009 Jul;1790(7):600-5.

[34] Ortiz-Estrada, G., Luna-Castro, S., Pina-Vazquez, C., Samaniego-Barron, L., Leon-Sicairos, N., Serrano-Luna, J., et al. Iron-saturated lactoferrin and pathogenic protozoa: could this protein be an iron source for their parasitic style of life? *Future Microbiol.* 2012 Jan;7:149-64.

[35] Masson, P.L., Heremans, J.F., Schonne, E. Lactoferrin, an iron-binding protein in neutrophilic leukocytes. *J Exp Med.* 1969 Sep 1;130(3):643-58.

[36] Masson, P.L., Heremans, J.F. Lactoferrin in milk from different species. *Comp Biochem Physiol B.* 1971 May 15;39(1):119-29.

[37] Bennett, R.M., Mohla, C. A solid-phase radioimmunoassay for the measurement of lactoferrin in human plasma: variations with age, sex, and disease. *J Lab Clin Med.* 1976 Jul;88(1):156-66.

[38] Legrand, D., Vigie, K., Said, E.A., Elass, E., Masson, M., Slomianny, M.C., et al. Surface nucleolin participates in both the binding and endocytosis of lactoferrin in target cells. *Eur J Biochem.* 2004 Jan;271(2):303-17.

[39] Russell, M.W., Bobek, L.A., Brock, J.H., Hajishengallis, G., Tenovuo, J., Jiri, M., et al. Chapter 5 - Innate Humoral Defense Factors. *Mucosal Immunology (Third Edition).* Burlington: Academic Press; 2005. p. 73-93.

[40] van der Velden, W.J., Blijlevens, N.M., Donnelly, J.P. The potential role of lactoferrin and derivatives in the management of infectious and inflammatory complications of hematology patients receiving a hematopoietic stem cell transplantation. *Transpl Infect Dis.* 2008 Apr;10(2):80-9.

[41] Van Snick, J.L., Masson, P.L., Heremans, J.F. The involvement of lactoferrin in the hyposideremia of acute inflammation. *J Exp Med.* 1974 Oct 1;140(4):1068-84.

[42] Finkelstein, R.A., Sciortino, C.V., McIntosh, M.A. Role of iron in microbe-host interactions. *Rev Infect Dis.* 1983 Sep-Oct;5 Suppl 4:S759-77.

[43] Wilson, M.E., Britigan, B.E. Iron acquisition by parasitic protozoa. *Parasitol Today.* 1998 Sep;14(9):348-53.

[44] Letendre, E.D. Iron metabolism during infection and neoplasia. *Cancer Metastasis Rev.* 1987;6(1):41-53.

[45] Otto, B.R., Verweij-van Vught, A.M., MacLaren, D.M. Transferrins and heme-compounds as iron sources for pathogenic bacteria. *Crit Rev Microbiol.* 1992;18(3):217-33.

[46] Jurado, R.L. Iron, infections, and anemia of inflammation. *Clin Infect Dis.* 1997 Oct;25(4):888-95.

[47] Weinberg, E.D, Miklossy J. Iron withholding: a defense against disease. *J Alzheimers Dis.* 2008 May;13(4):451-63.

[48] Garcia-Montoya, I.A., Cendon, T.S., Arevalo-Gallegos, S., Rascon-Cruz, Q. Lactoferrin a multiple bioactive protein: An overview. *Biochim Biophys Acta.* 2011 Jun 25.

[49] Arnold, R.R., Brewer, M., Gauthier, J.J. Bactericidal activity of human lactoferrin: sensitivity of a variety of microorganisms. *Infect Immun.* 1980 Jun;28(3):893-8.

[50] Arnold, R.R., Russell, J.E., Champion, W.J., Brewer, M., Gauthier, J.J. Bactericidal activity of human lactoferrin: differentiation from the stasis of iron deprivation. *Infect Immun.* 1982 Mar;35(3):792-9.

[51] Nibbering, P.H., Ravensbergen, E., Welling, M.M., van Berkel, L.A., van Berkel, P.H., Pauwels, E.K., et al. Human lactoferrin and peptides derived from its N terminus are highly effective against infections with antibiotic-resistant bacteria. *Infect Immun.* 2001 Mar;69(3):1469-76.

[52] Drago-Serrano, M.E., De la Garza, M., Campos-Rodríguez, R. *Lactoferrin as an Adjunctive Agent in the Treatment of Bacterial Infections Associated with Diabetic Foot Ulcers.* In: Dinh, T., editor. Global Perspective on Diabetic Foot Ulcerations: InTech; 2011.

[53] Singh, P.K., Parsek, M.R., Greenberg, E.P., Welsh, M.J. A component of innate immunity prevents bacterial biofilm development. *Nature.* 2002 May 30;417(6888):552-5.

[54] Farnaud, S., Evans, R.W. Lactoferrin--a multifunctional protein with antimicrobial properties. *Mol Immunol.* 2003 Nov;40(7):395-405.

[55] Legrand, D., Mazurier, J., Metz-Boutigue, M.H., Jolles, J., Jolles, P., Montreuil, J., et al. Characterization and localization of an iron-binding 18-kDa glycopeptide isolated from the N-terminal half of human lactotransferrin. *Biochim Biophys Acta.* 1984 May 31;787(1):90-6.

[56] Gifford, J.L., Hunter, H.N., Vogel, H.J. Lactoferricin: a lactoferrin-derived peptide with antimicrobial, antiviral, antitumor and immunological properties. *Cell Mol Life Sci.* 2005 Nov;62(22):2588-98.

[57] Tomita, M., Wakabayashi, H., Shin, K., Yamauchi, K., Yaeshima, T., Iwatsuki, K. Twenty-five years of research on bovine lactoferrin applications. *Biochimie.* 2009 Jan;91(1):52-7.

[58] Bellamy, W., Takase, M., Wakabayashi, H., Kawase, K., Tomita, M. Antibacterial spectrum of lactoferricin B, a potent bactericidal peptide derived from the N-terminal region of bovine lactoferrin. *J Appl Bacteriol.* 1992 Dec;73(6):472-9.

[59] Bellamy, W., Takase, M., Yamauchi, K., Wakabayashi, H., Kawase, K., Tomita, M. Identification of the bactericidal domain of lactoferrin. *Biochim Biophys Acta.* 1992 May 22;1121(1-2):130-6.

[60] Samuelsen, O., Haukland, H.H., Ulvatne, H., Vorland, L.H. Anti-complement effects of lactoferrin-derived peptides. *FEMS Immunol Med Microbiol.* 2004 Jun 1;41(2):141-8.

[61] Vaerman, J.P. [Effector mechanisms of IgA]. *Ann Biol Clin* (Paris). 1984;42(1):61-70.

[62] Leitch, E.C., Willcox, M.D. Lactoferrin increases the susceptibility of *S. epidermidis* biofilms to lysozyme and vancomycin. *Curr Eye Res.* 1999 Jul;19(1):12-9.

[63] Leitch, G.J., Ceballos, C. A role for antimicrobial peptides in intestinal microsporidiosis. *Parasitology.* 2009 Feb;136(2):175-81.

[64] Venkatesh, M.P., Rong, L. Human recombinant lactoferrin acts synergistically with antimicrobials commonly used in neonatal practice against coagulase-negative *staphylococci* and *Candida albicans* causing neonatal sepsis. *J Med Microbiol.* 2008 Sep;57(Pt 9):1113-21.

[65] Diarra, M.S., Petitclerc, D., Lacasse, P. Effect of lactoferrin in combination with penicillin on the morphology and the physiology of *Staphylococcus aureus* isolated from bovine mastitis. *J Dairy Sci.* 2002 May;85(5):1141-9.

[66] Lyons, T.E., Miller, M.S., Serena, T., Sheehan, P., Lavery, L., Kirsner, R.S., et al. Talactoferrin alfa, a recombinant human lactoferrin promotes healing of diabetic neuropathic ulcers: a phase 1/2 clinical study. *Am J Surg.* 2007 Jan;193(1):49-54.

[67] van der Strate, B.W., De Boer, F.M., Bakker, H.I., Meijer, D.K., Molema, G., Harmsen, M.C. Synergy of bovine lactoferrin with the anti-cytomegalovirus drug cidofovir in vitro. *Antiviral Res.* 2003 Apr;58(2):159-65.

[68] Seganti, L., Di Biase, A.M., Marchetti, M., Pietrantoni, A., Tinari, A., Superti, F. Antiviral activity of lactoferrin towards naked viruses. *Biometals.* 2004 Jun;17(3):295-9.

[69] Yamauchi, K., Wakabayashi, H., Shin, K., Takase, M. Bovine lactoferrin: benefits and mechanism of action against infections. *Biochem Cell Biol.* 2006 Jun;84(3):291-6.

[70] Tanaka, K., Ikeda, M., Nozaki, A., Kato, N., Tsuda, H., Saito, S., et al. Lactoferrin inhibits hepatitis C virus viremia in patients with chronic hepatitis C: a pilot study. *Jpn J Cancer Res.* 1999 Apr;90(4):367-71.

[71] Iwasa, M., Kaito, M., Ikoma, J., Takeo, M., Imoto, I., Adachi, Y., et al. Lactoferrin inhibits hepatitis C virus viremia in chronic hepatitis C patients with high viral loads and HCV genotype 1b. *Am J Gastroenterol.* 2002 Mar;97(3):766-7.

[72] Di Mario, F., Aragona, G., Bo, N.D., Ingegnoli, A., Cavestro, G.M., Moussa, A.M., et al. Use of lactoferrin for *Helicobacter pylori* eradication. Preliminary results. *J Clin Gastroenterol.* 2003a May-Jun;36(5):396-8.

[73] Di Mario, F., Aragona, G., Dal Bo, N., Cavestro, G.M., Cavallaro, L., Iori, V., et al. Use of bovine lactoferrin for *Helicobacter pylori* eradication. *Dig Liver Dis.* 2003b Oct;35(10):706-10.

[74] Yamauchi, K., Hiruma, M., Yamazaki, N., Wakabayashi, H., Kuwata, H., Teraguchi, S., et al. Oral administration of bovine lactoferrin for treatment of *tinea pedis*. A placebo-controlled, double-blind study. *Mycoses.* 2000;43(5):197-202.

[75] Teraguchi, S., Wakabayashi, H., Kuwata, H., Yamauchi, K., Tamura, Y. Protection against infections by oral lactoferrin: evaluation in animal models. *Biometals.* 2004 Jun;17(3):231-4.

[76] Zimecki, M., Mazurier, J., Spik, G., Kapp, J.A. Human lactoferrin induces phenotypic and functional changes in murine splenic B cells. *Immunology.* 1995 Sep;86(1):122-7.

[77] Haque, R., Mondal, D., Kirkpatrick, B.D., Akther, S., Farr, B.M., Sack, R.B., et al. Epidemiologic and clinical characteristics of acute diarrhea with emphasis on *Entamoeba histolytica* infections in preschool children in an urban slum of Dhaka, Bangladesh. *Am J Trop Med Hyg.* 2003 Oct;69(4):398-405.

[78] Stanley, S.L., Jr. Amoebiasis. *Lancet.* 2003 Mar 22;361(9362):1025-34.

[79] Tannich, E., Mirelman, D., Petri, W.A., Jr. Meeting report: EMBO workshop "Pathogenesis of amoebiasis: from genomics to disease", Institut Pasteur, Paris, May 19-22, 2003. *Protist.* 2003 Oct;154(3-4):293-8.

[80] Carrero, J.C., Cervantes-Rebolledo, C., Aguilar-Diaz, H., Diaz-Gallardo, M.Y., Laclette, J.P., Morales-Montor, J. The role of the secretory immune response in the infection by *Entamoeba histolytica*. *Parasite Immunol.* 2007 Jul;29(7):331-8.

[81] Roe, F.J. Toxicologic evaluation of metronidazole with particular reference to carcinogenic, mutagenic, and teratogenic potential. *Surgery.* 1983 Jan;93(1 Pt 2):158-64.

[82] Martinez-Palomo, A. The pathogenesis of amoebiasis. *Parasitol Today.* 1987 Apr;3(4):111-8.

[83] Dobias, L., Cerna, M., Rossner, P., Sram, R. Genotoxicity and carcinogenicity of metronidazole. *Mutat Res.* 1994 Jun;317(3):177-94.

[84] Barreda, A.R., Díaz, B.A., Chávez, M.B. Tratamiento. *Amibiasis Una perspectiva actual.* D.F., México: Ed. Mundo Médico. México; 2002.

[85] Leon-Sicairos, N., Reyes-Lopez, M., Ordaz-Pichardo, C., de la Garza, M. Microbicidal action of lactoferrin and lactoferricin and their synergistic effect with metronidazole in *Entamoeba histolytica*. *Biochem Cell Biol.* 2006 Jun;84(3):327-36.

[86] Ordaz-Pichardo, C., Shibayama, M., Villa-Trevino, S., Arriaga-Alba, M., Angeles, E., de la Garza, M. Anti-amoebic and toxicity studies of a carbamic acid derivative and its therapeutic effect in a hamster model of hepatic amoebiasis. *Antimicrob Agents Chemother.* 2005 Mar;49(3):1160-8.

[87] Upcroft, P., Upcroft, J.A. Drug targets and mechanisms of resistance in the anaerobic protozoa. *Clin Microbiol Rev.* 2001 Jan;14(1):150-64.

[88] Serrano-Luna, J.J., Negrete, E., Reyes, M., de la Garza, M. *Entamoeba histolytica* HM1:IMSS: hemoglobin-degrading neutral cysteine proteases. *Exp Parasitol.* 1998 May;89(1):71-7.

[89] Reyes-Lopez, M., Serrano-Luna, J.J., Negrete-Abascal, E., Leon-Sicairos, N., Guerrero-Barrera, A.L., de la Garza, M. Entamoeba histolytica: transferrin binding proteins. *Exp Parasitol.* 2001 Nov;99(3):132-40.

[90] Leon-Sicairos, N., Reyes-Lopez, M., Canizalez-Roman, A., Bermudez-Cruz, R.M., Serrano-Luna, J., Arroyo, R., et al. Human hololactoferrin: endocytosis and use as an iron source by the parasite *Entamoeba histolytica*. *Microbiology.* 2005 Dec;151(Pt 12):3859-71.

[91] Leon-Sicairos, N., Lopez-Soto, F., Reyes-Lopez, M., Godinez-Vargas, D., Ordaz-Pichardo, C., de la Garza, M. Amoebicidal activity of milk, apo-lactoferrin, sIgA and lysozyme. *Clin Med Res.* 2006 Jun;4(2):106-13.

[92] Lopez-Soto, F., Leon-Sicairos, N., Reyes-Lopez, M., Serrano-Luna, J., Ordaz-Pichardo, C., Pina-Vazquez, C., et al. Use and endocytosis of iron-containing proteins by *Entamoeba histolytica* trophozoites. *Infect Genet Evol.* 2009b Dec;9(6):1038-50.

[93] Lopez-Soto, F., Leon-Sicairos, N., Nazmi, K., Bolscher, J.G., de la Garza, M. Microbicidal effect of the lactoferrin peptides lactoferricin17-30, lactoferrampin265-284, and lactoferrin chimera on the parasite *Entamoeba histolytica*. *Biometals.* 2010 Jun;23(3):563-8.

[94] León-Sicairos, N., Martínez-Pardo, L., Sánchez-Hernández, B., De la Garza, M., Carrero, J.C. Oral lactoferrin treatment resolves amoebic intracecal infection in C3H/HeJ mice. *Biochemistry and Cell Biology.* 2012;In press.

[95] Ordaz-Pichardo, C., Leon-Sicairos, N., Hernandez-Ramírez, V.I., Talamás-Rohana, P., De la Garza, M. Effect of bovine lactoferrin in a therapeutic hamster model of hepatic amoebiasis. *Biochemistry and Cell Biology.* 2012;In press.

[96] Eckmann, L. Mucosal defences against *Giardia*. *Parasite Immunol.* 2003 May;25(5):259-70.

[97] Gillin, F.D., Reiner, D.S., Wang, C.S. Killing of *Giardia lamblia* trophozoites by normal human milk. *J Cell Biochem.* 1983;23(1-4):47-56.

[98] WHO. World Malaria Report 2010. 2010 [updated 2010; cited]; Available from: WHO | World Malaria Report 2010 - *World Health Organization.*

[99] Farrow, R.E., Green, J., Katsimitsoulia, Z., Taylor, W.R., Holder, A.A., Molloy, J.E. The mechanism of erythrocyte invasion by the malarial parasite, *Plasmodium falciparum*. Semin Cell Dev Biol. 2011 Dec;22(9):953-60.

[100] Newbold, C., Craig, A., Kyes, S., Rowe, A., Fernandez-Reyes, D., Fagan, T. Cytoadherence, pathogenesis and the infected red cell surface in *Plasmodium falciparum*. Int J Parasitol. 1999 Jun;29(6):927-37.

[101] Tilley, L., Dixon, M.W., Kirk, K. The *Plasmodium falciparum*-infected red blood cell. *Int J Biochem Cell Biol.* 2011 Jun;43(6):839-42.

[102] Kharazmi, A., Jepsen, S., Valerius, N.H. Polymorphonuclear leucocytes defective in oxidative metabolism inhibit in vitro growth of *Plasmodium falciparum*. Evidence against an oxygen-dependent mechanism. *Scand J Immunol.* 1984 Jul;20(1):93-6.

[103] Fritsch, G., Sawatzki, G., Treumer, J., Jung, A., Spira, D.T. *Plasmodium falciparum*: inhibition in vitro with lactoferrin, desferriferrithiocin, and desferricrocin. *Exp Parasitol.* 1987 Feb;63(1):1-9.

[104] Eda, S., Eda, K., Prudhomme, J.G., Sherman, I.W. Inhibitory activity of human lactoferrin and its peptide on chondroitin sulfate A-, CD36-, and thrombospondin-mediated cytoadherence of *Plasmodium falciparum*-infected erythrocytes. *Blood.* 1999 Jul 1;94(1):326-32.

[105] Kassim, O.O., Ako-Anai, K.A., Torimiro, S.E., Hollowell, G.P., Okoye, V.C., Martin, S.K. Inhibitory factors in breastmilk, maternal and infant sera against in vitro growth of *Plasmodium falciparum* malaria parasite. *J Trop Pediatr.* 2000 Apr;46(2):92-6.

[106] Tenter, A.M., Heckeroth, A.R., Weiss, L.M. *Toxoplasma gondii*: from animals to humans. *Int J Parasitol.* 2000 Nov;30(12-13):1217-58.

[107] Carruthers, V.B. Host cell invasion by the opportunistic pathogen *Toxoplasma gondii. Acta Trop.* 2002 Feb;81(2):111-22.

[108] Montoya, J.G., Liesenfeld, O. *Toxoplasmosis. Lancet.* 2004 Jun 12;363(9425):1965-76.

[109] Omata, Y., Satake, M., Maeda, R., Saito, A., Shimazaki, K., Yamauchi, K., et al. Reduction of the infectivity of *Toxoplasma gondii* and *Eimeria stieda*i sporozoites by treatment with bovine lactoferricin. *J Vet Med Sci.* 2001 Feb;63(2):187-90.

[110] Tanaka, T., Omata, Y., Saito, A., Shimazaki, K., Yamauchi, K., Takase, M., et al. *Toxoplasma gondii*: parasiticidal effects of bovine lactoferricin against parasites. *Exp Parasitol.* 1995 Dec;81(4):614-7.

[111] Tanaka, T., Omata, Y., Saito, A., Shimazaki, K., Igarashi, I., Suzuki, N. Growth inhibitory effects of bovine lactoferrin to *Toxoplasma gondii* parasites in murine somatic cells. *J Vet Med Sci.* 1996 Jan;58(1):61-5.

[112] Tanaka, T., Omata, Y., Isamida, T., Saito, A., Shimazaki, K., Yamauchi, K., et al. Growth inhibitory effect of bovine lactoferrin to *Toxoplasma gondii* tachyzoites in murine macrophages: tyrosine phosphorylation in murine macrophages induced by bovine lactoferrin. *J Vet Med Sci.* 1998 Mar;60(3):369-71.

[113] Steijns, J.M., van Hooijdonk, A.C. Occurrence, structure, biochemical properties and technological characteristics of lactoferrin. *Br J Nutr.* 2000 Nov;84 Suppl 1:S11-7.

[114] Dzitko, K., Dziadek, B., Dziadek, J., Dlugonska, H. *Toxoplasma gondii*: inhibition of the intracellular growth by human lactoferrin. *Pol J Microbiol.* 2007;56(1):25-32.

[115] Jensen, O.L., Gluud, B.S., Birgens, H.S. The concentration of lactoferrin in tears of normals and of diabetics. *Acta Ophthalmol (Copenh).* 1986 Feb;64(1):83-7.

[116] Adams, L.B., Hibbs, J.B., Jr., Taintor, R.R., Krahenbuhl, J.L. Microbiostatic effect of murine-activated macrophages for *Toxoplasma gondii*. Role for synthesis of inorganic nitrogen oxides from L-arginine. *J Immunol.* 1990 Apr 1;144(7):2725-9.

[117] Gazzinelli, R.T., Eltoum, I., Wynn, T.A., Sher, A. Acute cerebral toxoplasmosis is induced by in vivo neutralization of TNF-alpha and correlates with the down-regulated expression of inducible nitric oxide synthase and other markers of macrophage activation. *J Immunol.* 1993 Oct 1;151(7):3672-81.

[118] Seabra, S.H., de Souza, W., DaMatta, R.A. Toxoplasma gondii partially inhibits nitric oxide production of activated murine macrophages. *Exp Parasitol.* 2002 Jan;100(1):62-70.

[119] Guillermo, L.V., DaMatta, R.A. Nitric oxide inhibition after *Toxoplasma gondii* infection of chicken macrophage cell lines. *Poult Sci.* 2004 May;83(5):776-82.

[120] Pfefferkorn, E.R. Interferon gamma blocks the growth of *Toxoplasma gondii* in human fibroblasts by inducing the host cells to degrade tryptophan. *Proc Natl Acad Sci U S A.* 1984 Feb;81(3):908-12.

[121] Tanaka, T., Omata, Y., Narisawa, M., Saito, A., Shimazaki, K., Igarashi, I., et al. Growth inhibitory effect of bovine lactoferrin on *Toxoplasma gondii* tachyzoites in murine macrophages: role of radical oxygen and inorganic nitrogen oxide in *Toxoplasma* growth-inhibitory activity. *Vet Parasitol.* 1997 Jan;68(1-2):27-33.

[122] Lima, M.F., Beltz, L.A., Kierszenbaum, F. *Trypanosoma cruzi*: a specific surface marker for the amastigote form. *J Protozool.* 1988 Feb;35(1):108-10.

[123] Grey, A., Banovic, T., Zhu, Q., Watson, M., Callon, K., Palmano, K., et al. The low-density lipoprotein receptor-related protein 1 is a mitogenic receptor for lactoferrin in osteoblastic cells. *Mol Endocrinol.* 2004 Sep;18(9):2268-78.

[124] Bellamy, W.R., Wakabayashi, H., Takase, M., Kawase, K., Shimamura, S., Tomita, M. Role of cell-binding in the antibacterial mechanism of lactoferricin B. *J Appl Bacteriol.* 1993 Nov;75(5):478-84.

[125] Isamida, T., Tanaka, T., Omata, Y., Yamauchi, K., Shimazaki, K., Saito, A. Protective effect of lactoferricin against Toxoplasma gondii infection in mice. *J Vet Med Sci.* 1998 Feb;60(2):241-4.

[126] Kirchhoff, L.V. Epidemiology of *American trypanosomiasis* (Chagas disease). *Adv Parasitol.* 2011;75:1-18.

[127] Monteon-Padilla, V.M., Hernandez-Becerril, N., Guzman-Bracho, C., Rosales-Encina, J.L., Reyes-Lopez, P.A. *American trypanosomiasis* (Chagas' disease) and blood banking in Mexico City: seroprevalence and its potential transfusional transmission risk. *Arch Med Res.* 1999 Sep-Oct;30(5):393-8.

[128] Rassi, A., Jr., Rassi, A., Marin-Neto, J.A. Chagas disease. *Lancet.* 2010 Apr 17;375(9723):1388-402.

[129] Guedes, P.M., Silva, G.K., Gutierrez, F.R., Silva, J.S. Current status of Chagas disease chemotherapy. *Expert Rev Anti Infect Ther.* 2011 May;9(5):609-20.

[130] Sathler-Avelar, R., Vitelli-Avelar, D.M., Teixeira-Carvalho, A., Martins-Filho, O.A. Innate immunity and regulatory T-cells in human Chagas disease: what must be understood? *Mem Inst Oswaldo Cruz.* 2009 Jul;104 Suppl 1:246-51.

[131] Lima, M.F., Kierszenbaum, F. Lactoferrin effects on the interaction of blood forms of *Trypanosoma cruzi* with mononuclear phagocytes. *Int J Parasitol.* 1987b Aug;17(6):1205-8.

[132] Lima, M.F., Kierszenbaum, F. Lactoferrin effects of phagocytic cell function. II. The presence of iron is required for the lactoferrin molecule to stimulate intracellular killing by macrophages but not to enhance the uptake of particles and microorganisms. *J Immunol.* 1987a Sep 1;139(5):1647-51.

[133] Mehlhorn, H., Schein, E. Redescription of Babesia equi Laveran, 1901 as Theileria equi Mehlhorn, Schein 1998. *Parasitol Res.* 1998 Jun;84(6):467-75.

[134] Battsetseg, B., Xuan, X., Ikadai, H., Bautista, J.L., Byambaa, B., Boldbaatar, D., et al. Detection of *Babesia caballi* and Babesia equi in *Dermacentor nuttalli* adult ticks. *Int J Parasitol.* 2001 Apr;31(4):384-6.

[135] Rodriguez Bautista, J.L., Ikadai, H., You, M., Battsetseg, B., Igarashi, I., Nagasawa, H., et al. Molecular evidence of *Babesia caballi* (Nuttall and Strickland, 1910) parasite transmission from experimentally-infected SCID mice to the ixodid tick, *Haemaphysalis longicornis* (Neuman, 1901). *Vet Parasitol.* 2001 Dec 13;102(3):185-91.

[136] Jongejan, F., Uilenberg, G. The global importance of ticks. *Parasitology.* 2004;129 Suppl:S3-14.

[137] Schein, E. Equine babesiosis. In: Ristic, M., editor. *Babesiosis of Domestic Animals and Man.* Boca Raton, FL: CRC Press; 1988. p. 197-208.

[138] Friedhoff, K.T., Tenter, A.M., Muller, I. Haemoparasites of equines: impact on international trade of horses. *Rev Sci Tech.* 1990 Dec;9(4):1187-94.

[139] Avarzed, A., De Waal, D.T., Igarashi, I., Saito, A., Oyamada, T., Toyoda, Y., et al. Prevalence of equine piroplasmosis in Central Mongolia. *Onderstepoort J Vet Res.* 1997 Jun;64(2):141-5.

[140] de Waal, D.T. Equine piroplasmosis: a review. *Br Vet J.* 1992 Jan-Feb;148(1):6-14.

[141] Schwint, O.N., Knowles, D.P., Ueti, M.W., Kappmeyer, L.S., Scoles, G.A. Transmission of *Babesia caballi* by Dermacentor nitens (Acari: *Ixodidae*) is restricted to one generation in the absence of alimentary reinfection on a susceptible equine host. *J Med Entomol.* 2008 Nov;45(6):1152-5.

[142] Zaugg, J.L., Lane, V.M. Efficacy of buparvaquone as a therapeutic and clearing agent of *Babesia equi* of European origin in horses. *Am J Vet Res.* 1992 Aug;53(8):1396-9.

[143] Ikadai, H., Tanaka, T., Shibahara, N., Tanaka, H., Matsuu, A., Kudo, N., et al. Inhibitory effect of lactoferrin on in vitro growth of Babesia caballi. *Am J Trop Med Hyg.* 2005 Oct;73(4):710-2.

[144] Wang, J.S., Tsai, S.F. Prevalence and pathological study on rabbit hepatic coccidiosis in Taiwan. *Proc Natl Sci Counc Repub China B.* 1991 Oct;15(4):240-3.

[145] Razavi, S.M., Oryan, A, Rakhshandehroo E, Moshiri A, Mootabi Alavi A. Eimeria species in wild rabbits (Oryctolagus cuniculus) in Fars province, Iran. *Trop Biomed.* 2010 Dec;27(3):470-5.

[146] Frayer, R., Speer, C.A., Dubey, J.P. The general biology of *cryptosporidium.* In: Frayer, R., editor. *Cryptosporidium* and *Cryptosporidiosis.* New York: CRC Press LLC; 1997. p. 1-41.

[147] Current, W.L., Garcia, L.S. *Cryptosporidiosis. Clin Microbiol Rev.* 1991 Jul;4(3):325-58.

[148] Taylor, M.A., Coop, R.L., Wall, R.L. *Veterinary parasitology.* 3rd ed.: Blackwell Publishing Company; 2007.

[149] Clavel, A., Arnal, A.C., Sanchez, E.C., Cuesta, J., Letona, S., Amiguet, J.A., et al. Respiratory cryptosporidiosis: case series and review of the literature. *Infection.* 1996 Sep-Oct;24(5):341-6.

[150] Brook, E.J., Hart, C.A., French, N.P., Christley, R.M. Prevalence and risk factors for Cryptosporidium spp. infection in young calves. *Vet Parasitol.* 2008 Mar 25;152(1-2):46-52.

[151] Brook, E.J., Hart, C.A., French, N.P., Christley, R.M. Molecular epidemiology of Cryptosporidium subtypes in cattle in England. *Vet J.* 2009 Mar;179(3):378-82.

[152] Chen, X.M., Keithly, J.S., Paya, C.V., LaRusso, N.F. *Cryptosporidiosis.* N *Engl J Med.* 2002 May 30;346(22):1723-31.

[153] Verdon, R., Keusch, G.T., Tzipori, S., Grubman, S.A., Jefferson, D.M., Ward, H.D. An in vitro model of infection of human biliary epithelial cells by *Cryptosporidium parvum*. *J Infect Dis.* 1997 May;175(5):1268-72.

[154] Carryn, S., Schaefer, D.A., Imboden, M., Homan, E.J., Bremel, R.D., Riggs, M.W. Phospholipases and Cationic Peptides Inhibit *Cryptosporidium Parvum* Sporozoite Infectivity by Parasiticidal and Non-Parasiticidal Mechanisms. *J Parasitol.* 2011 Jul 25.

[155] Castro, M. Treatment and prophylaxis of *Pneumocystis carinii* pneumonia. *Semin Respir Infect.* 1998 Dec;13(4):296-303.

[156] Wilkin, A., Feinberg, J. Pneumocystis carinii pneumonia: a clinical review. *Am Fam Physician.* 1999 Oct 15;60(6):1699-708, 713-4.

[157] Morris, A., Beard, C.B., Huang, L. Update on the epidemiology and transmission of *Pneumocystis carinii. Microbes Infect.* 2002 Jan;4(1):95-103.

[158] Chabe, M., Dei-Cas, E., Creusy, C., Fleurisse, L., Respaldiza, N., Camus, D., et al. Immunocompetent hosts as a reservoir of Pneumocystis organisms: histological and rt-PCR data demonstrate active replication. *Eur J Clin Microbiol Infect Dis.* 2004 Feb;23(2):89-97.

[159] Aliouat-Denis, C.M., Martinez, A., Aliouat el, M., Pottier, M., Gantois, N., Dei-Cas, E. The Pneumocystis life cycle. *Mem Inst Oswaldo Cruz.* 2009 May;104(3):419-26.

[160] Merali, S., Chin, K., Grady, R.W., Weissberger, L., Clarkson, A.B., Jr. Response of rat model of *Pneumocystis carinii* pneumonia to continuous infusion of deferoxamine. *Antimicrob Agents Chemother.* 1995 Jul;39(7):1442-4.

[161] Weinberg, G.A., Boelaert, J.R., Shanzer, A. Multiple iron chelators of diverse chemical classes are active against: *Pneumocystis carinii*. + 1112. *Pediatric Research.* 1996;39(4):188.

[162] Cirioni, O., Giacometti, A., Quarta, M., Scalise, G. In vitro activity of topoisomeraseinhibitors against *Pneumocystis carinii*. *J Antimicrob Chemother.* 1997 Oct;40(4):583-6.

[163] Cushion, M.T., Chen, F., Kloepfer, N. A cytotoxicity assay for evaluation of candidate anti-*Pneumocystis carinii* agents. *Antimicrob Agents Chemother.* 1997 Feb;41(2):379-84.

[164] Walzer, P.D., Runck, J., Steele, P., White, M., Linke, M.J., Sidman, C.L. Immunodeficient and immunosuppressed mice as models to test anti-*Pneumocystis carinii* drugs. *Antimicrob Agents Chemother.* 1997 Feb;41(2):251-8.

[165] Bartlett, M.S., Shaw, M.M., Smith, J.W., Meshnick, S.R. Efficacy of sulfamethoxypyridazine in a murine model of Pneumocystis carinii pneumonia. *Antimicrob Agents Chemother.* 1998 Apr;42(4):934-5.

[166] Cirioni, O., Giacometti, A., Barchiesi, F., Scalise, G. Inhibition of growth of *Pneumocystis carinii* by lactoferrins alone and in combination with pyrimethamine, clarithromycin and minocycline. *J Antimicrob Chemother.* 2000 Oct;46(4):577-82.

[167] Holmberg, S.D., Moorman, A.C., Von Bargen, J.C., Palella, F.J., Loveless, M.O., Ward, D.J., et al. Possible effectiveness of clarithromycin and rifabutin for

cryptosporidiosis chemoprophylaxis in HIV disease. *HIV Outpatient Study (HOPS) Investigators. Jama.* 1998 Feb 4;279(5):384-6.

[168] Franzen, C., Muller, A. Microsporidiosis: human diseases and diagnosis. *Microbes Infect.* 2001 Apr;3(5):389-400.

[169] Canning, E.U., Lom, J., Dykova, I. *The microsporidia of vertebrates.* Academic Press; 1986.

[170] Kotler, D.P., Orenstein, J.M. Clinical syndromes associated with microsporidiosis. *Adv Parasitol.* 1998;40:321-49.

[171] Leitch, G.J., Visvesvara, G.S., He, Q. Inhibition of microsporidian spore germination. *Parasitol Today.* 1993 Nov;9(11):422-4.

[172] Didier, E.S., Weiss, L.M. Microsporidiosis: current status. *Curr Opin Infect Dis.* 2006 Oct;19(5):485-92.

[173] Mathis, A., Weber, R., Deplazes, P. Zoonotic potential of the microsporidia. *Clin Microbiol Rev.* 2005 Jul;18(3):423-45.

[174] Anane, S., Attouchi, H. Microsporidiosis: epidemiology, clinical data and therapy. *Gastroenterol Clin Biol.* 2010 Sep;34(8-9):450-64.

[175] Keohane, E.M., Weiss, L.M. The structure, function, and composition of the microsporidian polar tube. In: Wittner, M., Weiss, L.M., editors. The *microsporidia* and *microsporidiosis*. Washington DC: ASM Press; 1999. p. 196-224.

[176] Hancock, R.E., Scott, M.G. The role of antimicrobial peptides in animal defenses. *Proc Natl Acad Sci U S A.* 2000 Aug 1;97(16):8856-61.

[177] Cunliffe, R.N. Alpha-defensins in the gastrointestinal tract. *Mol Immunol.* 2003 Nov;40(7):463-7.

[178] Ouellette, A.J. Paneth cell alpha-defensin synthesis and function. *Curr Top Microbiol Immunol.* 2006;306:1-25.

In: Parasites
Editor: Gilmar S. Erzinger
ISBN: 978-1-62257-692-0

Chapter 8

Experimental and Natural Infections of House Flies (*Musca domestica*) with the Larval Stage of *Habronema muscae* (Nematoda: Habronematidae) a Parasite of Equids

Rolf K. Schuster* and Saritha Sivakumar
Central Veterinary Research Laboratory, Dubai, UAE

Abstract

Musca domestica is not only a nuisance and a mechanical vector of pathogens, it also can act as a true intermediate host for nematodes. Although the role of house-flies in the life cycle of *Habronema muscae* has been known for 100 years, only a few experimental and epidemiological studies were carried out. A rare case of pulmonal habronemosis in a mare in Sharjah Emirate/ UAE in the autumn of 2008 gave cause to examine the situation in the horse's stable of origin. In order to identify the source of infection, larvae of a laboratory-bred, local *M. domestica* field strain were experimentally raised on the fecal samples of individual horses. Eight of the 12 horses were identified as carriers (3 stallions, 4 mares and 1 foal). Samples from 2 foals and 2 mares, as well as samples from 6 feral donkeys, collected from an adjacent territory to the farm provided negative results. Another experiment was set up to get information about prevalence and intensity, as well as the location, of the nematode larvae in the body of adult flies of different ages. The prevalence and intensity of infection of *Habronema* larvae decreased with the age of the flies. In the first 2 days, nematode larvae were found in the caput, thorax and abdomen whereas only the heads of flies contained the parasites later on. The fly population in the stable was monitored over a period of 12 months. With decreasing temperatures in autumn, the fly density in the horse barn rose and peaked at 2,400 specimens per m^2 that got glued on a sticky paper in one hour. Applying a pyrethroid insect spray into the horse

* E-mail address: moniezia@zedat.fu-berlin.de.

boxes only brought a short-lived relief. Out of 4,085 house-flies examined, 517 were positive for *Habronema larvae*. Most of the positive flies harbored less than 10 larvae. The maximum intensity in a single specimen was 28. The highest prevalence of 35.7 and 45.4 was reached in September and October, respectively, with no significant difference between male and female flies. Prevalence decreased in the following months. None of the 36 *Stomoxys calcitrans* caught in the stable contained Habronema larvae.

INTRODUCTION

Habronematosis is a parasitic infection of equides caused by nematodes *Habronema muscae*, *H. microstoma (syn. H. majus*[1]*)* and *Draschia*[2] *megastoma* of the Habronematidae family. Flies of the Muscidae family serve as intermediate hosts. The adult nematodes inhabit the stomach of the final hosts, causing a mainly asymptomatic gastric habronemosis. However, when flies are attracted to wounds in the skin, they can release habronematid larvae that cause granulomatous wounds, or so-called summer sores. Larvae deposited on the eye lids can cause conjunctivitis. Habronematosis has a global distribution and is found in many countries of Eurasia, Africa, the Americas and Australia [1], and most of the data on the occurrence of these spirurids was obtained from a necropsy of horse stomaches.

The history of habronematids dates back to the beginning of the 19th century when Rudolphi [2] gave the first description of worms found in the stomach of a horse necropsied at the veterinary school in Berlin and named them *Spiculoptera megastoma*. The deposited material was later examined by Gurlt [3]. Gurlt concluded that Rudolphi's material contains larger and smaller varieties of these nematodes. Schneider [4], who re-examined the material given to him by Gurlt, described the larger variety as *Filaria microstoma,* known today as *Habronema microstoma.*

In his revision of the nematodes, Diesing [5] created the genus *Habronema* (habros: graceful, delicate) for the nematode *Filaria muscae* discovered by Carter [6] in *Musca domestica* in India. Piana [7] found the same nematode in *M. domestica* in Italy. Up to 30% of the examined house-flies harbored this 2.68 mm long nematode in its larval stage. Linstow [8] described a similar larval stage in *Stomoxys calcitrans* in Germany.

The first data on the life cycle of habronematids was obtained by Ransom [9-10] in the USA. In Australia, Hill [11] confirmed the role of *M. domestica* as an intermediate host for *H. muscae* and discovered that this fly species plays also a major role in the ontogeny of *D. megastoma* whereas *H. microstoma* uses *Stomoxys calcitrans* as intermediate host. For the first time, Descazeaux [12], Saceghem [13] and Bull [14] linked habronematid larvae to summer sores, or cutaneous habronematosis, in equines.

[1] Some authors prefer the name *H. majus* and put *H. microstoma* in the category of its synonyms. However, no source for the first description of *H. majus* could be found.

[2] The genus *Draschia* was introduced by Chitwod & Wehr [15] without an explanation of its etymology. It is suggested that it was named in honor of the Austrian helminthologist Richard v. Drasche. In this case, *Drascheia* would be the correct genus name.

1. MATERIAL AND METHODS

1.1. The Farm

The 14 ha farm under investigation is situated on an oasis (N: 25° 11.125'; E: 055 ° 56.367') in the foothills of the Hajar Mountains next to Al Dhaid (Sharjah Emirate) at a distance of 56 km from the Persian Gulf and 42 km from the Gulf of Oman.

When investigations started in April 2009, there were 12 horses kept at the farm: 3 stallions, 6 mares and 3 foals. The horse facility consisted of a barn with 16 boxes and 3 separate pastures, 0.15 ha each. The temperature in the area was extremely hot. Temperature data for the period during the investigation is presented in Figure 1. The highest temperature of 48 °C was recorded on July 1, 2009 whereas the lowest measured temperature was 8 °C on February 9, 2010.

Mares and foals are on the pastures of the farm 24 hours a day during winter months, and spent the hottest times of the day in the barn from April to October. Stallions remain in the barn the whole year, except for daily exercises.

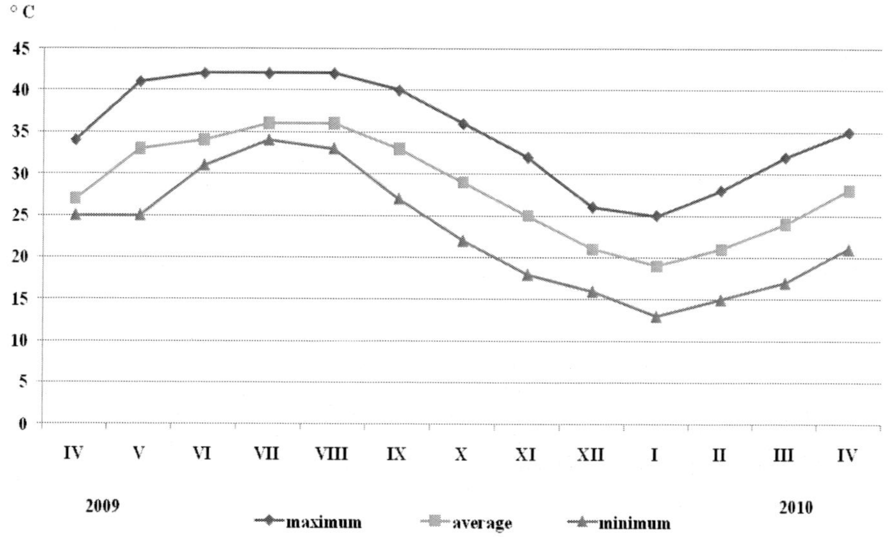

Figure 1. Monthly maximum, average and minimum temperatures during observation period from April 2009 – April 2010. (Source: Sharjah Airport).

Two of the stallions had a history of summer sores on their head and frontal extremities, respectively, in the previous year (Figure 2, Figure 3). The wounds were treated with iodine tincture and healed leaving hairless scars. A 15-year-old mare from the same farm, euthanised in October of 2008 due to acute respiratory problems, was diagnosed with pulmonal and cutaneous habronemosis (Figure 4, Figure 5). Since 210 out of 1,300 *M. domestica* examined were positive for *Habronema* larvae at the first visit to the farm and no medical treatment had been given to the horses, it was concluded that the focus of infection was still active. A molecular examination of larvae revealed the sole presence of *H. muscae* on this farm [16-17].

During the 12 months investigation period from April 2009 to March 2010, the farm was visited in monthly intervals between 7AM and 9AM in the morning in the second decade of the subsequent month. Flies were caught in the barn using hand nets and specially designed weir traps. At the same occasion, the fly density was measured by putting a sticky paper (Fa. Agrargigant, Germany) at the entrance of the barn for 1 hour. The fly density was expressed as the number of flies per m^2 per hour (Figure 6). Back in the laboratory, fly heads were examined for the presence of nematodes.

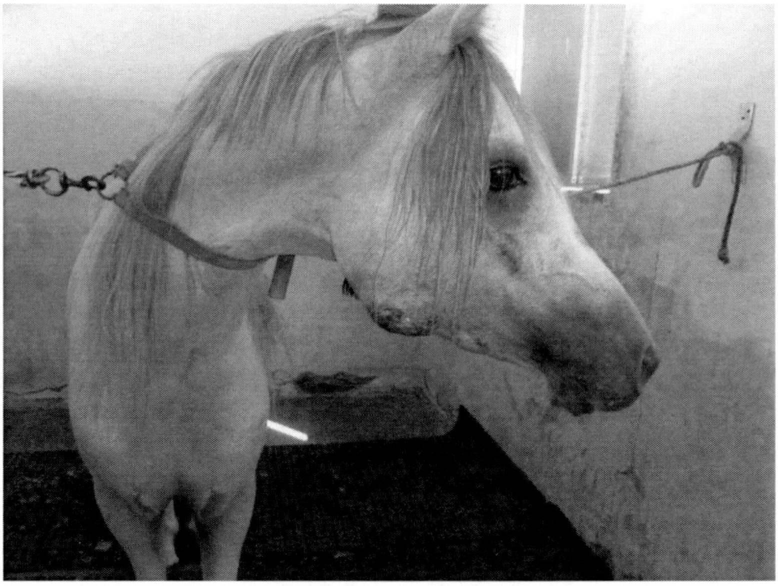

Figure 2. Summer sore on a horse: a granulomatous wound on the lower jaw.

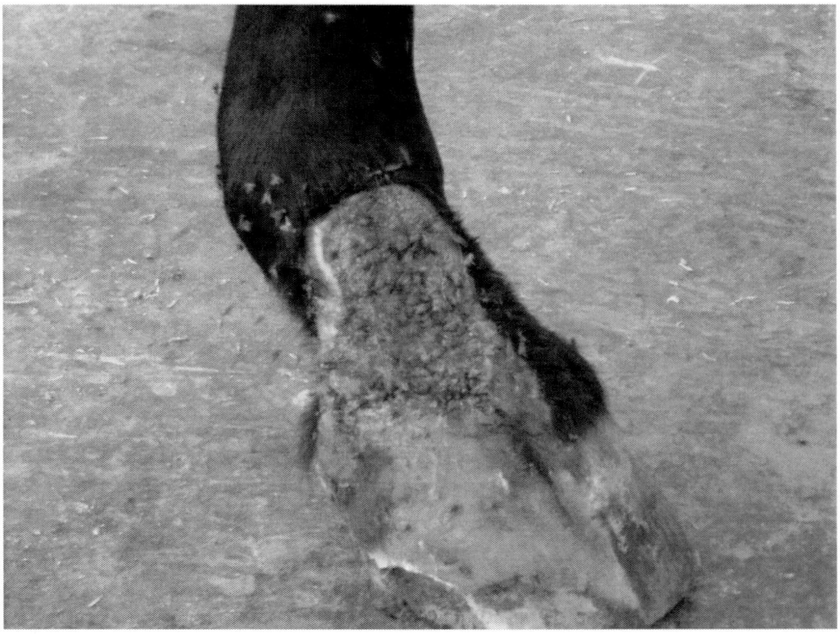

Figure 3. Healing summer sore in a horse. Despite iodine treatment, the wound attracts house flies.

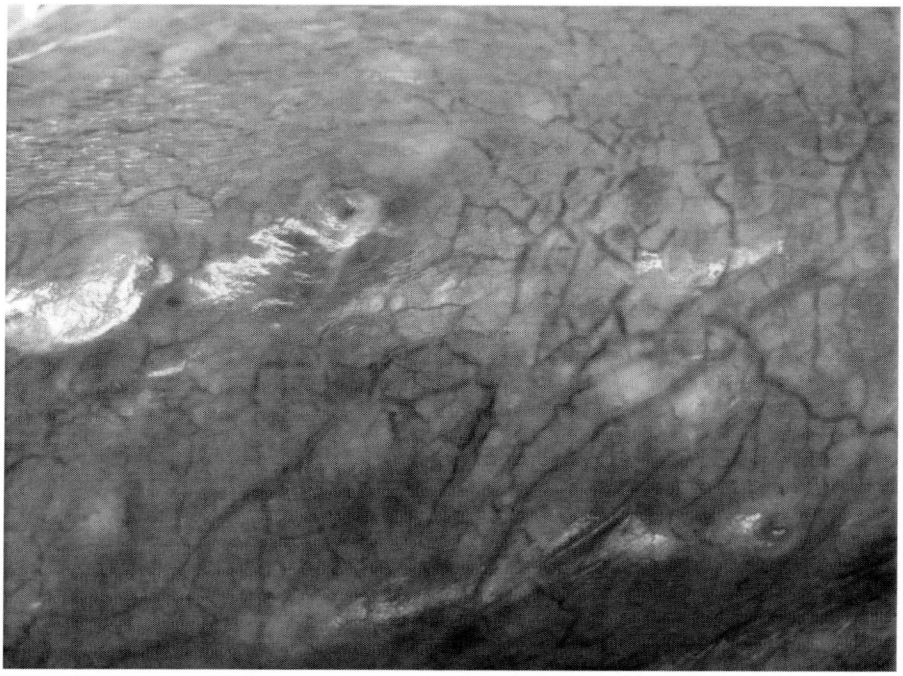

Figure 4. Pulmonal habronemosis : multible nodules containing parasite larvae can be seen in the lung tissue of a horse (by courtesy of Dr. J. Kinne, CVRL, Dubai).

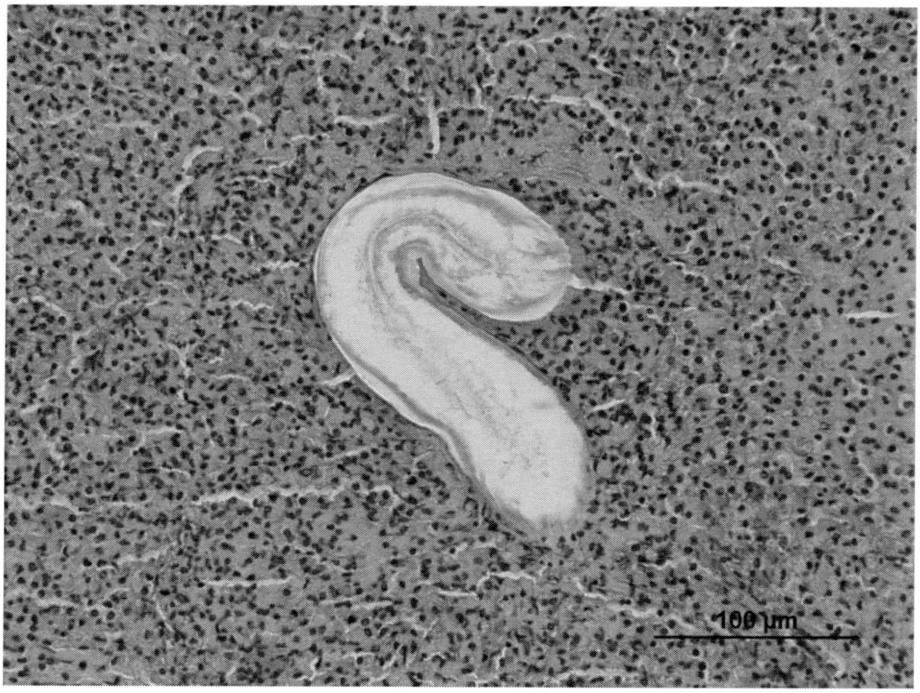

Figure 5. Pulmonal habronemosis : stained histocut of a horse's lung tissue with massive eosinophilic infiltration with a fraction of a *H. muscae* larva (by courtesy of Dr. J. Kinne, CVRL, Dubai).

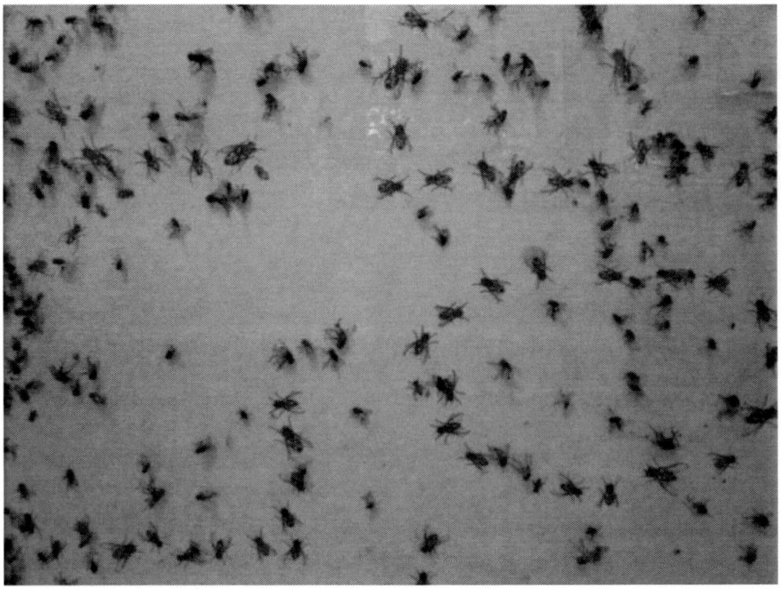

Figure 6. Sticky paper to monitor fly density.

2. Laboratory Investigations

2.1. Laboratory Breeding of *M. domestica*

A laboratory-bred field strain of *M. domestica* was the basis of the experimental work. The initial stock of *M. domestica* was obtained from the milking parlor of a camel milk farm in Dubai and consisted of approximately 100 adult flies. Laboratory breeding was performed in a similar way as described by Iglisch and Hollerbach [18]. The fly cage consisted of a wooden cage 700x300x250 mm with a glass window and site ventilation windows covered with fly gauze. A temperature of 26-30° C was assured by a 60 watt electrical light. Flies were fed sugar syrup and pasteurized milk. The stock nutrition medium for the fly larvae consisted of weatbran (9 parts), crushed dog biscuit (1 part) and pulverized alfalfa hay (1 part). Prior to feeding, some granules of active dry yeast and a teaspoon of molasses dissolved in 80 ml of water were added to 200 ml of a dry medium.

3. Experimental Infection of *M. domestica* with *H. muscae*

The aim of the first experiment was to test the technique for the experimental infection of flies and to identify the location site of *Habronema* larvae in the body of fly imagines. For this purpose, feces from the two stallions with summer sores were used since these horses were most likely infected, although we were unable to find eggs of the parasite with flotation or sedimentation methods.

Moistened horse feces (300 g) in 500 ml glass beakers were overlaid with a thin layer of nutrition medium and put into the fly cage containing 30 - 50 *M. domestica*. After 48 hours, when eggs and 1st fly instars were seen, the beakers were removed from the cage and kept at 26° C. A wet nutrition medium was added after another 3 days, and the beaker was closed by a layer of low-fat curt (Dmiati cheese). At day 7, larvae started to migrate out of the vessel. In order to prevent larvae from escaping, beakers were put into plastic buckets that contained dry sand. Pupation started 3 days later. Pupae were strained out and put into 11 glasses covered with gauze. Sugar syrup dropped on a piece of cotton wool served as food for the emerged flies. Flies of 1, 2, 3, 4 and 6 days of age were used to determine the success of infection and to determine the location of the nematode larvae in the body of the flies.

In order to identify the source of infection, a xenodiagnostic test with individual fecal samples from each of the 12 horses was set up in April 2009. Since feral donkeys were thought to be a reservoir of habronemosis, freshly-deposited feces from 6 donkeys next to the farm were included. The samples (100g each) were placed into laboratory glass beakers (250 ml) and mixed with 100 ml of tap water and were treated as described above.

4. Examination of *M. domestica* for *H. muscae* Larvae

To establish the success of experimental infection, but also to assess the natural *Habronema* prevalence, flies were killed by exposing them for 5 minutes at -18° C. In the first experiments the caput, thorax and abdomen of each were separated by a scalpel, carved by preparation needles and examined separately in a drop of normal saline under a stereoscopic microscope. Further investigations concentrated only on fly heads.

5. Statistical Evaluation

Prevalence, intensity of infection, abundance and subsequent 95%-confidence intervals were generated using the software package QP 2 [19].

6. Results

6.1. Location of *H. muscae* Larvae in the Body of *M. domestica*

The first exploratory investigation (trial 1 and 2) with the feces of the two stallions resulted in the hatching of 246 and 313 flies, respectively. The majority of *Habronema* larvae were situated in the caput of the flies, and only a few specimens harbored nematode larvae in the thorax and abdomen (Tab. 1, 2). Only 7 out of 127 (= 0.6%) and 14 out of 151 (= 0.9%) infected flies in trial 1 and 2, respectively, showed *Habronema* larvae exclusively in the thorax and abdomen. These flies were only 1 and 2 days old. Later, *Habronema* larvae were situated in the caput only (Figure 7, Figure 8).

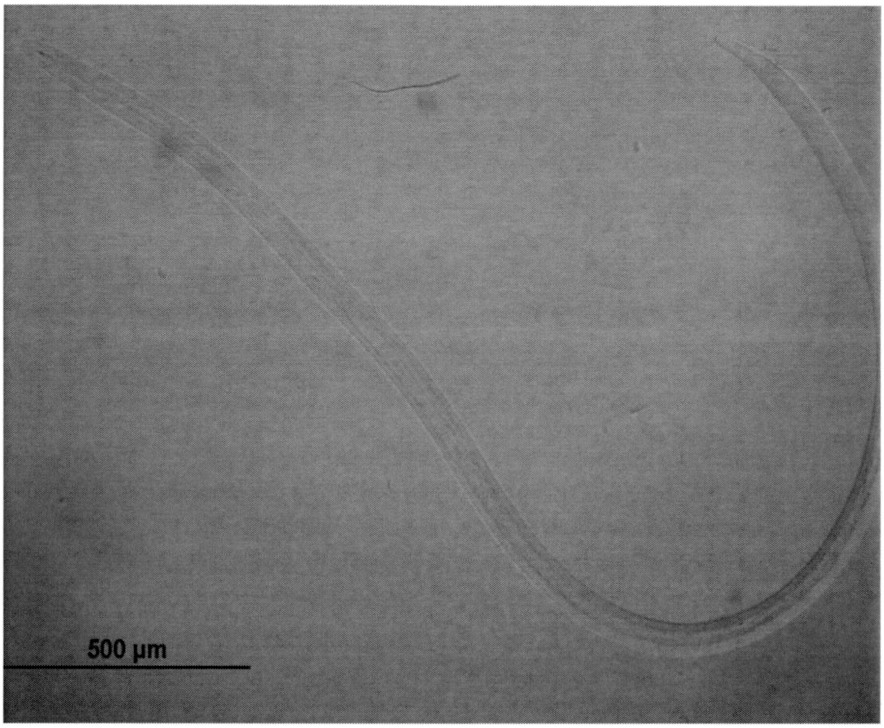

Figure 7. Isolated *H. muscae* larva.

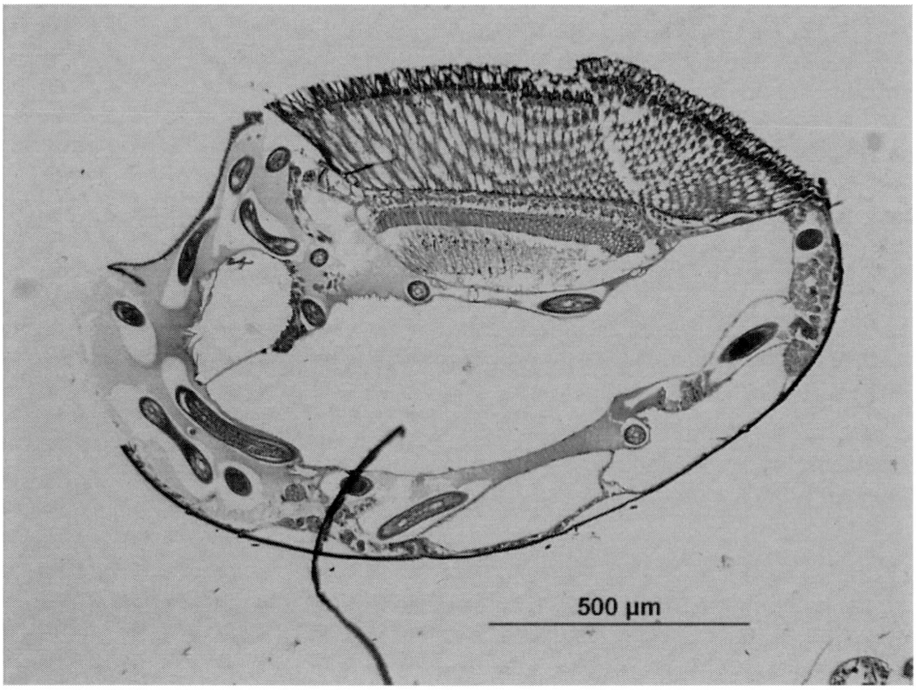

Figure 8. Histocut of *M. domestica* caput shows a section of *H. muscae* larva (by courtesy of Dr. J. Kinne, CVRL, Dubai).

Table 1. Location of *H. muscae* larvae in the body of experimentally infected male and female *M. domesica* of different age (trial 1)

Age (days)	Male flies					Female flies				
	exam-ined	infec-ted	with larvae in			exam-ined	infec-ted	with larvae in		
			caput	thorax	abdomen			caput	thorax	abdomen
1	13	11	9	6	5	17	13	10	5	6
2	32	20	19	1	1	26	16	15	2	2
3	21	14	14	0	0	17	11	11	0	0
4	22	9	9	0	0	25	8	8	0	0
6	30	8	8	0	0	43	18	18	0	0
all:	**118**	**61**	**59**	**7**	**6**	**128**	**66**	**62**	**7**	**8**

Table 2. Location of *H. muscae* larvae in the body of experimentally infected male and female *M. domesica* of different age (trial 2)

Age (days)	Male flies					Female flies				
	examined	infected	with larvae in			examined	infected	with larvae in		
			caput	thorax	abdo-men			caput	thorax	abdomen
1	23	18	13	6	5	23	17	10	5	6
2	17	12	11	1	1	15	7	7	2	2
3	42	25	25	0	0	33	17	17	0	0
4	41	18	18	0	0	38	9	9	0	0
6	31	8	8	0	0	50	20	20	0	0
all:	**154**	**81**	**75**	**7**	**6**	**159**	**70**	**63**	**7**	**8**

Table 3. Prevalence, intensity and abundance of *H. muscae* in *M. domesica* of different age

Age (days)	Trial 1			Trial 2		
	prevalence (%)	intensity	abundance	prevalence	intensity	abundance
1	80.1	2.38	1.9	76.1	2.66	2.02
	[61.40;92.29]	[1.79;3.04]	[1.33;253]	[61.23;87.42]	[2.17;3.23]	[1.52;2.52]
2	62.1	2.67	1.66	59.4	3.16	1.88
	[48.37;74.49]	[2.40;3.28]	[1.21;2.22]	[40.64;76.31]	[2.32;4.05]	[1.19;2.66]
3	65.8	2.4	1.58	56.1	2.42	1.25
	[48.64;80.37]	[1.80;3.00]	[1.08;2.18]	[44.05;67.46]	[1.83;2.71]	[0.93;1.61]
4	36.2 (S)	1.82	0.66 (S)	34.2 (S)	1.96	0.67 (S)
	[22.67;51.49]	[1.35;2.76]	[0.38;1.13]	[23.87;45.71]	[1.52;2.48]	[0.42;0.95]
6	34.2 (S)	1.4	0.48 (S)	34.6 (S)	1.46 (S)	0.51 (S)
	[23.53;46.29]	[1.16;1.64]	[0.32;066]	[24.34;45.96]	[1.21;1.79]	[0.35;0.70]

Results showed that *Habronema* prevalence and larval density decreased with the age of the flies (Tab 3). Flies examined at day 1 showed the highest larval prevalence of 80.1 and 76.1%, respectively. In 4-day-old flies, this parameter amounted only to 36.2 and 34.2 %, respectively. Similar data was recorded 2 days later. The decrease in prevalence and abundance of *Habronema* larvae in flies at day 4 and day 6 compared to day 1 was significant. The average number of *Habronema* larvae per infected fly varied between 1.46 and 3.16, and a maximum of 7 larvae was recorded.

6. Identifying the Source of Infection

Fecal samples of all 12 horses and 6 donkeys tested with routine coproscopical tests (flotation and sedimentation methods) were negative for parasites. Nonetheless, the experimental infection of laboratory-bred houseflies with feces collected at the farm revealed that 8 of the 12 horses were infected with stomach worms (Tab. 4). The prevalence of *Habronema* larvae in 3-day-old flies from positive samples in this experiment varied between 8.6 and 64.1%, suggesting a different egg density in the horse feces. A maximum number of 9 larvae/head were found. Flies bred on feces from 2 foals and one young mare were negative for nematode larvae. Feeding experiments with 6 samples from feral donkeys collected outside the farm also gave negative results.

Table 4. Identification of horses carrying *H. muscae* in xenodiagnostic trials

Horses			Flies		
			prevalence (%)	intensity	
No	sex	age(y)		average	Range
1	stallion	6	54.4	2.71	1-9
2	stallion	4	64.1	3.2	1-9
3	stallion	7	44.6	3.32	1-7
4	colt	> 1	0	0	0
5	mare	15	43.75	3.0	1-6
6	mare	15	53.9	2.21	1-8
7	mare	6	26.4	2.7	1-6
8	mare	3	0	0	0
9	mare	3	36.7	2.7	1-8
10	mare	2.5	0	0	0
11	filly	>1	8.6	2	1-5
12	filly	>1	0	0	0

6.2. Long-term Study of House Flies in the Horse Barn

House flies were present in the barn the whole year long. However, their density showed considerable variations. Starting with 455 flies per m^2 in 1 hour in April 2009, the fly density dropped without intervention during the hot summer months and reached a minimum of 55 and 46 specimens per m^2 in 1 hour in August and September, respectively (Tab. 5). With declining temperatures, the fly density started to rise in October and reached a peak of 2,400 per m^2 in 1 hour in November. Insecticide spraying reduced the density of adult flies in December to 430 per m^2 in 1 hour. A steady increase of the fly density was noticed in the following months, reaching another peak with 1,120 insects per m^2 in 1 hour at the end of our observation in March 2011.

During the 12 months investigation period, 4,121 muscid flies were caught and examined for the presence of *Habronema* larvae. The total number of *M. domestica* amounted to 4,085 with 2,177 male and 1,908 female specimens. In monthly catches, however, the sex ratio varied (Figure 9). A surplus of male flies was noticed from May to December 2009 and March and April 2010. At the beginning of our observation in April 2009, none of the 377 *M. domestica* were positive for *Habronema* larvae, but the infection rate rose steadily in the following months to reach a maximum of 35.7 and 45.4% in September and October 2009, respectively. In the following months, prevalence declined but the intensity of infection remained high (Tab. 5). As seen from Figure 10, the majority of the 517 infected flies harbored low larval burdens. However, the maximum number of larvae recovered from one fly was 28.

Stomoxys calcitrans was also present in the catch, but in low numbers and only from December to March 2010. All 36 stable flies were negative for nematode larvae.

Table 5. Prevalence and burden of *H. muscae* larvae in *M. domestica* in a horse stable in the UAE during the time period between April 2009 and March 2010. The fly pressure in the barn was measured with a sticky tape that was set for 1 hour during the netting

Month	Flies examined	Prevalence (%)	Prevalence 95% C.I.	Intensity Average	Intensity 95% C.I.	Fly density per m2/h
April	327	0	0;1.13	0	0	455
May	160	0.6	0.01;3.44	1	a	114
June	230	0.9	0.10;3.11	1.5	a	220
July	309	2.3	0.91;4.62	1.9	1;1.57	158
August	327	8.0	5.25;11.44	2.62	1.81;4.19	55
September	213	35.7	29.25;42;52	3.07	2.47;4.18	46
October	403	45.4	40.47;50;42	3.83	3.30;4.47	275
November	730	15.5	13.18;18.61	3.34	2.82;4.10	2398
December	287	2.8	1.12;5.42	4.13	2.13;7.50	427
January	355	19.4	15.44;23.95	5.14	3.91;6.84	598
February	506	2.0	0.95;3.61	5.2	1.80;14.60	801
March	238	8.4	5.20;14.12	3.2	2.2;4.30	1121

[a] Number of positives too small to generate 95% C.I.

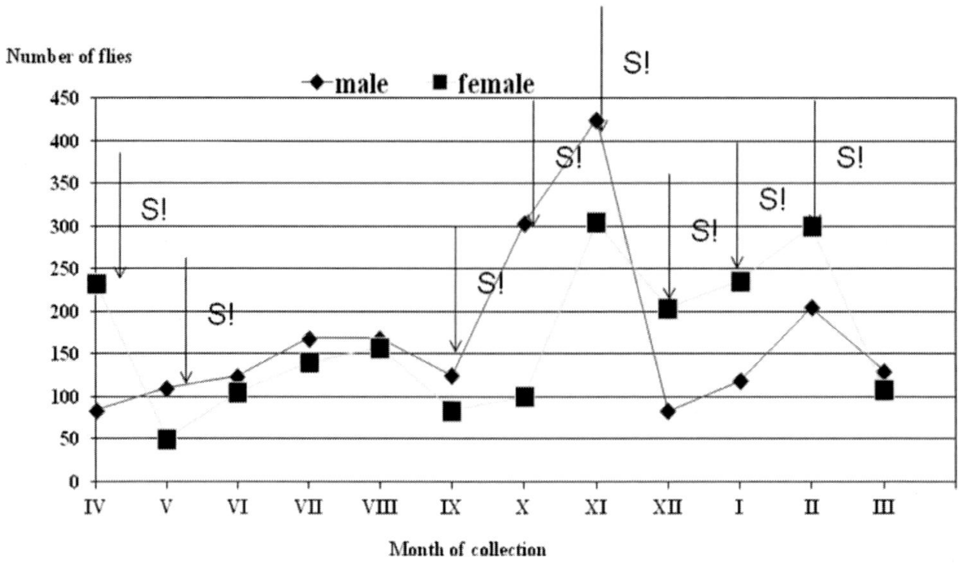

Figure 9. Number of male and female *M. domestica* in a net catch from a horse barn in the UAE. Arrows show a significant difference between sexes.

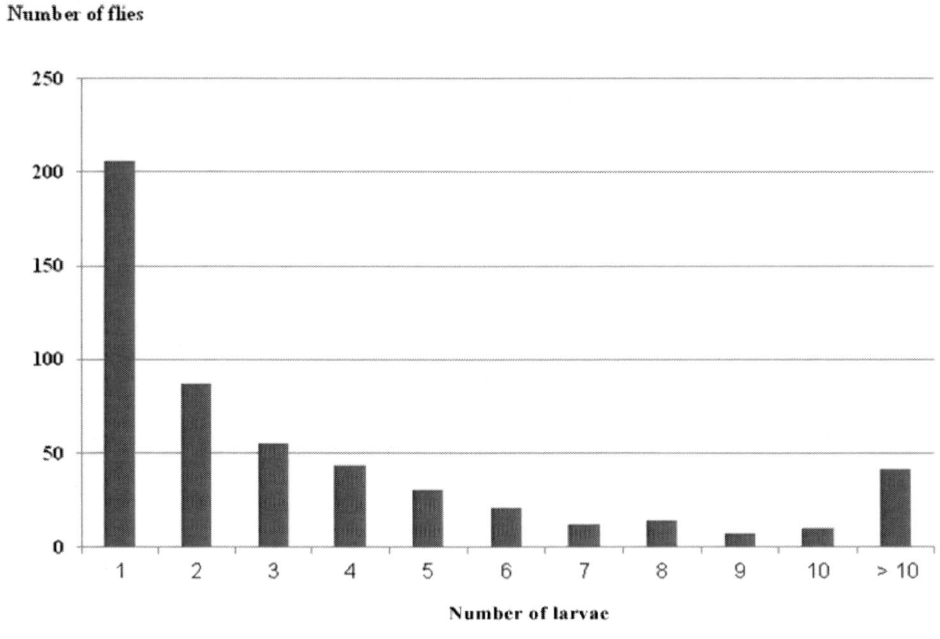

Figure 10. Burden of *H. muscae* larva in *M. domestica* caught in a horse barn.

7. DISCUSSION

Although veterinarians and horse owners are aware of summer sores, little is known about the epidemiology of habronematosis in equids. Recent publications about summer sores in horses exist from Brasil [20], the USA [21], UK [22], Belgium [23], Germany [24], Israel

[25] and the UAE [16]. Cutaneous habronematosis was also described in donkeys in Sudan [26] and South Africa [27]. Lesions can be found all over the body of equines with preposition sites at the prepuce, the urethral process of the penis, and the medial canthus of the eye, as well as the distal extremities, with conjunctival habronemosis being the most often diagnosed clinical form, whereas pulmonary and cerebral habronematosis seem to be very rare clinical forms [16, 28-29].

The stomach is the normal location site of adult habronematids and thus, gastric habronemosis is more common but usually does not cause obvious symptoms. Size, color and the morphology of structures surrounding the mouth opening are criteria for species determination of the adult worms [30-31]. Differences in the structure of the mouth opening and surrounding lips of 3 equine habronematid species have recently been investstigated by scanning an electron microscopy [32]. A diagnosis of gastric habronematosis is mainly based on necropsy since the ante-mortem detection of *Habronema* eggs and *Draschia* larvae in horse feces is difficult. In our own investigation on a farm with suspected gastric habronemosis in horses, we used a xenodiagnostic method to detect the infected horses.

Embryonated eggs of *H. muscae* excreted with the feces of the final host are ingested by *Musca* larvae and the 1st larval instars of the parasite hatch and penetrate the gut of its intermediate host. Contrary to larvae of *D. megastoma* that enter the Malpighian tubules, *Habronema* larvae develop in cells of the adipose tissue [33]. Nonetheless, development times of the larval stages are similar. *Habronema* larvae molt two times. A first molt takes place in the fly larva, and the second molt happens in the late pupal stage prior to the hatch of the adult insect [34]. Our own observations showed that 3rd -stage larvae quickly migrate from the abdominal cavity into the caput of the adult fly. Thus, larvae were found in abdomen and thorax only on days 1 and 2. In older flies, all larvae were only concentrated in the caput.

After discovering the habronematid's life cycle, a number of experiments associated with flies at their larval stages, grown on horse manure, were carried out in Australia [35]. The authors concluded that apart from *M. domestica, H. muscae* and *D. megastoma* can use a variety of other muscids as intermediate hosts, including *M. vetustissima, M. fergusoni, M. terraeregina, M. hilli* and *Pseudopyrellia* sp. Also, *Sarcophaga* spp. and blow flies can become infected, but their epidemiological role is limited. In detail, the authors examined native Australian flies and found that the majority of habronematid larvae in *M. fergusoni* can be found in the thorax and abdomen, whereas most of the larvae in *M. vetustissima* were present in the caput. Thus, larvae in *M. vetustissima* probably behave in the same way as in our experiments with *M. domestica*.

Under natural conditions, habronematid larvae are stimulated to leave the fly's mouth parts when the insect feeds on a moist, warm surface, such as around the nares or mouth of the definitive host [14, 36]. Since our experiments showed decreasing prevalence and larval burdens in aging flies, it can be concluded that larvae also leave the fly in the absence of a suitable mammalian host. Similar results were found after an artificial infection of 2nd instars of *M. domestica* with a suspension of *H. muscae* eggs [37]. The authors demonstrated the negative effect of the nematode larvae on the longevity of the adult flies and found that freshly emerged flies showed a prevalence of 98% and an abundance of 8.3, whereas the infection rate and larval burden in older specimens was 80% and 3.3, respectively. It is also suggested that horses become infected when swallowing flies with food or drinking water [38-39].

Feral donkeys that roam around in the oasis where the farm is located were suspected to be the source of the infection. However, all 6 tested donkeys were tested negative for *H. muscae* in xenodiagnostic experiments. Out of the 12 horses, 8 were infected with *H. muscae* and 2 of the 3 stallions showed granulomatous cutaneous wounds on the lower jaw and the coronary groove of the right frontal leg, respectively. The experimental infection of flies from the feces of 2 foals and 2 young mares probably failed because of the limited contact to infected flies. These animals were kept on the pasture during the colder period of the year when the fly density in the barn peaked. It was shown in a previous paper that the *H. muscae* prevalence of *M. domestica* in that barn was 20-times higher than outside [16].

There is only limited information about long-term studies on *Habronema* infection in flies. Subsequent investigations were carried out only in non-black-soil districts in Russia [31]. The authors showed that *Habronema* larvae were present in house flies between May and October with the highest prevalence in mid-summer.

Under Arabian conditions, the prevalence peak shifted to autumn because of the hot subtropical climate. It was experimentally shown that adult house flies choose areas of 30 °C in temperature gradients of 20-30 °C and 30-40 °C at a constant humidity [40], but the optimum temperature for the development of *M. domestica* was 25-27 °C [18,41-42]. After hot and dry summer months with low fly activity in the barn and hardly any flies on the pastures of the farm, *Musca* density started to increase in October when the average temperature dropped below 30 °C. Until October, the majority of flies underwent their larval development in horse manure in the barn, and this led to an increase in *Habronema* prevalence in adult flies that peaked in October when nearly every second fly was infected. The prevalence decrease in the following months might be attributed to a larger number of uninfected specimens that underwent their development in other sites on the farm and retire into the barn with decreasing temperatures. Furthermore, insecticide spraying in the boxes of the barn can be responsible for the decrease of the number of infected flies.

The average number of *Habronema* larvae per fly in both experimental and natural infections was low. Only 9 specimens harbored more than 20 larvae. Similar data was published in sources from Australia and Brazil [35, 37].

It is noteworthy that apart from *M. domestica,* another muscid fly, *S. calcitrans*, was present on the farm. However, stable flies occurred in low densities in the catch only after the first rains in December, and all specimens were negative for nematode larvae.

References

[1] Collobert-Laugier, C., Lamedey, C., Brisseau, N., Moussu, C., Hamet, N., (2000). Prevalence of stomach nematodes (*Habronema* spp., *Draschia megastoma* and *Trichostrongylus axei*) in horses examined post mortem in Normandy. *Rev. Med. Vet. 151*, 151 – 156.

[2] Rudolphi C.A. (1819). *Entozoorum synopsis cui accedunt mantissa duplex et indices locupletissimi*, Berolini: Sumptibus A. Rückeri.

[3] Gurlt M.F. (1831). *Lehrbuch der pathologischen Anatomie der Haussäugethiere nebst einem Anhange, welcher die Beschreibung der bei Haussäugethieren vorkommenden Eingeweidewürmer enthält.* Berlin, Verlag G. Reimer.

[4] Schneider, A. (1866) *Monographie der Nematoden*. Berlin: Verlag von Georg Reimer.

[5] Diesing, K. M. (1861). Kleine helminthologische Mittheilungen. *Sitzungsberichte der kaiserlichen Akademie der Wissenschaften, Wien. 43*, 269-282.

[6] Carter, H.J. (1861). On bi-sexual nematoid worm which infests the common house-fly (*Musca domestica*) in Bombay. *Ann.Mag.Nat. Hist*., 4, 29-33.

[7] Piana, G. P. (1896). Osservazioni sul *Dispharagus nasutus*, Rud. dei polli e sulle larve nematoelmintiche delle mosche e dei Porcellioni, *Atti Soc. Ital. d. Sci. Nat., 36*, pp. 239-262.

[8] Linstow, v. O.F.B. (1887). Beobachtungen an neuen und bekannten Helminthen. *Arch. Naturgesch., 53*, 194 - 205.

[9] Ransom, B.H. (1911). The life history of a parasite nematode – *Habronema muscae. Science, 34*. 690 – 692.

[10] Ransom, B.H. (1913). The life-history of Habronema muscae (Carter), a parasite of the horse transmited by the hous-fly. *U.S. Dept Agric. Bureau Animal Ind. Bull, Washington, (163)*, 1- 36.

[11] Hill, G.F. (1918). Relationship of insects to parasitic diseases. *Proc. Roy. Soc. Victoria,31*, 11 – 107.

[12] Descazeaux, J. (1915). Contribution a l' etude de l' esponja ou plaies d' ete des equides du Bresil. *Rev. Gen. Med. Vet. 25*, 431-433.

[13] Saceghem, v. R. (1917). Contribution a l'etude de la dermatite granuleuse chez des equides. *Bull. Soc. Pathol. Exot. 10*, 726-729.

[14] Bull, L.B. (1919). A contribution to the study of habromemiasis: a clinical, pathological and experimental Investigation of granulomatous condition of the horse – habronemic granuloma. *Trans. Proc. Royal Soc. S. Austr., 43*, 85-141.

[15] Chitwod, B.G., Wehr, E.E. (1932) The value of cephalic structures as characters in nematode classification, with special reference to the superfamily Spiruroidea. *Z. Parasitenkd. 7*, 294-335.

[16] Schuster, R.K., Sivakumar, S., Kinne, J., Babiker, H., Traversa, D., Buzzel, G.R. (2010). Cutaneous and pulmonal habronemosis transmitted by *Musca domestica* in a stable in the United Arab Emirates. *Vet. Parasitol. 174*, 170-174.

[17] Buzzell, G., Tariq, S., Traversa, D., Schuster, R. (2011). Morphology of the infective larval stage of the equid parasite *Habronema muscae* (Spirurida: Habronematidae), from houseflies (*Musca domestica*). *Parasitol. Res. 108*, 629-632.

[18] Iglisch, I. & Hollerbach, D. (1985). Erhaltungszucht mit Überschuss von Musca domestica L. (Diptera, Muscidae). *Anz. Schädlingskde, Pflanzenschutz, Umweltschutz, 58,* 81 – 86.

[19] Rozsa L Reiczigel J Majoros G (2000) Quantiifying parasites in samples of hosts. *J. Parasitol., 86*, 228 – 232.

[20] Sanavria, A., Fernandes, C.G. Reixoto, P.F., Fabris, V.E., Morais, M.C. (2000). Cutaneous habronemosis in equines: casuistic in four physiogeographic areas from the state of Rio de Janeiro. *Rev.Bras. Med. Vet. 22*, 104-107.

[21] Pusterla, N., Watson, J.L., Wilson, D., Afolter, V.K., Spier, S.J. (2003). Cutaneous and ocular habronemiasis in horses: 63 cases (1988-2002). J. *Am. Vet. Med. Assoc., 222*, 978-982.

[22] Down, S.S., Hughes, I., Henson, F.M.D. (2009). Cutaneous habronemiasis in a 9-year-old Arabic gelding in the United Kingdom. *Equine Vet. Educ. 21*, 4-8.

[23] Gasthuys, F.M.R., van Heerden, M., Vercruysse, J. (2004). Conjunctival habronemiosis in a horse in Belgium. *Vet Rec. 154*, 157-158.

[24] Toth, J. & Hollerrieder, J. (2008). Neubildungen im Augenbereich beim Pferd. *Tierärztl. Praxis (Grosstiere), 36*, 39-44.

[25] Yarmut, Y, Brommer, H., Weisler, S., Shehlah, M., Komarovsky, O., Steinmann, A. (2008). Ophthalmic and cutaneous habronemiasis in a horse: case report and review of the literature. *Israel J. Vet. Med., 63,* 87-90.

[26] Mohamed, F.H., Abu Samra, M.T., Ibrahin, K., Idris, S.O. (1990). Cutaneous habronemiasis in horses and domestic donkeys (*Equus asinus asinus*). Rev. d'Elev. Med. Vet. Pays Trop., 42, 535-540.

[27] Matthee, S., Krecek, R.C., Milne, A.A. (2000). Prevalence and biodiversity of helminth parasites in Donkeys from South Africa. *J. Parasitol. 86,* 756- 762.

[28] Bain, A.M., Rofe, J.C., Hotson, I.K., Murphy, S. (1969). *Habronema megastoma* larvae associated with pulmonary abscesses in a foal. *Austr. Vet. J. 45*, 101-102.

[29] Mayhew, I.G., Lichtenfels, J.R., Greiner, E£.C., MacKay, R.J., Enloe, C.W., (1982). Migration of a spiruroid nematode through the brain of a horse. *J. Am. Vet. Med. Ass. 180*, 1306-1311.

[30] Lichtenfels, J.R. (1975). Helminths of domestic equids. *Proc. Helminthol. Soc. Washington, 42*, 1-92.

[31] Ivashkin, V.M., Dvojnos, G.M. (1984). *Key to helminthes of horses*. Kiew, Naukova dumka,163 pp. (In Russian).

[32] Naem, S. (2007). The comparative morphology of three equine habronematid nematodes: SEM observations. *Parasitol Res., 101*, 1303-1310.

[33] Roubaud, E., Descazeaux, J. (1922). Evolution de l' *Habronema muscae*, Carter, chez la Mouche domestique et de l' *H. microstomum*, Schneider chez le Stomoxe. *Bull. Soc. Pathol. Exot. 15*, 572-574.

[34] Gorshkov, I.P. (1948). Biology of *Drashea megastoma* from horses. In: Papers on Helminthologydedicated to K.I. Skrjabin on the 40th anniversary of his scientific activity and on the 25th anniversary of the All-Union Institute of Helminthology, 98-108 (in Russian).

[35] Johnston, T. H. & Bankroft , M.J. (1920). The life history of *Habronema* in relation to *Musca domestica* and native flies in Queensland. *Proc.Royal Soc. Queensland, 32*, 61-82.

[36] Anderson, R.C. (2006). *Nematode parasites of Vertebrates. Their development and transmission.* 2nd Ed. Eastbourne, CPI Antony Rowe.

[37] Amado, S., Gulias-Gomes, C.C., Milward-de-Azevedo, E.M.V. (2000). Longevity of *Musca domestica* L. (Diptera-Muscidae) parasitized by *Habronema muscae* Carter (Nematoda: Habronematidae). *Parasitol dia 24*, 35-39.

[38] Ransom B.H. (1921). Relations of insects to the parasitic worms. In: Pierce, W.D. (Ed.) *Sanitary Entomology,* (50-96) Boston: Gorham Press.

[39] Skrjabin, K.I. & Ershov, V.S. (1933). *Helminths of equines.* Moskva: Selchozgiz, 408 pp (in Russian).

[40] Dakshinamurty, S. (1948). The common house-fly, *Musca domestica*, L. and its behavior to temperature and humidity. *Bull Entomol. Res. 39*, 339-357.

[41] Fletcher, M.G., Axtell, M.G., Stinner, R.E. (1990). Longevity and fecundity of *Musca domestica* (Diptera: Muscidae) as a function of temperature. . *Med. Entomol. 27*, 922-926.

[42] Crosskey, R.W., Lane, R.P. (1995). House-flies, blow-flies and their allies (calyptrateDiptera). In: Lane, R.P. & Crosskey, R.W. (eds.) *Medical insects and arachnids* (reprinted 1st ed. pp. 403-428) London: Chapman & Hall.

In: Parasites
Editor: Gilmar S. Erzinger

ISBN: 978-1-62257-692-0

Chapter 9

COMMUNITY ECOLOGY OF METAZOAN PARASITES IN THE RED ROCKFISH *SEBASTES MINIATUS* FROM BAJA CALIFORNIA, MÉXICO (EAST PACIFIC)

María Amparo Rodríguez-Santiago[1,*], Jorge Adrián Rosales-Casián[2] and Fernando García-Vargas[3,*]

[1]Facultad de Ciencias Marinas, Universidad Autónoma de Baja California, Km 106 carretera Tijuana-Ensenada, Ensenada, BC, México

[2]División de Oceanología, Departamento de Ecología Marina, Centro de Investigación Científica y de Educación Superior de Ensenada BC, Ensenada, BC, México

[3]Centro de Investigación en Alimentación y Desarrollo A.C. Unidad Mazatlán en Acuicultura y Manejo Ambiental, Laboratorio de Parasitología, Mazatlán, Sinaloa, México

ABSTRACT

Rockfishes (family *Scorpaenidae*) of the genus *Sebastes,* constitute one of the most important groundfish resources in the northern Pacific Ocean. The rockfishes comprise a large proportion of catches in the nearshore recreational fishery, and the commercial catch off California. Also, at Baja California, the rockfish group makes up a substantial part of the commercial nearshore fishing, and from the party boat activity. The parasite fauna of *Sebastes miniatus* from northern Baja California, Mexico was quantitatively described for the first time. Twelve parasite species were recorded: 5 ectoparasites (2 monogeneans, and 3 parasitic copepods) and 7 endoparasites (2 digeneans, 1 cestode, 3 nematodes, and 1 acanthocephala). Five of these parasites constituted new genera records to the genus *Sebastes* and nine were new geographic records. The most abundant species were the endoparasites *Bothriocephalus* sp., *Hysterothylacium* sp. and *Anisakis* sp. and the specific richness ranged from 1 to 8 parasite species per host. The most important

* E-mail address: amparor@cicese.mx.

parasite species in terms of prevalence were *Microcotyle sebastis* (93%) and *Anisakis* sp. (92%). The mean abundance of parasites found in *S. miniatus* showed significant variations (KW, H = 11.2068, $p < 0.05$) over the year, with maximum values (31.7 individuals/host) in August, and minimum (0.39 individuals/host) in February. *Botriocephalus* sp. showed the highest mean intensity of infection with 190.4 parasites/host, followed by *Anisakis* sp. (127.2 parasites/host) and *Hysterothylacium* sp. (46.6 parasites/host). Larval stages of *Anisakis*, *Pseudoterranova* and *Hysterothylacium* were particularly important due to their high abundance and prevalence and because these nematodes can represent a human health risk (anisakiasis).

Introduction

Currently, few parasite studies in marine fishes from the Mexican Pacific coasts are available, and particularly in Baja California. The identification of parasite species and the community description at northern Baja California has been realized in the California halibut *Paralichthys californicus* [1], in six rockfish species of the genus *Sebastes* [2] and in the ocean whitefish, *Caulolatilus princeps* [3].

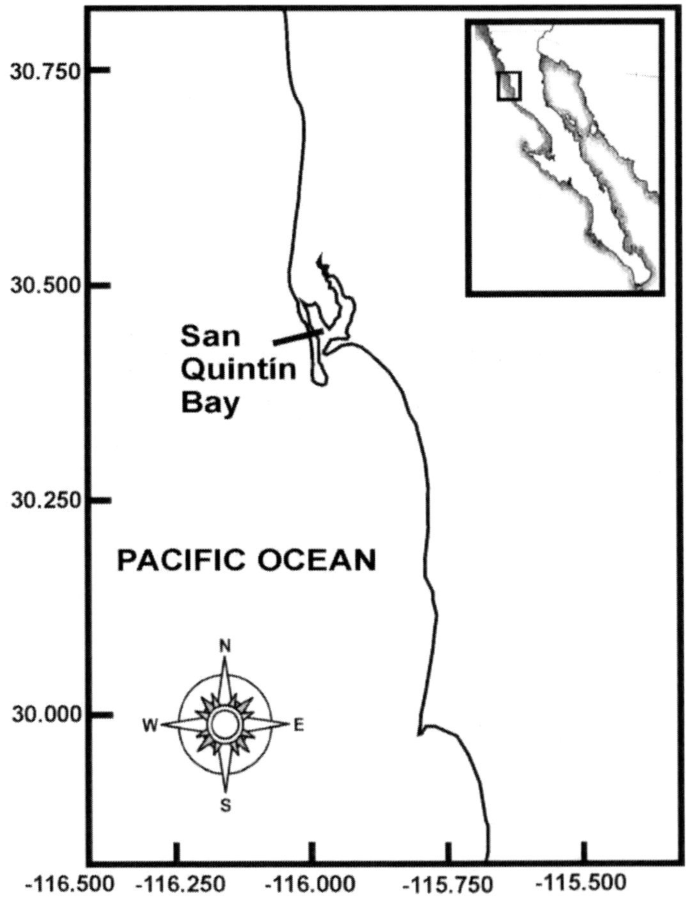

Figure 1. Location of the study area at San Quintín, Baja California, Mexico.

The rockfishes from the Pacific coast of North America are conformed at least by 65 species; however the greatest diversity (56 species) is found within the Southern California Bight [4]. This fish group comprised a large proportion in the coastal fisheries catch of Baja California, Mexico [5, 6. 7], and California USA [8, 9, 10].

The vermilion rockfish, red rockcod or simply named "red", *Sebastes miniatus* Jordan & Gilbert 1880, is a common target fish species in the recreational and commercial fisheries of Baja California [11]. Also, it is a highly prized fish by party and private vessel anglers throughout California with the majority of catches occurring south of Monterey Bay [4]. This rockfish fish species is found from Prince William Sound, south Alaska to central Baja California. They inhabit rocky reefs, kelp forests and canyons at depths of 15–467 m, but are commonly found at 50-150 m [4, 12].

By its wide latitudinal distribution this species represents a potential intermediate host for many parasites, and can be the prey of a final host like the California sea lion [13]. However, in spite of its economical importance, no one study describing the parasite community of *S. miniatus* at Baja California has been realized. Some studies on *S. miniatus* have been focused on growth, reproduction and feeding, and very few individual metazoan parasite species have been listed [4].

Therefore, the aim of the present work was 1) to identify the metazoan parasites in the red rockfish *S. miniatus* from the coasts of San Quintín B.C. Mexico, 2) to determine the community descriptors of parasites such as prevalence, abundance and intensity as well as its variability over an annual cycle and 3) to examine the possible relationships between parasite abundance and host size, condition factor, and water temperature.

1. Fish and Parasite Sampling

The data set comprises the parasite communities of the red rockfish *Sebastes miniatus*. The samples were obtained (bimonthly) from catches of the recreational fishing at San Quintín, Baja California, México (30o33''37' N; 115o56''33' W) (Figure 1). A total of 210 individuals (210-610 mm total length) were analyzed during 2005. The sea surface temperature was provided by the boat captains, and temperatures at depths of 50-140 m were obtained from a monitoring station close to Isla San Martin by the IMECOCAL program of CICESE [14].

All red rockfish specimens were measured for total length using an ictiometer (mm), and weighed (g) with a digital balance Accu-Lab. The Fulton's Condition Factor (KLP) was calculated for each specimen as K= [W/TL3] 10,000; where: W = weight (g) and TL = total length (mm) [15]. This factor is used to describe the physiological condition of the individual fish.

In the field, each specimen was examined to detect external parasites in skin and fins. After that they were dissected and the samples were transported to a laboratory for an examination under a stereoscopic microscope to find parasites in their internal structures (gill cavity, intestinal cecum, mesentery, coelomic cavity, stomach wall, stomach, intestine and gall bladder).

Helminthes were fixed in AFA (acetic acid-formaldehyde-alcohol) solution by 2-24 h. Then, they were preserved in ethyl alcohol (70%) and stained with *Gomori's trichromic* stain.

Nematodes were fixed in Berland's liquid, preserved in ethyl alcohol (70%), and cleared with a solution of phenol-ethanol (Lent's solution) and mounted on slides covered with glycerin-gelatin. Copepods were first fixed in ethyl alcohol (70%) and then they were cleared using a solution of glycerin-alcohol, and mounted on slides covered with glycerin-gelatin [16].

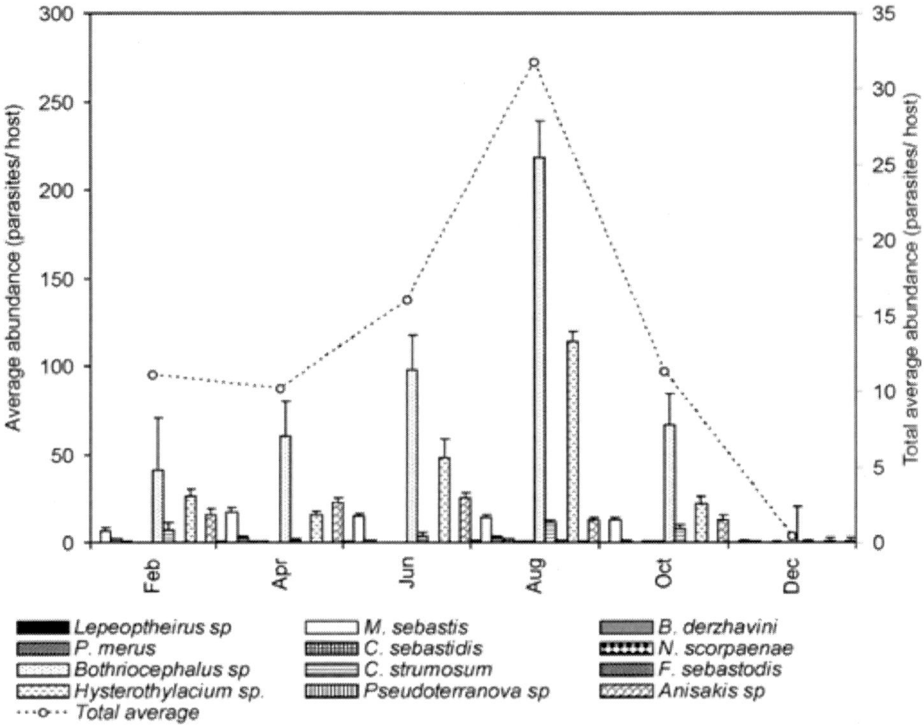

Figure 2. Average abundance (parasites/host ± SE) of parasites on *Sebastes miniatus*.

After obtaining the morphological data, identification of helminthes was realized with keys proposed by Yamaguti [17,18,19], Bravo-Hollis [20, 21, 22], Mendoza-Garfías and Pérez-Ponce de León [23], Vidal-Martinez et al [16] and Anderson, et al [24] for nematodes. To identify the parasitic copepods, keys of Cressey and Boyle-Cressey [25, 26], Kabata [27, 28, 29], and Boxshall [30] were used. Parasitological material was deposited in the Laboratorio de Ecología Pesquera, of the Centro de Investigación Científica y de Educación Superior de Ensenada CICESE, Ensenada, Baja California, México.

2. Community Descriptors

Species richness, prevalence (%), abundance (parasite number/host) and intensity (parasite number/infected hosts) of parasites were determined according to Margolis et al [31]. To assess significant variations in parasite abundance over time, a non-parametric analysis of variance of Kruskall-Wallis (KW) was performed [32]. Spearman rank correlations were used to assess relationships between the parasite abundance and the host size and weight, fish condition, and seawater temperature.

The sea surface temperature at the San Quintín fishing area averaged (± SE) of 16.0 ± 0.3°C. The highest temperature (18.2 ± 0.5°C) was in August and the lowest (14.9 ± 0.3°C) in February. At the fishing depth of 50-140 m, the average annual temperature was 10.9 ± 0.09°C with higher values in October (11.5 ± 0.20°C) and lower in June (10.2 ± 0.08°C)

3. CHARACTERIZATION OF PARASITES

A total of 29,640 parasites were found in the 210 red rockfish individuals, belonging to 12 parasite species. Cestodes were the most abundant group with up to 50% of the total (Table 1). The identified species were 7 endoparasites (2 digeneans, 1 cestode, 3 nematodes and 1 acanthocephalan) and 5 ectoparasites (2 monogeneans and 3 copepods) (Table 2).

Table 1. Groups and total number of parasites recorded in *Sebastes miniatus* caught on the coasts of San Quintín, B.C., Mexico

		Endoparasites				Ectoparasites		
	Hosts	Digeneans	Nematodes	Cestodes	Acanthocephales	Monogeneans	Copepods	Total
Feb	15	7	627	622	105	128	3	1,507
Apr	28	25	1,071	1,688	49	557	26	3,444
Jun	62	23	4,589	6,073	231	997	40	12,015
Aug	35	16	1,465	2,401	302	499	11	4,729
Oct	29	5	1,011	1,940	229	395	46	3,655
Dec	41	17	1,560	2,155	206	406	54	4,439
Total	210	93	10,323	14,879	1,122	2,982	180	29,789
%		0.3	34.9	50.3	3.8	10.1	0.6	100.0

Groups of endoparasites (digeneans, nematodes, cestodes, acantocephales), and ectoparasites (monogeneans and copepods).

The most infested organs were the stomach, pyloric cecum and intestine, with a presence of seven species (Table 2). Larvae of the nematode *Hysterothylacium* sp. showed lower organ specificity. They were found in most organs within the host (mesentery, stomach, pyloric cecum, and intestine). Nevertheless, trematodes and cestodes were mainly restricted to the digestive tract (mainly in the stomach). Also, the acanthocephalan species were found only in the mesentery. Ectoparasites were abundant in the gill cavity, except *Benedenia derzhavini,* which was also frequent in the skin and fins (Table 2).

The species with higher values of prevalence were *M. sebastis* (92.6%), *Anisakis* sp. (91.8%) and *Hysterothylacium* sp. (60.1%). These were, therefore, the most important species integrating the parasite community on *S. miniatus* and were regarded as core species (prevalences of 60% or more). In the case of *B. derzhavini, Bothriocephalus* sp. and *C. strumosum* they were termed as secondary species (prevalences upper of 20% and lower than 60%). The rest of the species were regarded as rare or satellite species (prevalences ≤ 20%).

From late spring to autumn (April-October), *Anisakis* sp. showed a prevalence of 100% and a decrease (80%) in early winter (December) (Figure 4). Similarly, *M. sebastis* sp. exhibited prevalences of up to 90% along the year. The other species such as *B. derzhavini, C.*

strumosum, and *C. sebastidis* showed an abrupt decrease in prevalence from spring to summer (Figure 4).

Table 2. Parasite species found in *Sebastes miniatus* (n = 210). Prevalence (%P), mean abundance (MA), mean intensity (MI)

	Species	Organ	Total	%P	MA	MI
Digenea	*Fellidostomum sebastodis*	G	28	11.2	0.3 ± 0.2	1.1 ± 0.1
	Parahemiurus merus	S, I	65	17.9	0.6 ± 0.4	1.5 ± 0.3
Cestoda	*Bothriocephalus* sp.	S, I, IC	14,879	32.8	80.8 ± 33.4	190.4 ± 48.1
Nematoda	*Anisakis* sp.	CC, Stw, I, Me	3,942	91.8	15.0 ± 3.9	15.8 ± 4.1
	Hysterothylacium sp.	Me, S, I	6,373	60.1	37.9 ± 18.0	46.6 ± 18.1
	Pseudoterranova sp.	S, Me, I, IC	8	2.9	0.2 ± 0.2	0.6 ± 0.3
Acantho-cephala	*Corynosoma strumosum*	Me, CC, I	1,112	37.5	5.3 ± 1.8	13.3 ± 4.4
Monogenea	*Microcotyle sebastis*	GC	2,696	92.6	11.2 ± 2.7	11.6 ± 3.0
	Benedenia derzhavini	S, F, GC	286	40.9	1.7 ± 0.5	2.8 ± 0.6
Copepoda	*Lepeoptheirus* sp.	GC, S, F	67	15.8	0.5 ± 0.3	1.5 ± 0.2
	Clavellostis sebastidis	GC	60	18.8	0.5 ± 0.2	1.2 ± 0.4
	Neobranchia scorpaenae	GC	53	15.5	0.4 ± 0.2	1.1 ± 0.4

Location in the body of the host: F = fins, S = skin, GC = gill cavity, IC = intestinal cecum, Me = mesentery, CC = coelomic cavity, Stw = stomach wall, S = stomach, I = intestine, G = gall bladder.

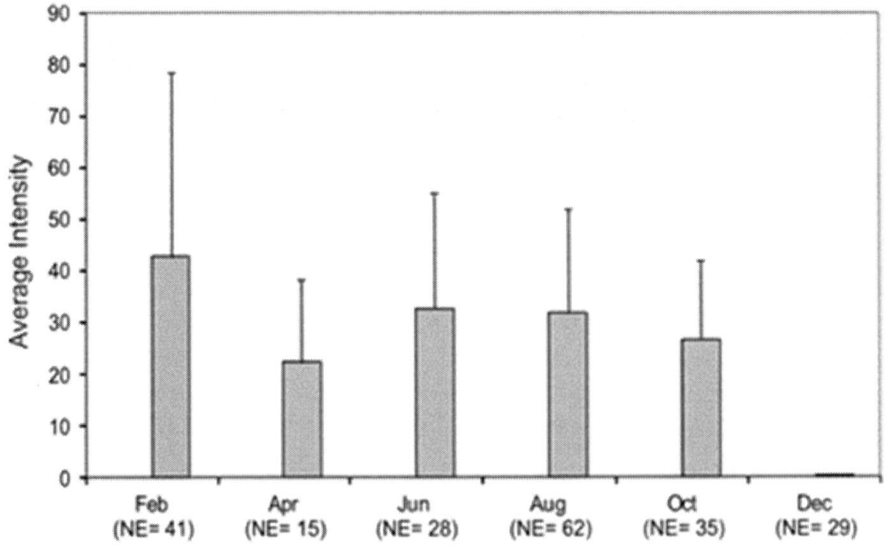

Figure 3. Average intensity of parasites in the red rockfish (*Sebastes miniatus*) during an annual cycle, in San Quintín Baja California, México. NE= number of examined fishes.

All the fish examined were parasitized with a minimum of 3 individuals of copepods to a maximum of 6,373 cestodes (Table 1). In general, the most abundant parasite species was *Bothriocephalus* sp. (14,879 individuals), followed by *Hysterothylacium* sp. (6,373 ind), *Anisakis* sp. (3,942 ind), and *Microcotyle sebastis* (2,696 ind) (Table 2).

The overall mean abundance of parasites showed significant variations (Kruskall Wallis, H = 11.206, p = 0.047) over the year (Table 1). The higher amount of parasites was detected in August and the lower in December (Figure 3). Six of the 12 species recorded (*M. sebastis, B. derzhavini, Parahemiurus merus, Clavellostis sebastidis, Corynosoma strumosum* and Anisakis sp.) showed also significant temporal variations in their abundance (Figure 3).

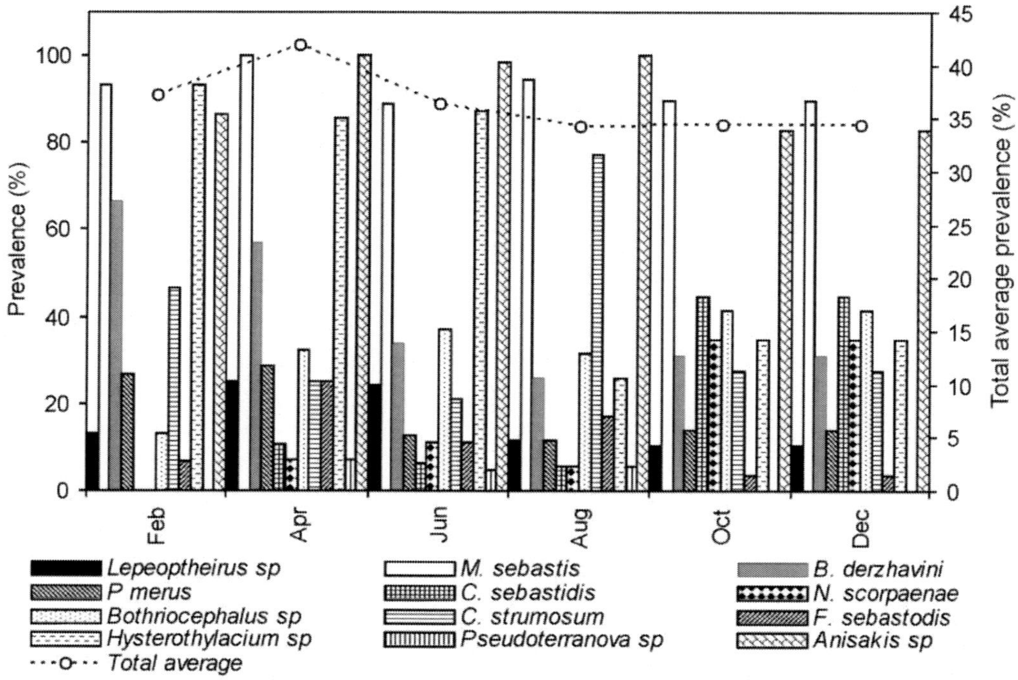

Figure 4. Prevalence (%) of parasites on specimens of *Sebastes miniatus* collected at San Quintín, Baja California, from February to December 2005.

The species with higher intensity was *Botriocephalus* sp. (average intensity = 190.4 ± 48.1) followed by *Anisakis* sp. (15.8 ± 4.1) and *Hysterothylacium* sp. (46.6 ± 18.1) (Table 2). Although there were not significant changes in the overall mean intensity of parasites through the year, it was higher in February and lower in December.

A significant positive correlation (Spearman rank) was found between the prevalence and the average abundance of parasites (R = 0.839, p < 0.05), indicating that the species with high values of prevalence were also the most abundant.

Non-significant correlations between abundance of parasites and total length (R = 0.1033, p = 0.134), weight (R = 0.196, p = 0.087), condition of fish (R = -0.0108, p = 0.876), or bottom temperature (R = -0.1035, p = 0.6004) were found.

4. Discussion

Rockfishes of the genus *Sebastes* sp. from the Southern California Bight (western coast of Baja California, México, and California, USA), are composed of 56 species and are a potential intermediate host for many parasites [33, 34] In the present study, a total of 12 parasite species were identified for *S. miniatus.* Three of them (*M. sebastis, Anisakis* sp. and *Pseudoterranova* sp.) had already been reported for this rockfish in the coasts of California, USA [4, 35], while the other 9 species constitute new parasites for this host and a new locality record.

The parasite species richness in this rockfish was relatively low in comparison with other congeneric species. In *S. nebulosus* (from southeastern Pacific) a richness of 26 parasite species was reported, mostly digeneans [36]. In *S. capensis* (from northern Chile) and *S. caurinus* (from the Atlantic and Pacific oceans) a richness of 16 and 50 parasite species was reported, respectively [37,38]. Also, in *S. serranoides* (a species that distributes from Oregon, USA to Baja California, Mexico) the species richness was of 36 species, which was positively related with the host age and size [39].

Although the parasite abundance in *S. miniatus* showed certain seasonality, it was not related to the characteristics of the fish or bottom temperature during our study period (2005). Such variations could be due to other factors like prey availability over the year. The diet of *S. miniatus* includes a variety of species (fishes, skids, octopuses, shrimps, copepods, isopods, polychaetes, etc.), many of them potential hosts for initial stages of cestodes [40]. Likewise, adult stages of parasites also inhabit predators like elasmobranches [41] and sea lions [13]. In this respect, *S. miniatus* is part of an intermediate food web permitting the parasites to reach their maturity in the final host [4, 42]. Parasites found in *S. miniatus* are possibly transmitted via ingestion of larval forms or by the presence of free-living stages of larvae in the plankton, which often determines the parasitic fauna of the host [43].

The lack of correlation between the parasite abundance and the host size could also be because specimens of *S. miniatus* were conditioned to a narrow size range (from 24 to 68 cm of total length). Therefore, we could not analyze the whole spectrum of sizes available in the stock. Similar findings have been documented in other studies, where there are no significant relationship between the host body size and abundance of parasites [43, 44, 45]. For example, in *S. capensis* from southern Chile, this lack of correlation was attributed to the small sample size or the narrow size range of fish collected in their study [46]. In a parasitological study conducted on 6 species of *Sebastes* from the North-Occidental coast of Baja California, the authors found that the host size was not a relevant variable to explain variations in the intensity of infection [2]. However, studies in *Sebastes fasciatus* (from the Scotia Shelf) revealed that the size of the host had little effect on parasite prevalence, except for *Anisakid* species, where smaller fish feed more heavily on the intermediate host of the parasite than do larger fish [47]. Thus, it is possible that other factors (e.g. the sexual maturity stage, changes in the diet, the ingest volume or the contact possibilities with other intermediate hosts) may be involved in this lack of correlation [48].

Some parasite species of *S. miniatus* showed microhabitat specificity for intestine, stomach and intestinal blind. This could be because they find there a high availability of nutrients and low competition for resources [45]. Thus, the specificity for certain microhabitats could also help determine the community structure of parasites in *S. miniatus*

[49]. This specificity for organ host was similar to those reported in *S. capensis* (from the Southeastern Pacific) [34].

Similar to other congeneric species (e.g. *S. capensis*) in the present study there was a predominance of certain groups of parasites such as the genera *Bothriocephalus, Hysterothylacium, Anisakis, Microcotyle*, and *Corynosoma* [34]. In *S. serranoides* larval cestodes were predominant [39]. In other rockfish species from Baja California (*S. constellatus, S. elongatus, S. rosaceus, S. umbrosus, S. helvomaculatus* and *S. chlorosticus*) the predominant group are the nematodes (*Anisakis*) (65-90% of prevalence), followed by digeneans (4.1-25% of prevalence) [2].

In this study nematodes and parasitic copepods were the most diverse groups (three species each one). However, in comparison with other studies this diversity was relatively low. For example in *S. alutus* from the coasts of California 8 species of nematodes and 12 copepod species were recorded [4]. Larvae of *Anisakis* sp. and *Hysterothylacium* sp. have been recorded from most of the fishes studied in the Mexican Pacific coasts as well as in many pelagic and demersal fishes from other areas of the world [50, 51]. The percentage of S. miniatus specimens infected by *Anisakis* larvae was relatively high (prevalence = 91.8%) in comparison with that reported in other congeneric species (*S. capensis,* 2.9-4.8%) [54]. Differences in parasite prevalences between *Sebastes* species and other fishes can probably be explained by differences in the diet and environmental conditions. Furthermore, previous studies have suggested that parasites may be present in a locality when hosts are available, and the abiotic conditions allow the survival of the infective stages [53].

Larvae of *Anisakis* sp. and *Hysterothylacium* sp. were particularly important because in addition to being abundant and causing diseases in the fish host, they may constitute a risk factor to human health (*anisakiasis*). These parasites can infect humans by eating marinated raw fish ('ceviche or cebiche') [54]; which is a seafood dish popular in the coastal regions of the Americas, especially Central and South America. Therefore we recommend avoiding consumption of raw or inadequately cooked fish.

In conclusion, the magnitude of infection (abundance, prevalence and intensity) in this host supports previous findings that demersal fishes are rich in the number of individual parasites [36], and that they have a parasitic fauna represented mainly by helminths (cestodes, nematodes and monogeneans). In this chapter, the importance of analyzing the parasitic fauna of commercially important fish species has been highlighted, in addition to contributing to the knowledge of marine biodiversity, which also helps to identify parasite species that could represent a potential risk for human health, mainly when fish is eaten raw or inadequately coked.

REFERENCES

[1] Castillo-Sánchez, E., Rosales-Casián, J.A., G. Pérez-Ponce de León, G. (1998). Helminth parasites of *Paralychtys californicus* (Osteichthyes: *Paralichthydae*) in Estero de Punta Banda, Todos Santos Bay and San Quintín Bay, Baja California, México. *Ciencias Marinas*, 24, 443-462.

[2] Alvarado-Villamar, M.R., Ruiz-Campos, G. (1992). Estudio comparativo del grado de infección de macroparásitos en seis especies de Sebastes (*Pisces Scorpenidae*) de la costa Noroccidental de Baja California, México. *Ciencias Mararinas,* 18, 79-92.

[3] Rodríguez-Santiago, M. A., Rosales-Casián, J. A. (2011). Parasite structure of the Ocean Whitefish *Caulolatilus* princeps from Baja California, México (East Pacific). *Helgoland Marine Research,* 65, 197–202.

[4] Love, M. S., Yoklavich, M., Thorsteinson, L. (2002). *The rockfishes of the Northeast Pacific*. University of California Press, Los Angeles, 405 p.

[5] Rodríguez-Medrano, M. C. (1993). Descripción y análisis de la pesca deportiva en Bahía de Todos Santos, Ensenada, B. C. *MSc Thesis,* Centro de Investigacion Cientifica y de Educacion Superior de Ensenada, Mexico, 88 p.

[6] Hernández-Hernández, A. (2002). *Composición específica de los peces escama comercializados en el Mercado de Mariscos del Puerto de Ensenada,* Baja California, México. Facultad de Ciencias Biológicas, Universidad Autónoma del Estado de Morelos. Cuernavaca, Morelos (México), 60 pp.

[7] Rosales-Casián J. A., González Camacho J. R. (2003). Abundance and importance of fish species from artisanal fishery on the Pacific coast of Northern Baja California. *Bulletin Southern California Academy of Sciences,* 102, 51-65.

[8] Eschmeyer, W. N., Herald, E. S., Hammann, H. (1983). *A field guide to Pacific Coast fishes of North America*. Houghton Mifflin, Boston, Massachusetts. 336 p.

[9] Love, M. S. (1996). *Probably more than you want to know about the fishes of the Pacific coast*. Really Big Press, Santa Barbara, California, 215 p.

[10] Stephens, J., Wendt, D., Wilson-Vandenberg, D., Carroll, J., Nakamura, R., Nakada, E., Rienecke, S., Wilson, J. (2006). *Rockfish resources of the South Central California coast: analysis of the resource from party boat data*, 2000-2005, CalCOFI Report 47, 140-155.

[11] Rodríguez-Santiago, M. A., Rosales-Casián, J. A. (2008). Abundance and size composition of vermilion rockfish *Sebastes miniatus* (Jordan and Gilbert 1880), from sport fishing catches of San Quintín, Ensenada, Baja California, Mexico. *Bulletin Southern California Academy of Sciences*, 107, 25-32.

[12] O'Connell, V. A., Gordon, D. A., Hoffmann, A., Hepler, K. (1992). Northern range extension of the vermilion rockfish (*Sebastes miniatus*). *California Department Fish and Game,* 78, 173.

[13] Lowry, M. S., Stewart, B. S., Heath, C. B., Yochem. P. K. (1991). Seasonal and annual variability in the diet of California sea lions *Zalophus californianus* at San Nicolas Island, California, 1981-1986. *Fishery Bulletin U.S,* 89, 331-336.

[14] García-Córdova, J., Robles-Pacheco, J. M., Gómez-Valdés, J. (2005). Informe de datos de CTD. Campaña IMECOCAL 0501/02. B/O Francisco de Ulloa. Enero 21-Febrero 10 de 2005. Informe Técnico, Departamento de Oceanografía Física, CICESE, 132 p.

[15] Ricker, W. E. (1975). Computation and interpretation of biological statistics of fish populations. *Bulletin of Fisheries Research Board of Canada,* 191, 1-382.

[16] Yamaguti, S. (1961). *Systema Helminthum.* III. Nematodes. Interscience Publishers, INC. New York, USA, 1261 pp.

[17] Yamaguti, S. (1963). *Systema Helminthum. Vol. IV: Monogenea and Aspidocotylea.* Interscience Publishers. New York, USA, 381 pp.

[18] Vidal-Martínez, V., Aguirre-Macedo, I., Sholz, T., González-Solis, D., Mendoza-Franco, F. (2002). *Atlas de los helmintos parásitos de cíclidos de México.* Instituto Politecnico Nacional, México D.F. 183 pp.

[19] Yamaguti, S. (1971). *Synopsis of digenetic trematodes of vertebrates.* Vol. I. Ed. Keigaku Publishing Co. Tokio, Japan, 1074 pp.

[20] Bravo-Hollis, M. (1967). *Helmintos del Pacífico mexicano XXV. Descripción de tres monogéneos del Golfo de California.* Anales del Instituto de Biología de la Universidad Nacional Autónoma de México, 37, 107-123.

[21] Bravo-Hollis, M. (1982). Helmintos de peces del Pacífico mexicano XXXVII. Sobre seis especies conocidas de monogéneos del Suborden *Microcotylinea Lebedev*, 1972. Anales del Instituto de Bilogía de la Universidad Nacional Autónoma de México, 52, 1-12.

[22] Bravo-Hollis, M. (1982). Helmintos de peces del Pacífico mexicano XXXIX. Estudio de Monogéneos del suborden *Microcotylinea Levedev*, 1972, con la presentación de una subfamilia y una especie nuevas. Anales del Instituto de Biología de la Universidad Nacional Autónoma de México, 52, 13- 26.

[23] Mendoza-Garfias, B., Pérez-Ponce de León, G. (1998). Microcotílidos (Monogenea: *Microcotylidae*) parásitos de peces marinos de la Bahía de Chamela, Jalisco, México. *Anales del Instituto de Biología de la Universidad Nacional Autónoma de México*, 69, 2, 139- 153.

[24] Anderson, R. C., Chabaud A.G., Willmott S. (1974-1983). CIH keys to the nematode parasites of vertebrates. *Commonwealth Agricultural Bureaux,* Farnham Royal, Bucks, England, Vol. 1-10.

[25] Cressey, R., Boyle-Cressey, H. (1980). Parasitic copepods of Mackerel and tuna-like fishes (*Scombridae*) of the world. *Smithsonian Contributions to Zoology*, 311, 54-61.

[26] Cressey, R., Boyle-Cressey, H. (1985). *Holobomolochus* (Copepoda: *Bomolochidae*) redefined, with descriptions of three new species from the eastern pacific. *Journal of Crusacean Biology,* 5, 717-727.

[27] Kabata, Z. (1979). *Parasitic Copepoda of British Fishes.* Ray Society, London, England, 468 p.

[28] Kabata, Z. (1992). Copepods Parasitic on Fishes: keys and notes for identification of the species. *Synopses of the British Fauna* (New Series) No. 47. Universal Book Services/W. Backhuys: Oegstgeest. ISBN 90-73348-16-1. VII, 264 pp.

[29] Kabata, Z. (1992b). Copepoda parasitic on Australian fishes, XV. Family *Ergasilidae (Poecilostomatoida). Journal of the Natural* History, 26, 47-66.

[30] Boxshall G. A. (2004). *An introduction to copepod diversity.* The Ray Society, London, United Kingdom, 940 p.

[31] Margolis, L., Esch, G. W., Holmes, J. C., Kuris A. M., Schad. G. A. (1982). The use of ecological terms in parasitology (report of an ad hoc committee of the American Society of Parasitologists). *Journal of Parasitolology,* 68, 131-133.

[32] Steel, R. G., Torrie, J. H. (1986). *Bioestadística: principios y procedimientos.* McGraw-Hill, México, 662 p.

[33] Love, M. S., Caselle, E., Snook, L. (2000). *Fish assemblages around seven oil platforms in the Santa Barbara Channel area.* Fishery Bulletin, 98, 96-117.

[34] González, M. T., Acuña, E. (1998). Metazoan parasites of the red rockfish *Sebastes capensis* off northern Chile. *Journal of Parasitology*, 84, 783–788.

[35] Love, M. S., Moser, M. (1983). *A checklist of parasites of California, Oregon and Washington marine and estuarine fishes.* NOAA Technical Report, NMFS SSRF-777, 576 pp.

[36] Holmes, J. C. (1990). Helminth communities in marine fishes. In: Esch, G., Bush, A., Aho, J. (Eds.) *Parasites communities: patterns and processes.* Chapman & Hall, London, 101-130 pp.

[37] Olsen, S. L. (1952). *Some nematodes parasitic in marine fishes.* Intitute of Marine Science. (University of Texas), 2, 174–215.

[38] González L. (1998). The life cycle of *Hysterothylacium aduncum* (Nematoda: *Anisakidae*) in Chilean marine farms. *Aquaculture,* 162, 173-186.

[39] Love, M. S., Shiner, K., Morris, P. (1984). Parasites of olive rockfish, Sebastes serranoides (Scorpaenidae) off central California. *U.S. Fishery Bulletin,* 82, 530-573.

[40] Rojas-Herrera, A. A., Mascaro, M., Chiappa-Carrara, X. (2003). Hábitos alimenticios de los peces *Lutjanus* peru y *Lutjanus guttatus* (Pisces: *Lutjanidae*) en Guerrero, Mexico. *Revista de Biología Tropical,* 52, 959-971.

[41] Schmidt, G. D. (1986). *CRC Handbook of tapeworm identification.* CRC Press, Inc. Florida, USA. 675 pp.

[42] Jensen, K. (2001). Four new genera and five new species of *lecanicephalideans* (Cestoda: *Lecanicephallidea*) from elasmobranchs in the Gulf o California, Mexico. *Journal of Parasitology,* 87, 845-861.

[43] Dogiel, V. A. (1964). *General Parasitology*. Oliver and Boyd, London, 164 pp.

[44] Rhode, K. C. Hayward, C., Gosper, D. (1994). A tropical assemblage of ectoparasites: gill and head parasites of *Lethrinus miniatus* (Teleostei: *Lethrinidae*). *International Journal of Parasitology,* 24, 1031-1053.

[45] Stromnes, E., Andersen, K. (1998). Distribution of whaleworm (*Anisakis simplex*, Nematoda, *Ascaridoidea)* L3 larvae in three species of marine fish; saithe (Pollachius virens L.), cod (*Gadus morhua L.)* and redfish (*Sebastes marinus L.*) from Norwegian waters. *Parasitology Researche,* 84, 281-285.

[46] Balboa, L., George-Nascimento, M. (1998). Variaciones ontogenéticas y entre años en las infracomunidades de parásitos metazoos de dos especies de peces marinos de Chile. Revista Chilena de Historia Natural, 71, 27–37.

[47] Scott, J. S. (1988). Helminth parasites of redfish (*Sebastes fasciatus*) from the Scotian Shelf, Bay of Fundy, and eastern Gulf of Maine. *Canadian Journal of Zoology,* 66, 617–621.

[48] Muñoz, G., Cribb, T. (2005). Infracomunity structure of parasites of *Hemigymnus melapterus* (Pisces: Labridae) from Lizard Island, Australia: the importance of habitat and parasite body size. *Journal of Parasitology,* 91, 38-44.

[49] Sasal, P., Trouve, S., Muller-Graf, C., Morand, S. (1999). Specificity and host predictability: a comparative analysis among monogenean parasites of fish. *Journal of Animal Ecology,* 68, 437-444.

[50] Aranda-Cruz, C. (2006). Platelmintos parasitos de la “Sarangola” *Microlepidotus brevipinnis Steindachner, 1869 (Haemulidae) de la Bahia de Chamela Jalisco.* MSc Thesis. Universidad Nacional Autonoma de México, 31-34 p.

[51] Laffon-Leal, S. M., Vidal-Martínez, V., Argora-Torres, G. (2000). Ceviche a potential source of human anisakiasis in Mexico? *Journal Helmintology,* 74, 151-154.

[52] Oliva, M. E., González, M. T. (2004). Metazoan parasites of *Sebastes capensis* from two localities in northern Chile as tools for stock identification. *Journal of Fish Biology,* 64, 170–175.

[53] Guidelli, G. M., Isacc, A., Takemoto, R. M., Pavanelli, G. C. (2003). Endoparasite infracommunities of *Hemisorubim platyrhynchos* (Valenciennes, 1840) (Pisces: *Pimelodidae*) of the Bahia River, up Paraná River floodplain, Brazil: specific composition and ecological aspects. *Brazilian Journal of Biology,* 63, 261-268.

[54] Domínguez Ortega, J., Alonso-Llamazares, A., Rodríguez, L., Chamorro, M., Robledo, T., Bartolomé, J.M., Martínez Cócera, C., (2001). Anaphylaxis due to hypersensitivity to *Anisakis simplex. International Archives of Allergy Immunology,* 125, 86–88.

In: Parasites
Editor: Gilmar S. Erzinger

ISBN: 978-1-62257-692-0

Chapter 10

The Forming and Functioning of Parasite Communities in Age Series of Freshwater Fishes

Zhargal Dugarov
Institute of General and Experimental Biology of Siberian Branch of Russian Academy of Sciences, Ulan-Ude, Russia

Abstract

The parasite communities are studied in age series of Baikal whitefish *Coregonus baicalensis* Dybowski, 1874 and roach *Rutilus rutilus* (L., 1758) of Lake Baikal basin from 0+ to oldest groups. The following definitions of parasite communities of different levels were used: 1) infracommunity; 2) component community. In addition, one more level of communities was considered, namely, the community of a particular host age group. It is considered a set of parasites infracommunities.

Parasites of Baikal whitefish and roach divided into three groups according to the pattern of changes in their abundance in the host age series: 1) parasites displaying an increase in infection rate with the host age; 2) parasites displaying a decrease in infection rate with the host age; and 3) parasites displaying a nonlinear correlation in abundance with roach age.

The number of species of parasites displays statistically significant positive correlations in both the Baikal whitefish and the roach on the infracommunities and the sets of infracommunities levels. Margalef and Menhinick's indices are in positive correlation with the Baikal whitefish age. But the Margalef index shows a significant positive correlation and Menhinick's index shows a significant negative correlation with the roach age in both aquatic bodies on two levels. It is discussed the reasons for it the opposite character of the correlations between these indices and the roach age.

The correlation between the indices reflecting the evenness of infracommunities of roach parasites and the host age is, in general, negative in both the Chivyrkuiskii Bay of Lake Baikal and the Selenga River Estuary at the infracommunities and the sets of infracommunities levels.

The opposite correlations between the Berger–Parker index and roach age were observed: the correlation for the Chivyrkuiskii Bay is positive and for the estuary of the Selenga River it is negative.

The species diversity of Baikal whitefish parasite communities increases naturally in the host age series in the Chivyrkuiskii Bay. The species diversity of roach decreased in the host age series in the Chivyrkuiskii Bay and, vice versa, increased in the Selenga River Estuary.

An analysis of the changes taking place in the parasite communities in connection with the fish age shows that it is necessary to use indices for measurement of biological diversity in an integrated manner, as well as to take into account the patterns of changes in the components contributing to these indices and the correlations between the indices.

INTRODUCTION

Dogel' (1958) characterized fishes as the most convenient objects for study of influence of age factor on fauna of parasites, first of all, because of availability of valuable property, namely, signs permitted exactly to determine the age of each individual [1]. He commenced studies on the dependence that parasitic fauna has on the age of the host [2]. This direction was then continued by many researchers, in particular, when studying the fish in lake Vrevo (Leningrad oblast, Russia) [3].

When the general concept of parasite communities was formed [4, 5] and the methods for assessing ecological diversity were developed [6], the number of papers on the parasite communities and the factors influencing them increased. Analysis of factors influencing the parasite's community structure formation, differences of parasites in various host groups and in the same hosts from different habitats, and fluctuations of the parasite community structure over various time intervals are reflected in numerous papers [7, 8, 9, 10, 11, 12]. Zander (2001) investigated the parasite communities of ecologically and taxonomically similar host species and proposed to use the term guild community as an intermediate level between component and supra community of parasites. The combination of several parasite guild communities may lead to an approximately accurate image of the parasite supra community within the whole ecosystem [13]. Later on the study of parasite supra community from the Flensburg fjord was carried out. 28 host species of the guilds snails, planktonic and benthic crustaceans, and fishes were investigated [14].

The determination of the structure of parasites communities in complete age series of hosts (from 0+ to oldest) is laborious process owing to necessity of parasitological investigation of required quantity of host individuals. The additional difficulty in this aspect is need of special efforts for catch of the youngest and oldest individuals. The regularities of forming and functioning of parasites communities in age series of Baikal whitefish and roach (from 0+ to oldest) were characterized in this paper.

MATERIALS AND METHODS

The regularities of forming and functioning of parasites communities are demonstrated in age series of Baikal whitefish *Coregonus baicalensis* Dybowski, 1874 and roach *Rutilus rutilus* (L., 1758) of Lake Baikal basin. The Baikal whitefish (313 ind.) was caught in

Chivyrkuiskii Bay of Lake Baikal. The studied sample's age groups (0+ to 16+) correspond to the age structure of the whitefish population in this bay. The roach was caught in the Chivyrkuiskii Bay of lake Baikal (0+ - 12+, 221 ind.) and in the Selenga River Estuary (0+ - 8+, 75 ind.).

The following definitions of parasite communities were used: infracommunity, comprising all the parasites (of all species) in a host individual, and component community, comprising all the infracommunities in a given host population [15, 16, 17, 18]. In addition to these types of communities, one more level of communities was considered when analyzing the distribution of parasites in host population, namely, the community of the age group of the host. In this hierarchical system, the parasite community of a particular host age group, referred to as an individual component community [19] or the parasite community of a host of a certain age [20], is considered *a set of parasite infracommunities* [21]. The sum of the parasite infracommunities of all the host age groups is referred to as the component community of the roach parasites of an aquatic body.

The considered characteristics of the roach parasite communities can be divided into four groups, namely: 1) species richness (number of parasite species, Margalef index, and Menhinick's index); 2) evenness (Simpson's measure of evenness, Camargo's evenness index, Smith and Wilson's evenness index, Shannon evenness measure, Brillouin evenness measure); 3) dominance (Berger–Parker index, Simpson domination index); and 4) species diversity based on the relative species abundance (Brillouin and Shannon indices). In addition to these characteristics, the number of parasite individuals was determined in the fish parasite communities, as was the fraction of the host individuals lacking parasites or infected with one parasite species. The chosen parameters include indices widely used when studying various groups of organisms [22], as well as parameters frequently used when studying parasite communities [16, 23]. The evenness indices were selected taking into account their comparative analysis, which has been done by several researchers [24, 25]. The Brillouin index, which is applied to collections and samples where all individuals can be counted, is optimal for assessing the species diversity of parasite infracommunities [26]. Mean values are given for all the parameters, and standard errors of the mean are given for some of them. The correlation between the parameters (the roach age and infection rate with individual parasite species, the roach age and parameters of parasite communities, and various parameters of parasite communities) were determined according to the nonparametric Spearman rank coefficient (r_S). The significance level (p) is shown when characterizing the Spearman correlation coefficient.

RESULTS

Parasite Fauna. Baikal Whitefish

A total of 28 species and subspecies of parasites from 11 classes and a group of protists of undefined status were registered in the Baikal whitefish of Chivyrkui Bay of Lake Baikal, namely Parasitomonada: *Hexamita truttae* (Schmidt); Myxosporea: *Chloromyxum coregoni* Bauer, *Henneguya zschokkei* (Gurley); Suctoria: *Capriniana piscium* (Butschli); Peritricha: *Trichodina nigra* Lom; Protozoa incertae sedis: *Dermocystidium sp.*; Monogenea:

Discocotyle sagittata (Leuckart); Cestoda: *Eubothrium crassum* (Bloch), *Diphyllobothrium dendriticum* (Nitzsch) (l.), *Diphyllobothrium ditremum* (Creplin) (l.), *Cyatocephalus truncatus* (Pallas), *Proteocephalus longicollis* (Zeder), *Triaenophorus crassus* Forel (l.); Trematoda: *Crepidostomum farionis* (Mueller), *Phyllodistomum umblae* (Fabricius), *Diplostomum spathaceum* (Rudolphi) (met.), *Ichthyocotylurus erraticus* (Rudolphi) (met.); Nematoda: *Cystidicola farionis* Fischer, *Contracaecum osculatum baicalensis* Mozgovoi et Ryjikov (l.), *Philonema sibirica* (Bauer); Palaeacanthocephala: *Pseudoechinorhynchus borealis* Linstow, *Metechinorhynchus salmonis salmonis* (Muller), *M. s. baicalensis* Bogolepova, *Metechinorhynchus truttae* (Schrank); Hirudinea: *Piscicola geometra* (L.); Crustacea: *Ergasilus sieboldi* Nordmann, *Salmincola extumescens* (Gadd), *Salmincola extensus* (Kessler). The taxonomic status of acantocephala of Baikal whitefish was brought by Baldanova, and Pronin [27].

Roach

Previous studies [28, 29, 30, 31, 32] have demonstrated that the parasite fauna of the lake Baikal roach comprises 37 species. According to our data, the roach from Lake Baikal and the Selenga River Estuary houses 49 parasite species belonging to 12 classes, namely Coccidiomorpha: *Eimeria leucisci* Schulman et Zaika; Myxosporea: *Myxidium rhodei* Leger, *M. pfeifferi* Auerbach, *Zschokkella costata* Kaschkowsky, *Chloromyxum fluviatile* Thelohan, *Myxosoma dujardini* Thelohan, *Myxobolus ellipsoides* Thelohan, *M. muelleri* Butschli, *M. bramae* Reuss, *M. pseudodispar* Gorbunova, *M. diversicapsularis* Sluchai in: Schulman, *M. macrocapsularis* Reuss, *Henneguya cutanea* Dogiel et Petruschewsky; Hymenostomata: *Ichthyophthirius multifiliis* Fouquet; Peritricha: *Apiosoma* sp. *Trichodina intermedia* Lom; Monogenea: *Dactylogyrus crucifer* Wagener, *D. tuba* Linstow, *D. sphyrna* Linstow, *D. suecicus* Nybelin, *D. similis* Wegener, *Gyrodactylus* sp., *Paradiplozoon homoion homoion* Bychowsky et Nagibina; Cestoda: *Caryophyllaeides fennica* (Schneider), *Proteocephalus torulosus* Batsch, *Triaenophorus nodulosus* Pallas, *Ligula intestinalis* L.; Trematoda: *Rhipidocotyle campanula* Dujardin, *Phyllodistomum folium* Olfers, *Allocreadium isoporum* Looss, *Diplostomum commutatum* Diesing, *D. rutili* Razmashkin, *D. helveticum* Dubois, *D. paracaudum* Iles, *D. mergi* Dubois, *D. volvens* Nordmann, *D. spathaceum* Rudolphi, *Tylodelphys clavata* Nordmann, *Ichthyocotylurus pileatus* Rudolphi; Nematoda: *Raphidascaris acus* Bloch; Eoacanthocephala: *Neoechinorhynchus rutili* Muller; Hirudinea: *Piscicola geometra* L.; Bivalvia: *Colletopterum anatinum* L., *C. piscinale* Nilsson, *C. nilssonii* Kuester; Crustacea: *Ergasilus briani* Markewitsch, *E. sieboldi* Nordmann, *Paraergasilus rylovi* Markewitsch, *Tracheliastes polycolpus* Nordmann.

Connection of Parasites Infection with the Host Age. Baikal Whitefish

A total of 21 taxonomic forms were analyzed to determine the correlation between the abundance of parasites and the host age, including acantocephalans (*P. borealis*, *M. s. salmonis*, *M. s. baicalensis*, *M. truttae* predominantly). Parasites of Baikal whitefish and roach are divided into three groups according to connection to the host age, the same division of parasites in age series of fishes first was marked by Dogel' [1].

1) Parasites with infection increasing together with host's age: *C. truncatus* (r_S =0.24, p <0.001), *Eubothrium crassum* (r_S =0.31, p <0.001), genus *Diphyllobothrium* cestodes (*D. dendriticum* + *D. ditremum*) (r_S =0.24, p <0.001), *Contracaecum osculatum baicalensis* (r_S =0.17, p <0.01).
2) Parasites with infestation decreasing with host's age: *P. longicollis* (r_S = –0.19, p < 0.001); *Ph. umblae* (r_S = –0.15, p < 0.01).
3) Parasites with relative abundance not dependent on host's age: *Dipl. spathaceum*, *Ichtyocotylurus erraticus*, acantocephalans, *Salmincola extumescens, Triaenophorus crassus, Crepidostomum farionis, Cystidicola farionis, Phylonema sibirica*, *Piscicola geometra*, *Ergasilus sieboldi*, and *Salmincola extensus*.

Roach

A total of 34 taxonomic forms were analyzed to determine the correlation between the abundance of parasites and the host age, including metacercariae of *Diplostomum* spp. (*D. commutatum* + *D. rutili* + *D. helveticum* + *D. paracaudum* + *D. mergi* + *D. volvens* +*D. spathaceum*) and the bivalves *Colletopterum* spp. (*C. anatinum* + *C. piscinale* + *C. nilssonii*).

Table 1. Species composition of parasites and the characteristics of a positive correlation between their number and host age

Parasite species	Chivyrkuiskii Bay		Selenga River Estuary	
	r_S	p	r_S	p
Myxidium rhodei	0.81	0	0.53	0
Myxobolus bramae	0.42	0	0.31	0.007
M. muelleri	0.49	0	–	–
M. pseudodispar	0.44	0	–	–
M. ellipsoides	0.15	0.023	–	–
M. diversicapsularis	0.47	0	–	–
Zschokkella costata	–	–	0.30	0.009
Ichthyophthirius multifiliis	0.14	0.044	–	–
Dactylogyrus crucifer	0.75	0	0.67	0
D. sphyrna	0.53	0	–	–
Paradiplozoon homoion homoion	0.19	0.006	–	–
Caryophyllaeides fennica	0.26	0	–	–
Rhipidocotyle campanula	0.26	0	–	–
Allocreadium isoporum	–	–	0.32	0.005
Diplostomum spp.	0.74	0	0.67	0
Tylodelphys clavata	0.35	0	0.36	0.001
Raphidascaris acus	0.35	0	–	–
Colletopterum spp.	–	–	0.29	0.011
Ergasilus briani	–	–	0.24	0.042

1) Parasites displaying an increase in infection rate with the host age (Table 1).
2) Parasites displaying a decrease in infection rate with the host age: *Trichodina intermedia*: for the Chivyrkuiskii Bay, r_S = –0.33, p = 0; *Apiosoma* sp.: for the Chivyrkuiskii Bay, r_S = –0.14, p = 0.044; *Ichthyocotylurus pileatus*: r_S = –0.24, p = 0.

3) Parasites displaying a nonlinear correlation in abundance with roach age: *Myxidium pfeifferi, Myxosoma dujardini, Chloromyxum fluviatile, Eimeria leucisci, Dactylogyrus suecicus, Gyrodactylus* sp., *Ligula intestinalis, Triaenophorus nodulosus, Phyllodistomum folium, Piscicola geometra, Paraergasilus rylovi*, and *Ergasilus sieboldi*.

During the first months of their life in the Chivyrkuiskii Bay of Lake Baikal, roach young fishes are most intensively infected with *Trichodina intermedia, Apiosoma* sp., and metacercariae of *Diplostomum* spp. *Trichodina* and *Apiosoma*, which are typical of young fishes and rarely met in the roach at ages of 2+ to 12+ in the Chivyrkuiskii Bay, can be regarded as so-called juvenile parasite species [3].

Infracommunities of Parasites and the Host Age. Baikal Whitefish

The analysis of Baikal whitefish parasite communities was involved 17 taxonomic forms including *Diphyllobothrium* genus cestodes (*D. dendriticum* + *D. ditremum*) and acantocephalans as jointed group mentioned above. Menhinick's index of species richness, Brillouin index of species diversity, Brillouin evenness measure, and the mean number of parasite species are in positive correlation with the Baikal whitefish age (Table 2). Spearman's rank correlation coefficient between the parasite infracommunity parameters and the Baikal whitefish age decreases in the following order: Menhinick's index > Brillouin evenness measure > Brillouin index > mean number of parasite species [33]. An increase in the mean number of parasite species per host individual was noted in the Ladoga whitefish [34] and other fish species [1, 35].

Menhinick's index is directly proportional to the mean number of parasite species and inversely proportional to the mean number of parasite individuals in one infracommunity. The dependence between the age of Baikal whitefish and the mean number of parasite individuals has a nonlinear character in one infracommunity ($r_S = 0.02$) in general; the mean number of parasite individuals in one infracommunity does not increase with age.

Brillouin index characterizes the relative abundance of species combining the species richness and evenness or uniformity of the distribution into a single value [22]. These features condition the interrelation of this index with a number of parameters of whitefish parasite infracommunities. Significant positive correlations of Brillouin index with the following parameters are noted: mean number of parasite species ($r_S = 0.87$, $p < 0.001$); Menhinick's index ($r_S = 0.56$, $p < 0.001$). These correlations reflect the positive correlation between the Brillouin index and the species richness. The significant negative correlation between the Brillouin index and the domination index of Berger–Parker ($r_S = -0.69$, $p < 0.001$) in the Baikal whitefish parasite infracommunities reflects the relation between the diversity and the evenness of communities. High evenness with equal species abundance is considered to be equivalent of high diversity [22]. Therefore, the higher the abundance of some dominating species, the lower the evenness and community diversity. A negative relation between the parasite community diversity and the domination of a certain species was noted by a number of researchers [16, 23, 36].

Table 2. The changes of the Baikal whitefish parasite infracommunities in relation to host age

Characteristic	Baikal whitefish age group 0+	1+	2+ - 3+	4+ - 5+	6+	7+	8+	9+	10+	11+	12+	13+	14+	15+ - 16+	r_S	P
Number of parasite species	3	7	9	7	8	8	11	9	11	11	12	8	8	9		
min-max	0 – 2	0 – 4	2 – 5	1 – 5	1 – 4	1 – 5	1 – 6	1 – 4	1 – 5	1 – 3	1 – 6	1 – 4	1 – 4	0 – 4		
X	0.68	1.68	3.25	2.47	2.38	2.20	2.90	2.24	2.39	2.04	2.75	1.96	2.29	1.89	0,14	0,017
s.e	0.12	0.24	0.27	0.24	0.20	0.26	0.25	0.21	0.22	0.13	0.23	0.17	0.22	0.19		
Menhinick's index *min-max*	0 – 1	0 – 1	0.3 – 0.5	0.2 – 1.2	0.2 – 1.0	0.2 - 1.0	0.4 -1.4	0.2 - 1.2	0.3 – 2.0	0.2 – 1.5	0.2 – 1.2	0.1 – 1.4	0.2 – 0.9	0.0 – 1.6		
X	0.25	0.48	1.10	0.63	0.54	0.53	0.64	0.50	0.69	0.57	0.64	0.69	0.56	0.75	0,27	0,000
s.e.	0.05	0.08	0.52	0.07	0.06	0.06	0.06	0.06	0.09	0.07	0.05	0.07	0.05	0.09		
Number of parasite individuals	276	425	980	341	450	391	861	673	448	470	882	445	372	321		
min-max	0 – 93	0 – 68	4 – 206	3 – 81	5 – 85	5 – 45	3 – 96	3 – 157	4 – 69	3 – 45	3 – 106	1 – 64	2 – 78	0 – 59		
X	9.86	19.32	61.25	22.73	28.13	19.55	28.70	32.05	19.48	20.43	31.50	17.12	21.88	11.46	-0,02	
s.e.	3.46	4.93	15.15	5.33	6.05	2.47	4.65	7.40	3.68	3.00	6.18	3.43	4.29	2.84		
Berger-Parker index *min-max*	0 - 1	0 – 1	0.3 – 1.0	0.4 – 1.0	0.5 – 1.0	0.4 – 1.0	0.3 – 1.0	0.5 – 1.0	0.3 – 1.0	0.4 – 1.0	0.4 – 1.0	0.5 – 1.0	0.4 – 1.0	0 – 1		
X	0.60	0.70	0.75	0.74	0.77	0.79	0.72	0.82	0.76	0.77	0.72	0.79	0.75	0.70	-0,03	
s.e.	0.09	0.08	0.05	0.05	0.04	0.04	0.03	0.04	0.04	0.04	0.03	0.04	0.05	0.06		
Brillouin index *min-max*	0.0 – 0.4	0.0 – 0.6	0.2 – 1.3	0.0 – 1.2	0.0 – 0.9	0.0 – 1.2	0.0 – 1.1	0.0 – 1.0	0.0 – 1.0	0.0 – 1.0	0.0 – 1.0	0.0 – 0.9	0.0 – 0.9	0.0 – 0.9		
X	0.02	0.23	0.52	0.63	0.42	0.40	0.55	0.36	0.40	0.38	0.50	0.32	0.46	0.31	0,17	0,003
s.e.	0.02	0.05	0.08	0.07	0.06	0.08	0.06	0.06	0.06	0.05	0.04	0.06	0.09	0.06		
Brillouin evenness measure *min-max*	0.0 – 0.7	0.0 – 1.0	0.1 – 1.0	0.0 – 1.0	0.0 – 1.0	0.0 – 0.9	0.0 – 1.0	0.0 – 1.0	0.0 – 1.0	0.0 – 1.0	0.0 – 1.0	0.0 – 1.0	0.0 – 1.6	0.0 – 1.0	0,23	0,000
X	0.03	0.35	0.55	0.62	0.56	0.48	0.61	0.46	0.54	0.59	0.67	0.50	0.58	0.49		
s.e.	0.03	0.08	0.08	0.08	0.09	0.08	0.05	0.07	0.08	0.07	0.05	0.08	0.11	0.08		
Number of studied fishes, spec.	28	22	16	15	16	20	30	21	23	23	28	26	17	28		
Number of infested fishes, spec.	17	18	16	15	16	20	30	21	23	23	28	26	17	26		

Roach

The analysis of roach parasite communities was involved 30 taxonomic forms from Chivyrkuiskii Bay and 17 ones from Selenga River estuary including jointed groups of *Diplostomum* spp. and *Colletopterum* spp. mentioned above. The number of species of roach parasites and the number of parasite individuals in one infracommunity display statistically significant positive correlations in both the Chivyrkuiskii Bay of Lake Baikal and the Selenga River Estuary. However, the Margalef and Menhinick's indices of species richness demonstrate oppositely directed correlations with the host age in both aquatic bodies: the former shows a significant positive correlation, whereas the latter shows a significant negative correlation (Tables 3, 4).

The correlation between the indices reflecting the evenness of infracommunities of roach parasites and the host age is, in general, negative in both the Chivyrkuiskii Bay of Lake Baikal and the Selenga River Estuary. The Smith and Wilson evenness index decreases in a statistically significant manner with the roach age in both aquatic bodies. The correlation between the Simpson evenness index and the fish age in the Chivyrkuiskii Bay is negative. As for the Selenga River Estuary, there is no correlation. There is no correlation between the Camargo's index of evenness and the roach age in Chivyrkuiskii Bay, and this correlation in the Selenga River Estuary is negative (Tables 3, 4). Roach groups of younger ages (0+ to 3+ in the Chivyrkuiskii Bay and 0+ in the Selenga River Estuary) contain individuals displaying the maximal values of the evenness indices (1.0) of their parasite infra communities. This is observed when infracommunities contain only one parasite species or two species in equal abundance. "Evenness" describes the variability in species abundances.

A community in which all species have approximately equal numbers of individuals (or similar biomasses) would be rated as extremely even [26]. In the older roach groups, the species richness of parasites increases while the abundance of parasite species considerably differs, which leads to a decrease in the evenness indices in the parasite infracommunities.

The correlation between the Brillouin index of species diversity in infracommunities and the host age is significantly positive in both aquatic bodies (Tables 3, 4). Most characteristics of species diversity in the infracommunities of roach parasites in the ChivyrkuiskiiBay of Lake Baikal and the Selenga River Estuary similarly correlate with the host age. The only exception is the Berger–Parker index, which displays a positive correlation with the host age in the Chivyrkuiskii Bay and a negative correlation in the estuary of the Selenga River. *Trichodina intermedia* was more frequently dominant in the parasite infracommunities of roach underyearlings in the Chivyrkuiskii Bay, *Diplostomum* spp. metacercariae were more frequently dominant in two-year-old individuals, and *Myxidium rhodei* more frequently dominated in fish 3 years old and older (Table 3). In the Selenga River Estuary, *M. rhodei* was dominant in the parasite infracommunities of roach of all ages (Table 4). In general, this Myxosporidia was the dominant species in most of the roach parasite infracommunities of all age groups.

Table 3. Changes in the characteristics of parasite infracommunities in the roach age groups in the Chivyrkuiskii Bay of Lake Baikal

Characteristic	Roach age group									r_S	p
	0+	1+	2+	3+	4+	5+	6+	7+	8+ – 12+		
Total number of parasites species	8	14	16	14	15	18	22	18	18	–	–
Average number of parasites species	1.05 0.17	2.45 0.16	3.69 0.59	5.08 0.35	4.69 0.36	6.14 0.17	6.85 0.29	7.08 0.22	8.22 0.81	0.86	0
Average number of parasites individuals	33.43 10.20	72.27 21.53	148.38 38.49	987.32 261.10	1598.06 288.54	1001.05 189.40	2157.42 615.46	2167.04 376.72	1344.22 506.05	0.80	0
Margalef index	0.17 0.05	0.50 0.06	0.65 0.13	0.66 0.06	0.53 0.06	0.79 0.03	0.86 0.05	0.85 0.04	1.07 0.09	0.64	0
Menhinick's index	0.34 0.07	0.56 0.05	0.45 0.09	0.25 0.03	0.15 0.02	0.25 0.02	0.25 0.03	0.21 0.02	0.30 0.03	–0.14	0.035
Simpson's measure of evenness	0.81 0.04	0.76 0.04	0.57 0.08	0.33 0.03	0.30 0.03	0.24 0.02	0.27 0.02	0.20 0.02	0.29 0.07	–0.76	0
Camargo's evenness index	0.18 0.04	0.59 0.05	0.49 0.08	0.30 0.02	0.31 0.02	0.25 0.02	0.27 0.02	0.21 0.02	0.29 0.06	–0.09	–
Smith and Wilson's evenness index	0.69 0.06	0.71 0.05	0.44 0.10	0.21 0.04	0.12 0.02	0.17 0.02	0.14 0.01	0.13 0.01	0.18 0.02	–0.65	0
Berger-Parker dominance index	0.52 0.07	0.70 0.04	0.74 0.05	0.82 0.03	0.88 0.04	0.83 0.03	0.75 0.04	0.85 0.03	0.70 0.08	0.16	0.021
Brillouin index	0.13 0.03	0.46 0.05	0.53 0.09	0.57 0.08	0.37 0.10	0.53 0.07	0.67 0.09	0.47 0.07	0.80 0.13	0.40	–
Dominant species*	*T. i.*	*D.* spp.	*M. r.*	*M. r.*	*M. r.*	*M. r.*	*M. r.*	*M. r.*	*M. r.*	–	–
Occurrence of dominant species, %	26.19	36.36	61.54	92.00	100	95.24	88.46	100	66.67	–	–
Number of examined fish, ind.	42	44	13	25	16	21	26	25	9	–	–

Note: Here and in Table 4, the mean value is shown above the line and the standard error of the mean is shown below the line; *T. i.* - *Trichodina intermedia*, *D.* spp. - *Diplostomum* spp., *M. r.* - *Myxidium rhodei*. *According to the occurrence in infracommunities.

Table 4. Changes in the characteristics of parasite infracommunities in the roach age groups in the Selenga River Estuary

Characteristic	Roach age group							r_S	p
	0+	1+	2+	3+	4+	5+	6+ - 8+		
Total number of parasites species	6	8	9	8	15	12	13	-	-
Average number of parasites species	3.00 0.22	3.60 0.22	4.38 0.38	4.20 0.33	5.94 0.32	5.00 0.24	6.00 0.28	0.66	0.000
Average number of parasites individuals	56.43 15.24	138.80 56.83	271.38 77.51	221.10 41.37	324.25 48.02	754.89 189.98	594.13 122.25	0.70	0.000
Margalef index	0.55 0.05	0.61 0.05	0.64 0.08	0.62 0.06	0.88 0.06	0.63 0.04	0.81 0.04	0.38	0.001
Menhinick's index	0.49 0.07	0.44 0.06	0.32 0.05	0.32 0.04	0.36 0.03	0.21 0.03	0.28 0.02	-0.38	0.001
Simpson's measure of evenness	0.54 0.09	0.48 0.05	0.35 0.03	0.40 0.06	0.34 0.03	0.37 0.05	0.39 0.03	-0.18	-
Camargo's evenness index	0.55 0.08	0.48 0.04	0.36 0.03	0.40 0.05	0.35 0.02	0.36 0.04	0.38 0.03	-0.25	0.032
Smith and Wilson's evenness index	0.43 0.10	0.38 0.04	0.23 0.03	0.26 0.08	0.23 0.02	0.14 0.02	0.17 0.02	-0.66	0.000
Berger-Parker dominance index	0.77 0.06	0.75 0.05	0.82 0.04	0.79 0.05	0.69 0.03	0.72 0.06	0.61 0.04	-0.37	0.001
Brillouin index	0.48 0.05	0.60 0.05	0.55 0.10	0.56 0.07	0.86 0.06	0.70 0.10	1.20 0.23	0.56	0.000
Dominant species	*M. r.*	*M. r.*	*M. r.*	*M. r.*	*M. r.*	*M. r.*	*M. r.*	-	-
Occurrence of dominant species, %	85.71	60.00	100	100	93.75	88.89	86.67	-	-
Number of examined fish, ind.	7	10	8	10	16	9	15	-	-

Sets of Parasite Infracommunities and Host Age. Baikal Whitefish

The Margalef index of species richness and the Shannon index both have positive correlations with the age of Baikal whitefish. The Simpson index of domination is related to the host age by a significant negative correlation. The decrease of this index (from 0.93 in 0+ whitefish to 0.28 in 15+ to 16+ whitefish) evidences a decrease in the dominant species share (*P. longicollis* and *C. truncatus,* by abundance) in the sets of parasite infracommunities of the host's age series (Table 5). The negative correlation ($r_S = -0.94$) between Shannon index and Simpson's index of domination is significant in the Baikal whitefish sets of parasite infracommunities ($p < 0.001$). The increase in Shannon index in host's age series is related to two factors: increase of parasite species richness and decrease in domination of the most abundant species in the Baikal whitefish sets of parasite infracommunities.The minimum number of parasite species (three) was observed in 0+ individuals of the Baikal whitefish; their numbers increase in the following age groups, reaching a maximum in 12+ fishes. Baikal whitefish individuals from 13+ to 16+ demonstrate both a lower mean number of parasite species and an average number of parasite individuals in the infracommunities (Table 2).

Table 5. The changes of the Baikal whitefish sets of parasite infracommunities in relation to host age.

Characteristic	Baikal whitefish age group														r_S	*p*
	0+	1+	2+ - 3+	4+ - 5+	6+	7+	8+	9+	10+	11+	12+	13+	14+	15+ - 16+		
Total number of parasite species	3	7	9	7	8	8	11	9	11	11	12	8	8	9	0.53	
Total number of parasite ind.	276	425	980	341	450	391	861	673	448	470	882	445	372	321		
Share of hosts without parasites or with one parasite species	0.93 ± 0.05	0.41 ± 0.10	0 ± 0	0.07 ± 0.06	0.13 ± 0.08	0.30 ± 0.10	0.13 ± 0.06	0.24 ± 0.09	0.22 ± 0.09	0.17 ± 0.08	0.04 ± 0.04	0.35 ± 0.09	0.18 ± 0.09	0.39± 0.09		
Margalef index	0,36	0,99	1,16	1,03	1,15	1,17	1,48	1,23	1,64	1,63	1,62	1,15	1,18	1,39	0.66	0.001
Dominant species	*P.l.*	*P.l.*	*P.l.*	*P.l.*	*C.t.*	*C.t.*	*C.t.*	*C.t.*	*C.t.*	*C.t.*	*C.t.*	*C.t.*	*C.t.*	*C.t.*		
Dominant species occurrence, %	50.0 ± 9.4	9.1 ± 10.5	62.5± 12.1	26.7± 11.4	62.5± 12.1	95.0 ± 4.9	80.0± 7.3	85.7 ± 7.6	69.6 ± 9.6	69.6 ± 9.6	71.4 ± 8.5	61.5 ± 9.5	64.7 ± 11.6	53.6 ± 9.4		
Simpson domination index	0,93	0,75	0,63	0,38	0,45	0,53	0,42	0,51	0,40	0,51	0,43	0,50	0,48	0,28	-0.56	0.039
Shannon index	0,17	0,61	0,80	1,18	1,05	1,05	1,20	0,99	1,24	1,02	1,15	1,13	1,04	1,50	0.56	0.037
Shannon evenness measure	0,15	0,31	0,36	0,61	0,51	0,50	0,50	0,45	0,52	0,42	0,46	0,54	0,50	0,68	0.50	
Number of studied fish, ind.	28	22	16	15	16	20	30	21	23	23	28	26	17	28		

Note: P.l. – *Proteocephalus longicollis*; C.t. – *Cyathocephalus truncatus*.

A shift of dominant species in the age range of the host takes place in the Baikal whitefish parasite communities. *P. longicollis* cestode dominates in the younger age groups of the Baikal whitefish (0+ to 3+), in 4+ to 5+ fish the relative abundance of *P. longicollis* is insignificantly higher than that of *C. truncatus* cestode. *C. truncatus* dominates in the following age groups (6+ to 16+).

A significant negative correlation between Simpson index and Baikal whitefish age may be observed only in the sets of parasite infracommunities level. One dominant species is found in the parasite component communities of each age group of the Baikal whitefish. The dominant species shift during the transition from 4+ to 5+ (*P. longicollis*) to 6+ (*C. truncatus*) is not an obstacle for significant negative correlation between the age of the Baikal whitefish and the domination index in the sets of parasite infracommunities. A different character of domination was observed in the parasite infracommunities of the majority of age groups: the share of infracommunities where *P. longicollis* dominated fluctuated in the host's age range 0+ to 5+ with two peaks (50.0 and 62.5%); share of infracommunities where *C. truncatus* dominated, first increased from 62.5 to 95%, then decreased to 53.6% in the host's age range 6+ to 16+ (Table 5). The number of species dominating in the infracommunities was up to 6 in single age groups of the Baikal whitefish. The lack of a significant correlation between the domination index and the host age in the Baikal whitefish parasite infracommunities is due to domination of a number of parasites species in the infracommunities of the majority of the host's age groups.

Roach

The total number of parasite species in the sets of infracommunities in the roach positively correlates with the host age in both aquatic bodies. An increase in the number of species in the parasite communities with the host age was recorded in the common minnow *Phoxinus phoxinus* from the Pechora River [20].

The Margalef index for the sets of parasite infracommunities displays a trend for a positive correlation with the host age in both aquatic bodies, whereas the Menhinick's index displays a trend for a negative correlation (this correlation is statistically significant for the Selenga River Estuary) (Tables 6, 7).

The correlation between the evenness indices for the sets of parasite infracommunities and the host age is negative in both aquatic bodies. The largest values of the Spearmen rank coefficient were recorded for the Simpson's measure of evenness; the values of this coefficient for the Camargo's evenness index are lower (the correlations for the Chivyrkuiskii Bay are statistically significant); and the Smith and Wilson's evenness index displayed the lowest values (Tables 6, 7).

The evenness indices in the sets of infracommunities of roach parasites decrease with the host age for the same reason as in the infracommunities of roach parasites: a considerable difference in the abundances of parasite species is observed on the background of an increase in the species richness.

Table 6. Changes in the characteristics of sets of parasite infracommunities in the roach age groups in the Chivyrkuiskii Bay of Lake Baikal

Characteristic	Roach age group									r_S	p
	0+	1+	2+	3+	4+	5+	6+	7+	8+ - 12+		
Total number of parasites species	8	14	16	14	15	18	22	18	18	0.85	0.004
Share of hosts without parasites or with one parasite species	0.64	0.16	0.08	0.04	0	0	0	0	0	-0.91	0.001
Margalef index	0.966	1.612	1.983	1.285	1.379	1.708	1.912	1.560	1.808	0.38	-
Menhinick's index	0.214	0.248	0.364	0.089	0.094	0.124	0.093	0.077	0.164	-0.57	-
Simpson's measure of evenness	0.200	0.278	0.139	0.086	0.077	0.068	0.061	0.064	0.079	-0.80	0.010
Camargo's evenness index	0.211	0.283	0.146	0.099	0.084	0.085	0.071	0.073	0.102	-0.70	0.036
Smith and Wilson's evenness index	0.097	0.109	0.131	0.076	0.083	0.102	0.071	0.077	0.099	-0.37	-
Berger-Parker dominance index	0.776	0.382	0.602	0.908	0.928	0.903	0.857	0.930	0.833	0.55	-
Shannon index	0.736	0.911	1.080	0.439	0.346	0.492	0.604	0.360	0.726	-0.48	-
Dominant species	*T.i.*	*M.b.*	*M.r.*	*M.r.*	*M.r.*	*M.r.*	*M.r.*	*M.r.*	*M.r.*	-	-
Number of examined fish, ind.	42	44	13	25	16	21	26	25	9	-	-

Note. T.i. - *Trichodina intermedia*, M.b. - *Myxobolus bramae*, M.r. - *Myxidium rhodei*.

Table 7. Changes in the characteristics of the sets of parasite infracommunities in the roach age groups in the estuary of the Selenga River

Characteristic	Roach age group							r_S	p
	0+	1+	2+	3+	4+	5+	6+ - 8+		
Total number of parasites species	6	8	9	8	15	12	13	0.83	0.021
Share of hosts without parasites or with one parasite species	0	0	0	0	0	0	0	-	-
Margalef index	0.836	0.967	1.041	0.909	1.637	1.247	1.319	0.79	0.036
Menhinick's index	0.302	0.215	0.193	0.170	0.208	0.146	0.138	-0.89	0.007
Simpson's measure of evenness	0.238	0.197	0.145	0.172	0.125	0.133	0.171	-0.68	-
Camargo's evenness index	0.259	0.222	0.165	0.191	0.182	0.149	0.195	-0.54	-
Smith and Wilson's evenness index	0.156	0.098	0.120	0.118	0.107	0.078	0.111	-0.50	-
Berger-Parker dominance index	0.830	0.786	0.872	0.846	0.711	0.780	0.642	-0.68	-
Shannon index	0.628	0.768	0.548	0.609	1.003	0.779	1.174	0.68	-
Dominant species	*M.r.*	*M.r.*	*M.r.*	*M.r.*	*M.r.*	*M.r.*	*M.r.*	-	-
Number of examined fish, ind.	7	10	8	10	16	9	15	-	-

Note. *M.r.* - *Myxidium rhodei*.

Discussion

Infracommunities of Parasites and Host Age

An increase in Menhinick's index with whitefish age is linked to growth of the mean number of parasite species number in one infracommunity in the first place. The impact of the mean number of parasite individuals upon Menhinick's index is leveled by the absence of a rectilinear link of its number with whitefish age.Opposite correlations between two indices of species richness (Margalef and Menhinick's) and the roach age were observed: a positive for the former and negative for the latter. This fact is explainable based on the properties of both indices: they are directly proportional to the total number of species in the communities and inversely proportional to the total number of individuals of all species in the community (the Margalef index is proportional to the natural logarithm of the total number of individuals and the Menhinick's index is proportional to the total number of individuals to the power of 1/2) [22]. The opposite character of the correlations between these indices and the host age is explained by the difference in the changes in the values of the natural logarithm of the total number of parasite individuals and the total number of parasite individuals raised to the power of 1/2. The values of the mean number of species in an individual infracommunity of roach parasites (the numerator in the formula for the Margalef index) increase in general synchronously with the increase in the mean value of the natural logarithm of the total number of parasite individuals of all species in one infracommunity (the denominator in the formula for the Margalef index) in the host age series in both aquatic bodies. The former value exceeds the latter only in roach groups of older ages (8+ to 12+) in the Chivyrkuiskii Bay of Lake Baikal (Figure 1).

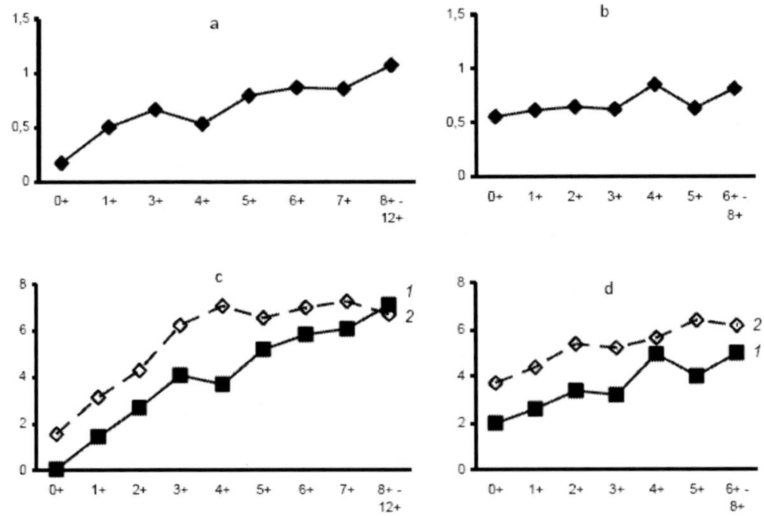

Figure 1. Changes in the averages of the Margalef index (a, b) and its components (c, d) in parasite infracommunities in the roach age groups in the Chivyrkuiskii Bay of lake Baikal (a, c) and the Selenga River estuary (b, d): (*1*) the mean number of parasite species minus one and (*2*) the mean value of the natural logarithm of the total number of parasite individuals of all species; the abscissa shows the roach age.

The consequence of a synchronous increase in the values of the numerator and denominator of the Margalef index is a positive correlation between this index in the infracommunities of roach parasites and the host age in the Chivyrkuiskii Bay and the Selenga River Estuary. The rate of increase in the mean number of species in one infracommunity (the numerator in the formula for the Menhinick's index) is considerably (several times) lower than the rate of increase in the mean value of the total number of all individuals of all species raised to the power of 1/2 in the infracommunity (the denominator in the formula for the Menhinick's index) of parasites in the host age series in the Chivyrkuiskii Bay and the Selenga River Estuary (Figure 2). This results in a negative correlation between the Menhinick's index and the roach age in both aquatic bodies.

Opposite correlations between the Berger-Parker index in two aquatic bodies and roach age were observed: the correlation for the Chivyrkuiskii Bay is positive and for the estuary of the Selenga River it is negative. One characteristic specific feature of *M. rhodei* is that there is a higher abundance in the fish in natural and artificial lakes than in rivers [37].

The spatial distribution pattern of this Myxosporidia species in roach was detected earlier: the highest infection level was recorded in the estuary front of the Selenga River (Lake Baikal) (avandelta), the medium level was recorded in the estuary, it then decreased in the fore estuary, and the minimal infection level was found in the Selenga river [38] (Figure 3). The population size of *M. rhodei* and its ratio to the populations of all other species in the infracommunities of roach parasites determine the overall pattern of the correlation between the Berger–Parker dominance index and the host age in both aquatic bodies.

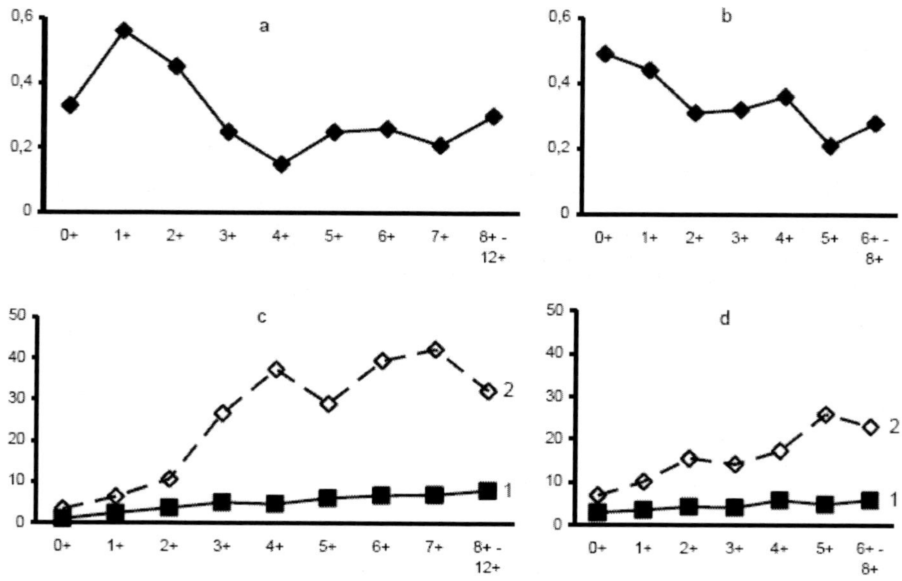

Figure 2. Changes in the averages of the Menhinick's index (a, b) and its components (c, d) in parasite infracommunities in the roach age groups in the Chivyrkuiskii Bay of lake Baikal (a, c) and the estuary of the Selenga River (b, d): (1) the mean number of parasite species and (2) the mean value of the total number of parasite individuals of all species raised to the power of 1/2; the other designations are the same as in figure 1.

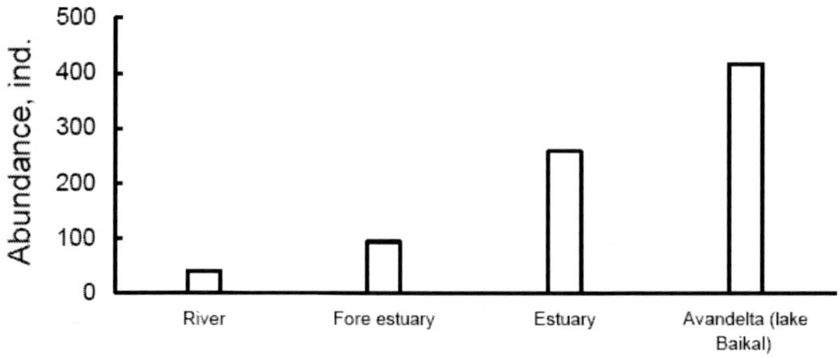

Figure 3. The changes of roach infection with *Myxidium rhodei* in transect «Selenga River - Fore estuary - Estuary – Avandelta (Lake Baikal)».

The opposite correlations between the Berger–Parker index and the roach age in two aquatic bodies can depend on both the properties of *M rhodei* spores themselves (which belong to the slowly descending type [37] and the abundance of oligochaetes (intermediate hosts for Myxosporidia) in the lake (Chivyrkuiskii Bay) and river (Selenga River Estuary) ecosystems.

Sets of Parasite Infracommunities and Host Age

The parasite dominant species shift that takes place with Baikal whitefish aging is related to the host's shift in feeding. Baikal whitefish young fishes (0+ to 3+) feed on plankton organisms including crustaceans, which are intermediate hosts of proteocephalidean cestodes, dominating in the diet. As a consequence *Proteocephalus longicollis* cestoda dominates in young fishes. Benthic organisms such as gamarrids, which are intermediate hosts of cyatocephalidean cestodes, start to dominate in the diet in the following age groups (4+ and older) [39, 40]. Hence *Cyatocephalus truncatus* cestoda dominates in age groups 6+ and older, and at the same time abundances of *P. longicollis* and *Cyatocephalus truncatus* are almost equal in age groups of Baikal whitefish 4+ - 5+.

A common characteristic of the diversity indices (Brillouin and Shannon) relation to host age on different community levels (infracommunities and sets of infracommunities) of the Baikal whitefish parasites is the presence of a significant positive correlation between them and Baikal whitefish age. Differences are manifested in the character of the fluctuations of these indices on different levels of parasites communities. Fluctuations of the Brillouin index in the parasites infracommunities in the age series of the Baikal whitefish correspond to the species richness parameter dynamics (the mean number of parasite species in one infracommunity and the Menhinick's index) and the dynamics of the mean number of parasite individuals in one infracommunity: increase from 0+ to 12+ and decrease in the following age groups 13+ to 16+. The values of the relatively stable Berger–Parker domination index do not affect the Brillouin index dynamics in the parasite infracommunities (Table 2) significantly. Fluctuations of Shannon index in the parasites sets of infracommunities in the Baikal whitefish age series (increase from minimum in the 0+ to maximum in 15+ to 16+ age) correspond to, first, the dynamics of Margalef's index of parasite species richness (increasing with the host's age) and, second, the opposite vector of Simpson domination index dynamics (decreasing with the host's age) (Table 5). In general, differences in parasite diversity indices dynamics in the age series of the Baikal whitefish at various levels of communities (infracommunities and sets of infracommunities) are determined by the differences in the domination indices dynamics (at increasing of indices of species richness in connection with host age).

Brillouin's index of parasite species diversity (infracommunities) decreases significantly from age 4+ to 5+ to age 6+ along with the dominant species (*P. longicollis* to *C. truncatus*) shift. At the same time, the Menhinick's index of species richness decreases insignificantly and the Berger–Parker index stays on the same level. The share of infracommunities where dominant species prevailed increases significantly (from 26.7% in 4+ to 5+ whitefish to 62.5% in 6+ hosts) (Tables 2, 5). The Brillouin index happened to be the most sensitive to rapid increase in dominant species occurrence in the sets of parasite infracommunities during the shift of Baikal whitefish age groups from 4+ and 5+ to 6+.The opposite correlations between the Margalef and Menhinick's indices of the species richness of parasite communities and roach age, which were observed for infracommunities, are preserved for the sets of infracommunities. The reason for this is the same as for the individual infracommunities: the rate of increase in the number of parasite species is higher than the rate of increase in the values of the natural logarithm of the mean number of parasite individuals of all species and lower than the rate of increase in the mean number of parasite individuals raised to the power of 1/2 in the sets of infracommunities (Figures 4, 5).

The correlation between the Shannon index of species diversity for sets of roach parasite infracommunities and the host age is characterized by a negative value of the Spearman rank coefficient in the Chivyrkuiskii Bay and a positive value in the Selenga River Estuary (Tables 6, 7).

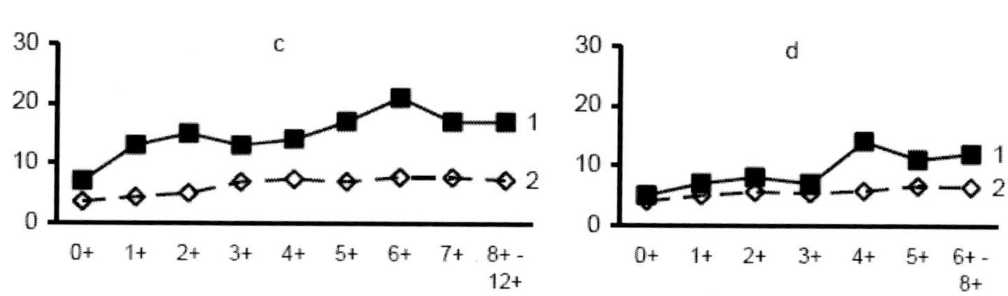

Figure 4. Changes in the values of the Margalef index (a, b) and its components (c, d) in sets of parasite infracommunities in the roach age groups in the Chivyrkuiskii Bay of Lake Baikal (a, c) and the estuary of the Selenga River (b, d): (1) the total quantity of parasite species minus one and (2) the natural logarithm of the total quantity of parasite individuals of all species; the other designations are the same as in figure 1.

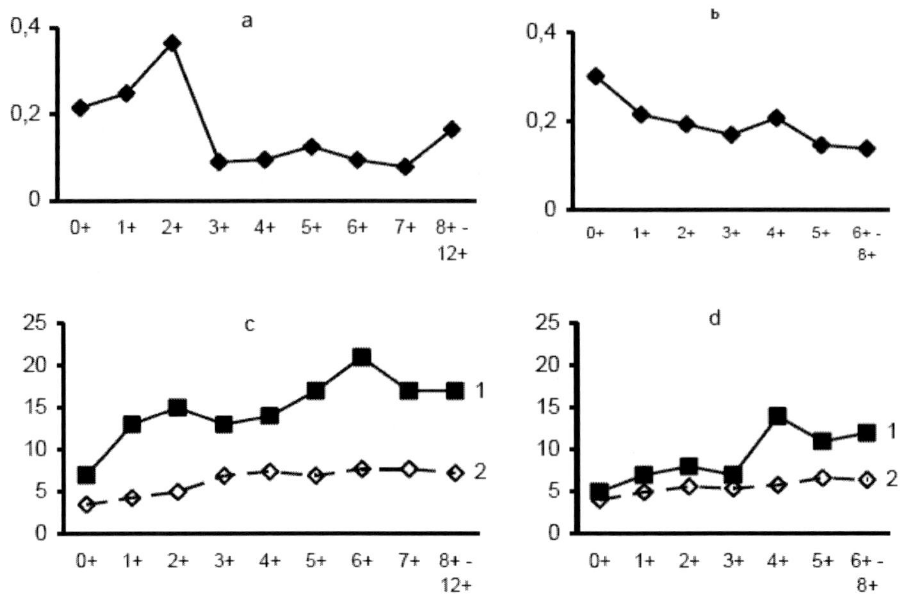

Figure 5. Changes in the values of the Menhinick's index (a, b) and its components (c, d) in sets of parasite infracommunities in the roach age groups in the Chivyrkuiskii Bay of Lake Baikal (a, c) and the estuary of the Selenga River (b, d): (*1*) the total number of parasite species and (*2*) the total number of parasite individuals of all species raised to the power of 1/2; the other designations are the same as in figure 1.

When analyzing the reasons for this dependence, it is necessary to take into account two things. First, the indices based on relative species abundances, in particular, the Shannon and Brillouin indices, concurrently take into account the species richness and evenness [22]. Second, a strong negative correlation between the Shannon and Brillouin indices of species diversity, on the one hand, and the Berger–Parker dominance index, on the other, was empirically found by the example of the communities of common minnow parasites [16]. Only the Berger–Parker dominance index displays opposite patterns in these two aquatic bodies. The Shannon index negatively correlates with the Berger–Parker index in both the Chivyrkuiskii Bay ($rS = -0.967$, $p < 0.001$, $n = 9$) and the Selenga River Estuary ($rS = -1.0$, $p < 0.001$, $n = 7$); i.e., the changes in the Shannon index in the parasite communities in the roach age series are opposite the changes in the Berger–Parker index in both aquatic bodies. It was shown earlier [36] that the component communities of the tropical fish *Cichlasoma urophthalmus* (Gunther, 1862) display a high level of species richness (in number of species) and a high number of parasite individuals per individual host, whereas the species diversity (according to the Brillouin index) is low, which reflects the significant prevalence of one species, the trematode *Oligogonotylus manteri* (Watson, 1976).

So, species richness in parasite communities of two fish species in lake Baikal basin increases in connection with the host age. It is demonstrated by increase of number of parasite species in fish age series. The opposite character of changes of Margalef and Menhinick's indices in roach age series in Lake Baikal basin connects with the properties of indices themselves. Older, larger yellow perch, Perca flavescens, from Garner Lake, Alberta, Canada tend to have more species of parasites combined with larger parasite infrapopulations [41].

Host diet and age contributed significantly to the rate of parasite species accumulation of yellow perch (*Perca flavescens*) from four small Canadian Shield lakes [35].

An other nature of connection between host age and parasite diversity was revealed of marine fishes. When investigating the parasite communities of three corall-reef fish species (*Stegastes nigricans, Dascyllus aruanus and Cephalopholis argus*) from Tiahura reef, French Polynesia species richness and Brillouin's diversity index increased significantly with age and length of both *S. nigricans* and *C. argus*. No significant relationship was found between body size and age of *D. aruanus* and parasite diversity [42]. As a result of the study of metazoan parasite fauna of the bigeye flounder, *Hippoglossina macrops*, from Northern Chile was revealed that species richness, mean intensity and prevalence of ectoparasites and endoparasites were not correlated with host age [43]. In contrast to freshwater fishes the relationships between the age of marine fishes and changes of their parasite diversity appear less clear.

The correlation between the indices reflecting the evenness of infracommunities of roach parasites and the host age is, in general, negative in both the Chivyrkuiskii Bay of Lake Baikal and the Selenga River Estuary at the infracommunities and the sets of infracommunities levels.

Opposite correlations between the Berger–Parker index in two aquatic bodies and roach age were observed: the correlation for the Chivyrkuiskii Bay is positive and for the estuary of the Selenga River it is negative. It is shown that ecological and biological features of dominant species, *Myxidium rhodei*, extensively determine the opposite character of changes of Berger-Parker index in host age series in two various water bodies.

The changes in the Brillouin and Shannon diversity indeces of the parasite communities in the Baikal whitefish and roach age series are opposite the ones of the domination level of ones in both aquatic bodies of Lake Baikal basin.

CONCLUSION

Regular changes in the structure of the parasite communities in the Baikal whitefish and roach age series in two aquatic bodies of the Lake Baikal basin were detected: the species richness increased, whereas the evenness decreased.

Presented data of opposite character of changes of Berger-Parker index in parasite communities of age series of roach in two different aquatic bodies show considerable significance of the ecological and biological features of most abundant species, *Myxidium rhodei*, in domination level.

It is marked the negative interrelation between domination level and diversity indices in parasite communities of fishes in Baikal basin.

An analysis of the changes taking place in the parasite communities in connection with the fish age shows that it is necessary to use indices for measurement of biological diversity in an integrated manner, as well as to take into account the patterns of changes in the components contributing to these indices and the correlations between the indices.

REFERENCES

[1] Dogel', V. A. (1958). The Parasite fauna and the environment. Some problems of ecology of parasites of freshwater fishes. In V.A. Dogel', G.K. Petrushevskii, and J.I. Poljanskii (Eds.), *Osnovnye problemy parazitologii ryb (The main problems of fish parasitology)* (pp. 9–54). Leningrad: Izdatel'stvo Leningradskogo universiteta. (In Russian).

[2] Dogel', V. A. (1936). Age-related changes in the parasitic fauna of the eel in relation to the problem of its migration. *Uchenije zapiski LGU: serija biologicheskaja, No. 3*, 114–122. (In Russian).

[3] Strelkov, Yu. A., Chernysheva, N. B., and Yunchis, O. N. (1981). Regularities of formation of the parasite fauna of young freshwater fishes. *Trudi Zoologicheskogo Instituta of Academy of Sciences of USSR*, *108*, 23-30. (In Russian).

[4] Holmes, J. C.; and Price, P. W. (1980). Parasite communities: the roles of phylogeny and ecology. *Systematic Zoology, 29*, 203-213.

[5] Goater, T. M., Esch, G. W., and Bush, A. O. (1987). Helminth parasites of sympatric salamanders: Ecological concepts at infracommunity, component and compound community levels. *American Midland Naturalist*, *118*, 289-300.

[6] Magurran, A. E. (1988) *Ecological diversity and its measurement*. Princeton, NJ: Princeton University Press.

[7] Bush, A. O., Aho, J. M., and Kennedy, C. R. (1990). Ecological versus phylogenetic determinants of helminth parasite community richness. *Evolutionary Ecology, 4*, 1-20.

[8] Kennedy, C. R. (1993). The dynamics of intestinal helminth communities in eels *Anguilla anguilla* in a small stream: long-term changes in richness and structure. *Parasitology*, *107*, 71-78.

[9] Kennedy, C.R., and Guegan, J.-F. (1994). Regional versus local parasite richness in British freshwater fish: saturated or unsaturated parasite communities. *Parasitology*, *109*, 175-185.

[10] Poulin, R. (1995). Phylogeny, ecology, and the richness of parasite communities in vertebrates. *Ecological Monographs*, *65*, 283–302.

[11] Valtonen, E.T., Pulkkinen, K., Poulin, R., and Julkunen M. (2001). The structure of parasite component communities in brackish water fishes of the Northeastern Baltic Sea. *Parasitology*, *122*, 471–481.

[12] Dorovskikh, G. N., and Golikova, E. A. (2004). Seasonal dynamics of the structure of component communities of parasites of the common minnow *Phoxinus phoxinus* (L.). *Parazitologiya, 38*, 413–425. (In Russian).

[13] Zander, C.D. (2001). The guild as a concept and a means in ecological parasitology. *Parasitology Research*, *87*, 484-488.

[14] Josten, N., Detloff, K. C., and Zander, C. D. (2009). Analysis of a parasite supra community from the Flensburg fjord. *Parasitology Research*, *104*, 449-461.

[15] Balashov, Yu.S., Terms and Concepts Used in the Study of Populations and Communities of Parasites, *Parazitologiya*, 2000, vol. 34, no. 5, pp. 361–370.

[16] Pugachev, O. N. (2000). Parasitic communities of the river minnow. *Parazitologiya*, *34*, 196–209. (In Russian).

[17] Bush, A.O., Lafferty, K.D., Lotz, J.M., and Shostak, A.W. (1997). Parasitology meets ecology on its own terms: Margolis et al. revisited. *Journal of Parasitology, 83*, 575–583.

[18] Holmes, J.C. and Price, P.W. (1986). Communities of Parasites. In D.J. Anderson, and J. Kikkawa (Eds.), *Community Ecology: Patterns and Processes* (pp.187-213). Oxford: Blackwell Scientific Publications.

[19] Timi, J.T., and Poulin, R. (2003). Parasite community structure within and across host populations of a marine pelagic fish: how repeatable is it? *International Journal for Parasitology, 33*, 1353–1362.

[20] Dorovskikh, G. N., and Stepanov, V. G. (2008). Change in the structure of component parasite communities with host age. *Russian Journal of Ecology, 39*, 215-220.

[21] Dugarov Zh. N., Pronin N. M., Sondueva L. D., Burdukovskaya T. G., Batueva M. D., and Pronina S. V. (2011). Dependence that the Community Structure of the Roach *Rutilus rutilus* (L.) Parasites Has on the Host Age. *Inland Water Biology, 4*, 78–89.

[22] Magurran, A. E. (1992). *Ekologicheskoe raznoobrazie i ego izmerenie (Ecological diversity and its measurement)*. Moscow: Mir. (In Russian: translation from English).

[23] Zhokhov, A.E. (2003). Seasonal dynamics of the structure of intestinal helminth community in ide (*Leuciscus idus* L.) from the Rybinsk Reservoir. *Russian Journal of Ecology, 34*, 413-417.

[24] Smith, B. and Wilson, J. B. (1996). A consumer's guide to evenness indices. *Oikos, 76*, 70–82.

[25] Payne, L.X., Schindler, D. E., Parrish, J. K., and Temple, S. A. (2005). Quantifying spatial pattern with evenness indices. *Ecological Applications, 15*, 507–520.

[26] Magurran, A. E. (2004). *Measuring biological diversity*. Oxford: Blackwell Publishing.

[27] Baldanova, D. R., and Pronin, N. M. (2001). *Skrebni (Tip Acanthocephala) Baikala: Morfologiya i ekologiya (Acanthocephalans (phylum Acanthocephala) of lake Baikal: morphology and ecology)*. Novosibirsk: Nauka. (In Russian).

[28] Dogel', V. A., Bogolepova, I.P., and Smirnova, I.V. (1949). The parasite fauna of lake Baikal and its zoogeographical significance. *Vestnik Leningradskogo universiteta, No. 7*, 13–34. (In Russian).

[29] Zaika, V.E. (1965). *Parazitofauna ryb ozera Baikal (Parasite Fauna of Fishes of Lake Baikal)*. Moscow: Nauka.

[30] Timoshkin, O. A., Sitnokova, T. Ya., Rusinek, O. T., et al. (2001). *Annotirovannyi spisok fauny Baikala i ego vodosbornogo basseina (Index of animal species inhabiting lake Baikal and its catchment area)*: In 2 volumes. Novosibirsk: Nauka.Vol. I: *Ozero Baikal* (Lake Baikal), Book 1. (In Russian).

[31] Timoshkin, O. A., Sitnokova, T. Ya., Rusinek, O. T., et al. (2004). *Annotirovannyi spisok fauny Baikala i ego vodosbornogo basseina (Index of animal species inhabiting lake Baikal and its catchment area)*: In 2 volumes. Novosibirsk: Nauka. Vol. I: *Ozero Baikal* (Lake Baikal), Book 2. (In Russian).

[32] Rusinek, O. T. (2007). *Parazity ryb ozera Baikal (fauna, soobshchestva, zoogeografiya, istoriya formirovaniya). (Fish parasites of Lake Baikal (fauna, communities, zoogeography and historical background)*. Moscow: KMK Scientific Press Ltd. (In Russian).

[33] Dugarov, Zh. N., and Pronin, N. M. (2010). Baikal whitefish *Coregonus baicalensis* Dybowski, 1874 parasite communities and host age. *Biology Bulletin, 37*, 630–637.

[34] Bauer, O. N., and Nikol'skaya, N. P. (1957). Dynamics of parasitic fauna of Ladoga whitefish and its epizootic significance. *Izvestija VNIIORKh, 42*, 227–242.

[35] Johnson, M. W., Nelson, P. A., and Dick, T. A. (2004). Structuring mechanisms of yellow perch (*Perca flavescens*) parasite communities: host age, diet, and local factors. *Canadian Journal of Zoology, 82*, 1291–1301.

[36] Salgado-Maldonado, G., and Kennedy, C. R. (1997). Richness and similarity of helminth communities in the tropical cichlid fish *Cichlasoma urophthalmus* from the Yucatan Peninsula, Mexico. *Parasitology, 114*, 581–590.

[37] Shul'man, S. S., Donets, Z. S., and Kovaleva, A. A. (1997). *Klass miksosporidii mirovoi fauny (Class Myxosporea of the world fauna): Vol. 1: Obshchaya Chast' (General Part).* St. Petersburg: Nauka. (In Russian).

[38] Pronin, N. M., Dugarov, Zh. N., Burdukovskaya, T. G., Sondueva, L. D., Vasiljeva O. S., and Pronina S. V. (2008). The spatial distribution of fish parasites. In A.K. Tulochonov, and A.M. Plusnin (Eds.), *Del'ta reki Selengi – estestvennyi biofil'tr i indikator sostoyaniya ozera Baikal (The Selenga River Delta: natural biofilter and indicator of the condition of lake Baikal)* (pp. 243-250). Novosibirsk: Publishing House of the Siberian Branch of the Russian Academy of Sciences. (In Russian).

[39] Sterlyagova, M. A. (1958). Biology and commercial catching of Baikal whitefish. In M. M. Kozhov, K. I. Misharin (Eds.), *Ryby i rybnoe khozyaistvo v basseine ozera Baikal (Fishes and fisheries in the lake Baikal basin)* (pp. 288-310). Irkutsk: Irkutskoe knizhnoe izdatel'stvo. (In Russian).

[40] Pronin, N. M., Matveev, A. N., Samusenok, V. P., et al. (2007). *Ryby ozera Baikal i ego basseina (Fishes of the lake Baikal and its basin).* Ulan-Ude: Publishing Buryat Scientific Center, Siberian Branch of Russian Academy of Sciences. (In Russian).

[41] Zelmer, D.A., and Arai, H.P. (1998). The contributions of host age and size to the aggregated distribution of parasites in yellow perch, *Perca flavescens*, from Garner Lake, Alberta, Canada. *Journal of Parasitology, 84*, 24–28.

[42] Lo, S. M., Morand, S., and Galsin, R. (1998). Parasite diversity/host age and size relationship in three coral-reef fishes from French Polynesia. *International Journal for Parasitology, 28*, 1695–1708.

[43] Gonzalez, M. T., Acuna, E., and Oliva, M. E. (2001) Metazoan parasite fauna of the bigeye flounder, *Hippoglossina macrops*, from Northern Chile. Influence of host age and sex. *Memorias do Instituto Oswaldo Cruz, 96*, 1049-1054.

INDEX

A

B

C

D

E

F

G

H

I

J

K

L

M

N

O

P

Q

R

S

T

U

V

W

Y

Z